Silicon Front-End Junction Formation Technologies

MATERIALS RESEARCH SOCIETY
SYMPOSIUM PROCEEDINGS VOLUME 717

Silicon Front-End Junction Formation Technologies

Symposium held April 2–4, 2002, San Francisco, California, U.S.A.

EDITORS:

Daniel F. Downey
Varian Semiconductor Equipment Associates
Gloucester, Massachusetts, U.S.A.

Mark E. Law
University of Florida
Gainesville, Florida, U.S.A.

Alain Claverie
CNRS
Toulouse, France

Michael J. Rendon
Motorola Advanced Products R&D Lab
Austin, Texas, U.S.A.

MRS

Materials Research Society
Warrendale, Pennsylvania

Single article reprints from this publication are available through
University Microfilms Inc., 300 North Zeeb Road, Ann Arbor, Michigan 48106

CODEN: MRSPDH

Published by:

Materials Research Society
506 Keystone Drive
Warrendale, PA 15086
Telephone (724) 779-3003
Fax (724) 779-8313
Web site: http://www.mrs.org/

Library of Congress Cataloging-in-Publication Data

Silicon front-end junction formation technologies, symposium held April 2–4, 2002,
San Francisco, California, U.S.A. / editors, Daniel F. Downey, Mark E. Law, Alain Claverie, Michael J. Rendon
p.cm.—(Materials Research Society symposium proceedings ; v. 717)
Includes bibliographical references and indexes.
ISBN 1-55899-653-2
I. Downey, Daniel F. II. Law, Mark E. III. Claverie, Alain IV. Rendon, Michael J.
V. Materials Research Society symposium proceedings ; v. 717
2002

Manufactured in the United States of America

CONTENTS

*Invited Paper

DEVICE ENGINEERING OPTIONS

DOPING ACTIVATION

*Invited Paper

EPITAXIAL TECHNIQUES, FLUORINE, DEFECT FUNDAMENTALS

*Invited Paper

DEFECT AND DIFFUSION MODELS

*Invited Paper

CHARACTERIZATION USING SURFACE ANALYSIS TECHNIQUES

CHARACTERIZATION TECHNOLOGIES

*Invited Paper

PREFACE

This volume contains papers presented at Symposium C, "Silicon Front-End Junction Formation Technologies", held April 2-4 at the 2002 MRS Spring Meeting in San Francisco, California. This was the fourth symposium covering topics related to Si front end processing. Unlike the previous three, this symposium expanded its focus to include more topics related to formation of ultra-shallow junctions. With the challenges presented by the requirements of the sub-100 nm node, the need for new activation technologies which yield minimal diffusion of the dopant while producing high activation are paramount. In addition to this, the metrology required to measure these shallow profiles in both one and two dimensions becomes more critical. This symposium was aimed at addressing these new requirements and potential solutions, by covering a variety of topics such as: alternate annealing technologies, device engineering options, dopant activation, epitaxial techniques, defect and diffusion models and characterization techniques.

In the first session, alternate annealing technologies were covered. There was an invited talk on low temperature shallow junction formation primarily focusing on SPE. This was followed by talks on new and promising techniques such as flash lamp milli-second annealing, microwave annealing, non-melt laser annealing and laser melt annealing. Many issues related to laser melt annealing were addressed such as the effects of various dopants, point defect formation, defect removal and dopant deactivation. The next session addressed device engineering options which could help optimize shallow junctions or provide alternate fabrication approaches. These options include SPE optimization, elevated source drains and SOI. The third session dealt with issues related to dopant activation such as boron interstitial clustering, dopant deactivation defects, the stress dependence and surface effects on solid solubility, diffusion effects and dopant segregation at interfaces. The next session addressed epitaxial techniques primarily employing SiGe. This session was kicked off by an invited talk about SiGe source drain technology. This was followed by three talks addressing various aspects of SiGe, SiGeC, and SiC epitaxy for junction formation. This session was closed with talks on the effects of fluorine on diffusion. The fifth session addressed defect and diffusion modeling. The first four talks modeled TED effects in Si while other talks in the session addressed self interstitial diffusion, interstitial clustering, evolution of defects, and dislocation loop formation. The sixth session was a joint session with Symposium B with an invited talk on SIMS analysis techniques for SiGe, SiON and USJ boron profiling. This was followed by the last session which addressed the challenges facing the characterization technologies in dopant profiling. Techniques such as scanning tunneling microscopy, carrier illumination, radiochemical neutron activation, TEM, electron holography, medium energy ion scattering and pulsed force scanning spreading resistance were discussed.

Finally we wish to acknowledge the financial support for this symposium which was provided by Applied Materials, Inc., Fujitsu Ltd., Intel Corporation, Verdant Technologies and Varian Semiconductor Equipment Associates.

Daniel F. Downey
Mark E. Law
Alain Claverie
Michael J. Rendon

June 2002

MATERIALS RESEARCH SOCIETY SYMPOSIUM PROCEEDINGS

Volume 686— Materials Issues in Novel Si-Based Technology, W. En, E.C. Jones, J.C. Sturm, S. Tiwari, M. Hirose, M. Chan, 2002, ISBN: 1-55899-622-2
Volume 687— Materials Science of Microelectromechanical Systems (MEMS) Devices IV, A.A. Ayon, S.M. Spearing, T. Buchheit, H. Kahn, 2002, ISBN: 1-55899-623-0
Volume 688— Ferroelectric Thin Films X, S.R. Gilbert, Y. Miyasaka, D. Wouters, S. Trolier-McKinstry, S.K. Streiffer, 2002, ISBN: 1-55899-624-9
Volume 689— Materials for High-Temperature Superconductor Technologies, M.P. Paranthaman, M.W. Rupich, K. Salama, J. Mannhart, T. Hasegawa, 2002, ISBN: 1-55899-625-7
Volume 690— Spintronics, T.J. Klemmer, J.Z. Sun, A. Fert, J. Bass, 2002, ISBN: 1-55899-626-5
Volume 691— Thermoelectric Materials 2001—Research and Applications, G.S. Nolas, D.C. Johnson, D.G. Mandrus, 2002, ISBN: 1-55899-627-3
Volume 692— Progress in Semiconductor Materials for Optoelectronic Applications, E.D. Jones, M.O. Manasreh, K.D. Choquette, D. Friedman, 2002, ISBN: 1-55899-628-1
Volume 693— GaN and Related Alloys—2001, J.E. Northrup, J. Neugebauer, S.F. Chichibu, D.C. Look, H. Riechert, 2002, ISBN: 1-55899-629-X
Volume 695— Thin Films: Stresses and Mechanical Properties IX, C.S. Ozkan, R.C. Cammarata, L.B. Freund, H. Gao, 2002, ISBN: 1-55899-631-1
Volume 696— Current Issues in Heteroepitaxial Growth—Stress Relaxation and Self Assembly, E. Stach, E. Chason, R. Hull, S. Bader, 2002, ISBN: 1-55899-632-X
Volume 697— Surface Engineering 2001—Fundamentals and Applications, W.J. Meng, A. Kumar, Y-W. Chung, G.L. Doll, Y-T. Cheng, S. Veprek, 2002, ISBN: 1-55899-633-8
Volume 698— Electroactive Polymers and Rapid Prototyping, Y. Bar-Cohen, D.B. Chrisey, Q.M. Zhang, S. Bauer, E. Fukada, S.C. Danforth, 2002, ISBN: 1-55899-634-6
Volume 699— Electrically Based Microstructural Characterization III, R.A. Gerhardt, A. Washabaugh, M.A. Alim, G.M. Choi, 2002, ISBN: 1-55899-635-4
Volume 700— Combinatorial and Artificial Intelligence Methods in Materials Science, I. Takeuchi, C. Buelens, H. Koinuma, E.J. Amis, J.M. Newsam, L.T. Wille, 2002, ISBN: 1-55899-636-2
Volume 702— Advanced Fibers, Plastics, Laminates and Composites, F.T. Wallenberger, N. Weston, K. Chawla, R. Ford, R.P. Wool, 2002, ISBN: 1-55899-638-9
Volume 703— Nanophase and Nanocomposite Materials IV, S. Komarneni, R.A. Vaia, G.Q. Lu, J-I. Matsushita, J.C. Parker, 2002, ISBN: 1-55899-639-7
Volume 704— Nanoparticle Materials, R.K. Singh, R. Partch, M. Muhammed, M. Senna, H. Hofmann, 2002, ISBN: 1-55899-640-0
Volume 705— Nanopatterning—From Ultralarge-Scale Integration to Biotechnology, L. Merhari, K.E. Gonsalves, E.A. Dobisz, M. Angelopoulos, D. Herr, 2002, ISBN: 1-55899-641-9
Volume 706— Making Functional Materials with Nanotubes, P. Nikolaev, P. Bernier, P. Ajayan, Y. Iwasa, 2002, ISBN: 1-55899-642-7
Volume 707— Self-Assembly Processes in Materials, S. Moss, 2002, ISBN: 1-55899-643-5
Volume 708— Organic and Optoelectronic Materials, Processing and Devices, S. Moss, 2002, ISBN: 1-55899-644-3
Volume 709— Advances in Liquid Crystalline Materials and Technologies, P.T. Mather, D.J. Broer, T.J. Bunning, D.M. Walba, R. Zentel, 2002, ISBN: 1-55899-645-1
Volume 710— Polymer Interfaces and Thin Films, C.W. Frank, 2002, ISBN: 1-55899-646-X
Volume 711— Advanced Biomaterials—Characterization, Tissue Engineering and Complexity, 2002, ISBN: 1-55899-647-8

Materials Research Society Symposium Proceedings

Volume 712— Materials Issues in Art and Archaeology VI, P.B. Vandiver, M. Goodway, J.R. Druzik, J.L. Mass, 2002, ISBN: 1-55899-648-6
Volume 713— Scientific Basis for Nuclear Waste Management XXV, B.P. McGrail, G.A. Cragnolino, 2002, ISBN: 1-55899-649-4
Volume 714E—Materials, Technology and Reliability for Advanced Interconnects and Low-k Dielectrics II, S. Lahiri, 2002, ISBN: 1-55899-650-8
Volume 715— Amorphous and Heterogeneous Silicon-Based Films—2002, J.R. Abelson, J.B. Boyce, J.D. Cohen, H. Matsumura, J. Robertson, 2002, ISBN: 1-55899-651-6
Volume 716— Silicon Materials—Processing, Characterization and Reliability, J. Veteran, D.L. O'Meara, V. Misra, P. Ho, 2002, ISBN: 1-55899-652-4
Volume 717— Silicon Front-End Junction Formation Technologies, D.F. Downey, M.E. Law, A. Claverie, M.J. Rendon, 2002, ISBN: 1-55899-653-2
Volume 718— Perovskite Materials, K. Poeppelmeier, A. Navrotsky, R. Wentzcovitch, 2002, ISBN: 1-55899-654-0
Volume 719— Defect and Impurity Engineered Semiconductors and Devices III, S. Ashok, J. Chevallier, N.M. Johnson, B.L. Sopori, H. Okushi, 2002, ISBN: 1-55899-655-9
Volume 720— Materials Issues for Tunable RF and Microwave Devices III, S.C. Tidrow, J.S. Horwitz, J. Levy, X. Xi, 2002, ISBN: 1-55899-656-7
Volume 721— Magnetic and Electronic Films—Microstructure, Texture and Application to Data Storage, P.W. DeHaven, D.P. Field, S.D. Harkness IV, J.A. Sutliff, J.A. Szpunar, L. Tang, T. Thomson, M.D. Vaudin, 2002, ISBN: 1-55899-657-5
Volume 722— Materials and Devices for Optoelectronics and Microphotonics, R.B. Wehrspohn, S. Noda, C. Soukoulis, R. März, 2002, ISBN: 1-55899-658-3
Volume 723— Molecularly Imprinted Materials—Sensors and Other Devices, K.J. Shea, M.J. Roberts, M. Yan, 2002, ISBN: 1-55899-659-1
Volume 724— Biological and Biomimetic Materials—Properties to Function, J. McKittrick, J. Aizenberg, C. Orme, P. Vekilov, 2002, ISBN: 1-55899-660-5
Volume 725— Organic and Polymeric Materials and Devices—Optical, Electrical and Optoelectronic Properties, G.E. Jabbour, N.S. Sariciftci, S.T. Lee, S. Carter, J. Kido, 2002, ISBN: 1-55899-661-3
Volume 726— Organic/Inorganic Hybrid Materials—2002, R.M. Laine, C. Sanchez, S. Yang, C.J. Brinker, 2002, ISBN: 1-55899-662-1
Volume 727— Nanostructured Interfaces, G. Duscher, J.M. Plitzko, Y. Zhu, H. Ichinose, 2002, ISBN: 1-55899-663-X
Volume 728— Functional Nanostructured Materials through Multiscale Assembly and Novel Patterning Techniques, Steven C. Moss, 2002, ISBN: 1-55899-664-8
Volume 729— BioMEMS and Bionanotechnology, L.P. Lee, J.T. Borenstein, R.P. Manginell, M. Okandan, P.J. Hesketh, 2002, ISBN: 1-55899-665-6
Volume 730— Materials for Energy Storage, Generation and Transport, G. Ceder, S.A. Ringel, R.B. Schwarz, 2002, ISBN: 1-55899-666-4
Volume 731— Modeling and Numerical Simulation of Materials Behavior and Evolution, V. Tikare, E.A. Olevsky, A. Zavaliangos, 2002, ISBN: 1-55899-667-2
Volume 732E—Chemical-Mechanical Planarization, S.V. Babu, R. Singh, N. Hayasaka, M. Oliver, 2002, ISBN: 1-55899-668-0
Volume 733E—Polymer Nanocomposites, S. Nutt, R. Vaia, W. Rodgers, G.L. Hagnauer, G.W. Beall, 2002, ISBN: 1-55899-669-9

Prior Materials Research Society Symposium Proceedings available by contacting Materials Research Society

Alternate Annealing Technologies

Mat. Res. Soc. Symp. Proc. Vol. 717 © 2002 Materials Research Society C1.1

Low Temperature Shallow Junction Formation For 70nm Technology Node And Beyond

John O. Borland
Varian Semiconductor Equipment Associates
4 Stanley Tucker Dr.
Newburyport, MA 01950, USA

ABSTRACT

Low temperature shallow junction formation is an attractive activation technique for 70nm technology node and beyond as it can easily be integrated into device structures that are formed using disposable spacer (reverse source drain extension formation) or low power CMOS devices using high-k/metal gate stack structures. Therefore, this paper will first review the shallow junction requirements as stated in the 2001 ITRS (international technology roadmap for semiconductors) and it's interpretation to ion implantation shallow junction formation for various dopant activation and annealing techniques. First high temperature (>1000°C) RTA spike, flash or sub-melt laser annealing techniques with oxide or oxynitride/polysilicon electrode gate stack structures will be discussed and its limitations to <8E19/cm^3 boron electrically active dopant level due to boron solid solubility limit in silicon satisfying only the 100nm technology node requirement (2003). Next, higher temperature laser melt annealing (1200°C to 1400°C) will be discussed and it's applicability beyond 70nm node technology (2006) to 25nm node (2016) where boron solid solubility limit is up to 5E20/cm^3. However, if high-k (HfO) dielectric/metal electrode gate stack structures are to be used starting at sub-100nm node in 2005 for low power CMOS then low temperature (<700°C) annealing must be used for shallow junction formation to prevent recrystallization and dielectric constant degradation. Using low temperature SPE (solid phase epitaxial regrowth) annealing techniques in the 550°C to 750°C for short anneal times of <5mins., shallow & abrupt junctions 8.0nm deep, <2.0nm/decade with up to 2.5E20/cm^3 boron electrical active dopant level can be achieved satisfying the 25nm technology node (2016) requirements.

INTRODUCTION

With the continued growth trends in portable/mobile devices replacing stationary desk top devices, more and more emphasis is being placed on improving low power CMOS devices. This is driving the switch from oxynitride type gate dielectrics with polysilicon electrodes to high-k (HfO_2) dielectrics with metal electrodes by 2005 to satisfy the 70-90nm technology node. High-k dielectric material have demonstrated orders of magnitude reduction in gate leakage with super thin equivalent gate oxide thicknesses but their control and deposition process are still very immature and therefore the focus of development at consortiums such as International-SEMATECH and SELETE. However, a fundamental high-k dielectric and metal electrode material issue is it's thermal stability and thermal resistance properties at elevated (high) temperatures above 700°C. Several groups have reported the thermal stability of amorphous high-k dielectric is limited to <700°C for HfO_2 and <800°C for ZrO_2 before recrystalization occurs degrading the dielectric constant [1]. Selection of the metal electrode material can also impose thermal budget limitations. To avoid these thermal budget constraints in the formation of shallow junctions the replacement gate process scheme was proposed several years ago [2,3]. Unfortunately, the replacement gate process has proven to be very difficult to manufacture and abandoned by most. The replacement gate process adds 2+ masking levels and 20% more processing steps [2]. Another issue is the key-hole structure resulting from the metal electrode deposition process due to the narrow aspect ratio [3]. Others have reported the difficulty of pre-cleaning the silicon surface prior to high-k dielectric deposition and this becomes extremely more difficult in the replacement gate structure where you have the narrow and steep side-wall geometries of the removable dummy gate structure to deal with. Therefore, with the conventional gate stack formation scheme for high-k dielectric with metal electrodes, the ion implanted shallow junction formation process will require a paradigm shift from high temperature annealing (>1000°C) to low temperature annealing (<700°C) for dopant activation and defect location engineered annealing for both SDE (source drain extension) and deep source drain formation. For this reason most of all the advanced logic manufacturers around the world are now investigating low temperature SPE in detail, implanting amorphous shallow junction structures using beam-line B_{11} and BF_2 species or plasma BF_3 and B_2H_6 species with various PAI (pre-amorphizing implant) conditions in the temperature range of 550°C to 700°C. Some of these results will be discussed below describing how to achieve: 1)low temperature (<700°C), 2)shallow & abrupt junctions, 3)low resistance and 4)high quality (acceptable leakage) SDE structure that satisfies the 70nm technology node and beyond.

2001 ITRS SHALLOW & ABRUPT JUNCTION ROADMAP

The 2001 ITRS SDE junction depth, sheet resistance and lateral abruptness is shown in table I [4]. However, another key SDE parameter that was not identified in the 2001 ITRS is the SDE under diffusion (gate overlap) value. Ghani et al from Intel reported in their paper on CMOS scaling, projections of transistor parameters including the critical SDE shallow junction for both vertical (X_j) and lateral (Y_j) dimensions[5]. The lateral Y_j coorelates to the amount of gate under diffusion of the SDE structure also know as gate overlap. They showed insufficient gate overlap can degrade I_{DSAT} however improving lateral junction abruptness from 7nm/decade down to 1.5nm/decade improves it. With 7nm/dec. abruptness their 180nm technology transistors needed >20nm gate overlap while with 3.5nm/dec. lateral abruptness they could reduce gate overlap by 12.5nm to >7.5nm. Similarly, Osburn et al also reported on lateral dopant abruptness and for extreme abruptness <1.0nm/dec. negative gate overlap (gate underlap) of 5.0nm is desirable as shown in figure 1 [6]. Reducing gate overlap improves SCE (short channel effects) not only for bulk CMOS but also for SOI-CMOS [7]. With the insertion point for high-k gate with metal electrodes in 2006 for the 70nm node low power CMOS devices and 2010 for the 50nm high performance MPU CMOS devices the USJ (ultra shallow junction) roadmap for dopant and activation methods based on gate stack structure is shown in figure 2 [8]. Note the emergence of high-k/metal gate stack structure option in 2006 with low temperature dopant activation methods and new plasma doping method. Also, by 2007 elevated source/drain structures will re-emerge for thin SOI applications and shallow source drain structures requiring low parasitic source drain resistance. The universal Rs-vs-Xj chart with technology node windows and boron dopant activation levels (solid solubility limits) for a box profile 1st created by Shishiguchi of NEC and modified by Osburn of NCSU, Muto of I-SEMATECH and Borland of VSEA is summarized in figure 3 [9-12]. From this chart, 70nm node requires box profile boron dopant level of 1E20/cm^3, while 50nm node requires a level of 1.5E20/cm^3 and 35nm node requires a level of 2E20/cm^3. Note the dramatic change between the 1999 and 2001 ITRS roadmap for Rs (sheet resistance) values. This came about from the device simulation studies reported by Kim et al of UCLA and Gossmann et al of Agere where they showed that a more relaxed specification window in Rs is acceptable in the 750 to 2000 ohms/square range rather than 200 to 300 ohms/square range [13,14].

Table 1. 2001 ITRS [4].

Year of Production	2001	2002	2003	2004	2005	2006	2007	2010	2013	2016
DRAM 1/2 Pitch [nm]	130	115	100	90	80	70	65	45	32	22
MPU/ASIC 1/2 Pitch [nm]	150	130	107	90	80	70	65	50	35	25
MPU Printed Gate Length [nm]	90	75	65	53	45	40	35	25	18	13
MPU Physical Gate Length [nm]	65	53	45	37	32	28	25	18	13	9
Equivalent physical oxide thickness for MPU/ASIC Tox[nm]	1.3-1.6	1.2-1.5	1.1-1.6	0.9-1.4	0.8-1.3	0.7-1.2	0.6-1.1	0.5-0.8	0.4-0.6	0.4-0.5
Gate Electrode Thickness [nm]	65-130	53-106	45-90	37-74	32-64	30-60	25-50	18-36	13-26	37517
Profile Control (side wall angle)	>89	>89	>89	90	90	90	90	90	90	90
Drain Extension Xj [nm]	27-45	22-36	19-31	15-25	13-22	12-19	10-17	7-12	5-9	4-6
Maximum Drain Extension Rs (PMOS) [Ohm/Sq]	400	460	550	660	770	830	760	630	940	1210
Extension Lateral Abruptness [nm/decade]	7.2	5.8	5.0	4.1	3.5	3.1	2.8	2.0	1.4	1.0

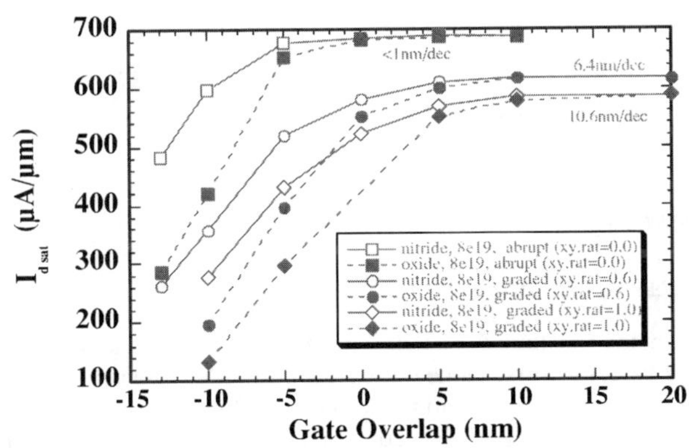

Figure 1. Lateral gradient (abruptness) effects on gate overlap requirements [6].

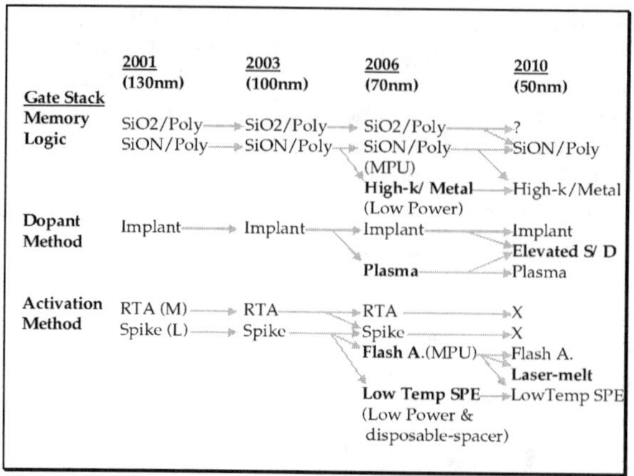

Figure 2. USJ dopant & activation roadmap [8].

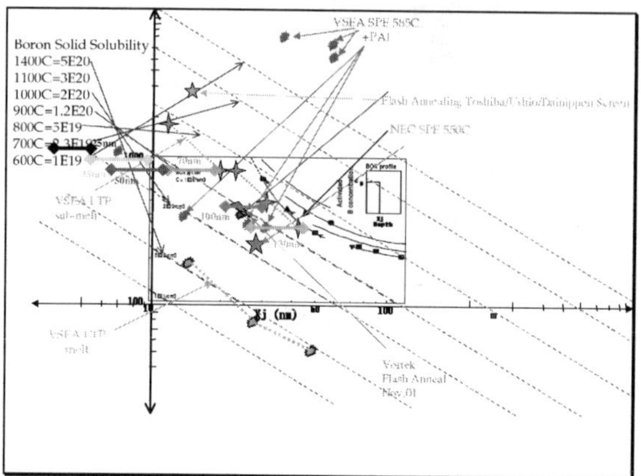

Figure 3. Rs-vs-Xj annealing results.

HIGH TEMPERATURE ANNEALING

RTA spike, flash and non-melt laser annealing

The most advanced annealing technique used in current state-of-the-art 130nm production manufacturing in 2002 is RTA spike annealing with controlled oxygen to reduce OED (oxidation enhanced diffusion) with beam-line implant energies down to 500eV in decel mode for B_{11} and 2keV for BF_2 requiring <0.3% energy contamination to achieve <22.0nm as-implanted Xj (junction depth) at $1E18/cm^3$ and in drift mode <0.05% energy contamination to achieve 14.3nm as-implanted Xj and abruptness of 3.0nm/decade as shown in figure 4 [15]. Development for 100nm technology is being done at energies as low as 200eV in decel mode for B_{11} and 1keV for BF_2 with 0.3% energy contamination to achieve as-implanted X_j <18.0nm. For

70nm R&D development, drift mode implantation is required with no energy contamination (<0.05%) at 200eV B_{11} is being used to achieve 9.0nm as-implanted X_j value at $1E18/cm^3$ level with 2.1nm/decade abruptness [16]. Using BF_2 higher implant energies are possible as well as better amorphization but there have been some reported concerns with F impurity affecting SiO_2 gate oxide quality [17]. However, use of BF_2 implantation can be extended through optimized oxynitride gate dielectric material [18]. Optimizing high temperature RTA annealing conditions can significantly impact junction depth (X_j). Shishiguchi et al showed the benefits of PAI on reducing TED (transient enhanced diffusion) for B_{11} implants >1keV [9]. Downey et al reported on the influence of controlled oxygen ambient during RTA processing to reduce OED (oxidation enhanced diffusion) [19]. Also, for sub-keV B_{11} implants, high ramp-up and cool-down rates both have been reported to reduce X_j while controlled oxygen showed minimal effects for sub-keV implants as reported by Matsuda et al [20]. However, sub-keV B_{11} implants using PAI of Si or Ge with RTA spike annealing produced deeper junctions than non-PAI. Sub-keV B_{11} implants also show surface dose sputtering limit as the energy is reduced below 500eV and the dose increases above $2E15/cm^2$ as reported by Jacobson et al [21].

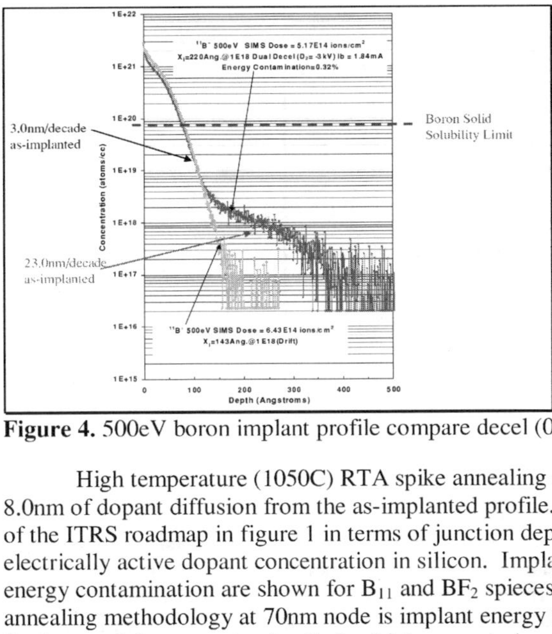

Figure 4. 500eV boron implant profile compare decel (0.3% energy contamination) to drift [15].

High temperature (1050C) RTA spike annealing on the average results in approximately 8.0nm of dopant diffusion from the as-implanted profile. Table 2 shows the direct interpretation of the ITRS roadmap in figure 1 in terms of junction depth Xj and Rs into implant energy and electrically active dopant concentration in silicon. Implant energies for both 0.3% and <0.05% energy contamination are shown for B_{11} and BF_2 spieces. Also, note that the limiter to this annealing methodology at 70nm node is implant energy for 0.3% energy contamination and fundamental dopant activation limit which seems to be at the solid solubility limit of boron in silicon, $<8E19/cm^3$ for a box like dopant profile corresponding to an equilibrium temperature of $<900^oC$ as shown in figure 3. To further reduce dopant diffusion/movement and maintain the as-implanted profile with abrupt junctions super fast ramp-up and ramp-down rate annealing techniques have been proposed called flash annealing where the times are on the msec. (milli-second) duration. Examples include arc-lamp annealing, sub-melt laser annealing and most recently xenon lamp annealing [22,23,24]. Ito et al reported using standard tungsten halogen lamps on the bottom of an RTA system to heat the wafer first up to $\sim 500^oC$ and then using xenon top lamps to achieve flash annealing up to 1050°C in <30msec. Using a 200eV B_{11} implant they achieved X_j=15nm and Rs=3K ohms/sq. (3E19/cm^3)as shown in figure 3 and they stated this is their target for 70nm node devices which is 3.5x higher than the target Rs value stated in the 2001 ITRS [24]. Figure 3 also shows the results from Vortek's arc lamp flash annealing and Verdant's laser non-melt flash annealing techniques achieving Rs values equivalent to 1E20/cm^3 and 8E19/cm^3 respectively. Note that the best high temperature annealing result is only 1E20/cm^3 equivalent to 900°C solid solubility limit of boron in silicon. Therefore, the best high temperature flash annealing technique can satisfy only two out of the four requirements for high-k USJ process integration: 1) shallow & abrupt junctions and 2) high quality junctions and falls short on 3) low Rs and 4) low temperature annealing requirement of <700°C.

Table 2. 2001 ITRS interpretation for high temperature RTA spike anneal assuming 8.0nm diffusion in as-implanted X_j.

	130nm	100nm	70nm	50nm	35nm
Year	01	03	06	10	13
Xj	27-45nm	19-31nm	12-19nm	7-12nm	5-9nm
Xj-8nm	(19-37)	(11-23)	(4-11)	(-1-4)	(-3-1)
Rs	400	550	830	830	940
Dopant Level/ cm3	5E19	8E19	1E20	1.5E20	2E20
Dose Range	0.5-1E15	0.5-1E15	0.5-1E15	BSS	BSS
B11 (0.3% E. C.)	<1keV	<100eV	-	-	-
(No E. C.)	0.7-1.3keV	300-800eV	<300eV	-	-
BF2	1.9-4.8keV	0.2-2.2keV	<200eV	-	-
(no E.C.)	3.5-6.5keV	1.5-4.0keV	<1.5keV	-	-
PLAD	1.2-2.5kV	0.4-1.7kV	<400V	-	-

Laser melt annealing

High temperature laser annealing has been studied for over 25+ years and led to the development of RTA lamp annealing in the early 80's. Laser melt annealing in the 1200°C to 1400°C temperature range has been recently studied again for shallow junction formation. A key element to laser melt annealing is the location of the Ge-PAI layer which defines the melt region and therefore junction depth both vertically and laterally. The Ge-PAI layer is used to define the junction and lower the melting temperature of silicon by 200°C from 1400°C to 1200°C. Upon surface melting, the implanted boron dopant atoms rapidly and uniformly redistributes itself throughout the melted Si-Ge amorphous region at the boron solid solubility limit level at 1400°C (5E20/cm^3). However, one of the issues with this technique is the creation of EOR (end-of-range) damage that are present beyond the amorphous/crystalline interface leading to degradation in junction leakage current in both the vertical as well as in the lateral channel directions. To improve leakage the electrical junction must diffuse beyond the EOR damage as reported by Talwar et al through an additional RTA anneal [25]. Using an additional 800°C 20 seconds RTA anneal after the laser melt annealing step they were able to drive the junction an additional 15.0nm deeper thus improving junction leakage however, this increased the junction depth by 33% from 45.0nm to 60.0nm and also resulted in dopant deactivation (solid solubility limit) effects. Takamura et al showed dopant deactivation in the 500°C to 900°C temperature range occurs within the first 10 sec. of an RTA anneal and can result in 30% to 75% dopant deactivation [25]. Nevertheless, table 3 summaries high temperature laser melt annealing results showing it's ability in achieving the high boron electrically active dopant level of 5E20/cm^3 for sub-25nm technology node but would require boron implant energies <100eV. As stated above, process integration and material compatibility are what will limit the production application of laser melt annealing. Also, post laser melt wafer thermal treatment resulting in dopant deactivation and solid solubility limit degradation in Rs value and the need to improve junction leakage by driving the junction deeper beyond the EOR defects which may negates all the advantages of laser melt annealing. The device level integration issues with laser melting must be overcome before this technique will be accepted in manufacturing and Talwar et al summarized their results with laser-melt annealing at the April 2001 Spring MRS meeting [26]. They include: 1) melting of the silicon under the STI (shallow trench isolation) structure which can result in the STI structure popping out, 2) polysilicon gate stack electrode melting and rounding and 3) optimization of lateral junction profile for gate over-lap and junction leakage control due to the lateral placement of the PAI layer.For high-k gate stack structures laser melting is not an option, therefore, limiting it's potential application to oxide or oxynitride /polysilicon gate stack structures only. For these reasons laser melt annealing has been abandoned by a number of leading logic device manufacturing companies and they have switched their attention to low temperature activation and annealing techniques since only two out of the four high-k gate stack integration shallow junction requirements can be achieved: 1) shallow & abrupt junctions and 2) low Rs and not meeting 3) the low temperature annealing requirement of <700°C and 4) the high quality junction based on low junction leakage current.

Table 3. 2001 ITRS interpretation for high temp. laser melt annealing (Xj=Ge-PAI EOR depth).

	130nm	100nm	70nm	50nm	35nm
Year	01	03	06	10	13
Xj	27-45nm	19-31nm	12-19nm	7-12nm	5-9nm
Rs	400	550	830	830	940
Dopant Level/ cm3	5E19	8E19	1E20	1.5E20	2E20
Dose Range	0.5-1E15	0.5-1E15	0.5-1E15	0.5-1E15	0.5-1E15
Ge-PAI	17-29keV	12-20keV	7-12keV	3-7keV	2-4.5keV
B11 (0.3% E. C.)	0.5-1.3keV	<500eV	-	-	-
(no E.C.)	1-1.7keV	0.6-1.1keV	300-600eV	150-300eV	80-200eV
BF2 (0.3% E.C.)	3-6.5keV	1.8-3.7keV	0.2-1.7keV	<200eV	-
(no E.C.)	5-8.3keV	3-5.5keV	1.5-3keV	0.75-1.5keV	0.4-1keV
PLAD	1.6-3kV	1.2-2keV	0.5-1.2kV	200-600V	100-300V

LOW TEMPERATURE ANNEALING (550°C TO 750°C) BY SPE TECHNIQUE

Low temperature SPE annealing has also been investigated over the last 25+ years but the need for a manufacturing solution when using high-k gate stack structures has created great interest in this technique over the last few year. With optimized SPE processing all four requirements for high-k USJ process integration can be realized: 1) low temperature <700°C annealing (600°C in 3.5 minutes), 2) shallow (8.0nm) & abrupt (<2.0nm/decade) junctions, 3) low Rs (2.5E20/cm^3) and 4) acceptable junction leakage (<20pA/um) for low power CMOS. The main focus on SPE research is optimizing acceptable level of junction quality through PAI optimization. Nevertheless, the top logic manufacturers in the US, Japan, Taiwan and Europe are all actively developing low temperature annealing processes for 70nm technology node driven by low power CMOS devices which will use high-k gate stack structures in order to reduce gate leakage [16]. At temperatures below 750°C dopant diffusion is not observed as reported by Osburn et al. comparing temperatures from 750°C up to 1000°C [27]. Using PAI of Si or Ge eliminates channeling thereby shallower junctions are realized with SPE techniques as shown in figure 5 for 600V and 2kV BF_3 plasma implantation, note this was not the case with high temperature RTA spike annealing as discussed earlier where PAI resulted in deeper junctions for sub-keV implants [9]. Also, abruptness of <2.0nm/dec. can be achieved as shown in figure 5 for 600V plasma implant. SPE recrystalization rates for various temperatures was reported by Jacobson where he observed the rate to be 0.1A/sec at 500°C, 1A/sec at 550°C, 10A/sec at 600°C, 100A/sec at 660°C, 1000A/sec at 730°C and 10,000A/sec at 830°C as shown in figure 6 [28]. Complete dopant activation by SPE at <600°C can be achieved in the 3 to 5 minute time frame for either PAI or non-PAI wafers as reported by Borland & Galewski [29]. Antimony SPE results were also reported by Ponmanev et al showing similar supersaturation levels to boron [30]. Figure 3 shows the low temperature SPE results from various organizations around the world using low temperature SPE annealing techniques showing dopant activation limit at the supersaturation level of 2.5E20/cm^3. Table 4 shows the extendability of SPE to sub-25nm technology node but would require implantation energies for boron below 100eV range in order to realize 5.0nm junctions.

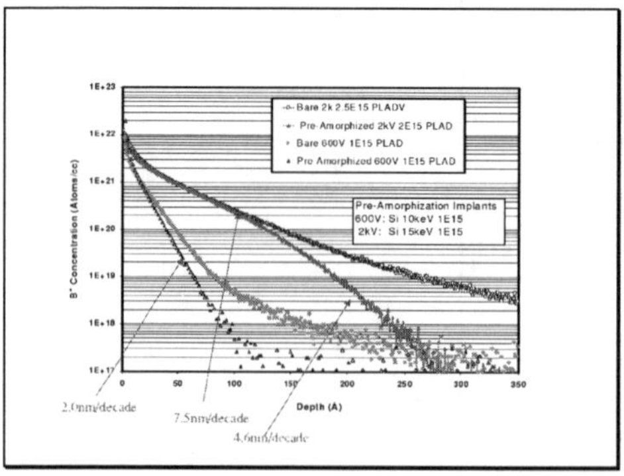

Figure 5. PAI for channeling reduction.

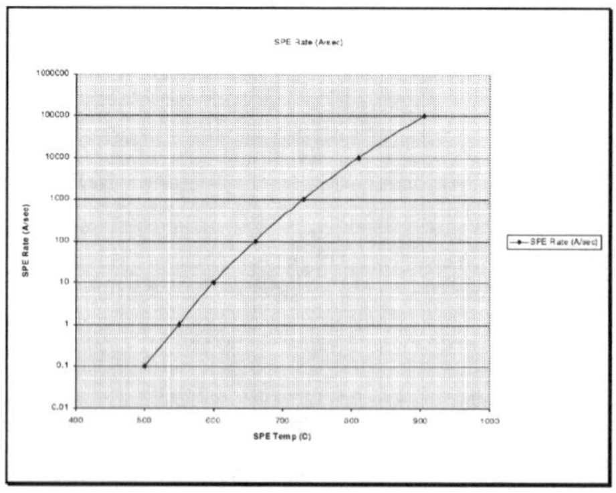

Figure 6. SPE recrystallization rate [28].

Table 4. 2001 ITRS interpretation for low temperature SPE (X_j=as-implanted junction depth).

	130nm	100nm	70nm	50nm	35nm
Year	01	03	06	10	13
Xj	27-45nm	19-31nm	12-19nm	7-12nm	5-9nm
Rs	400	550	830	830	940
Dopant Level/ cm3	5E19	8E19	1E20	1.5E20	2E20
Dose Range	0.5-1E15	0.5-1E15	0.5-1E15	0.5-1E15	5E15
Ge-PAI or	11-21keV	10-14keV	6-10keV	3-6keV	2.5-5keV
Si-PAI	9-16keV	7-10keV	4-7keV	2-4keV	2-3keV
B11 (0.3% E.C.)	0.5-1.3keV	<500eV	-	-	-
(no E.C.)	1-1.7keV	0.6-1.1keV	300-600eV	150-300eV	80-200eV
BF2 (0.3% E.C.)	3-6.5keV	1.8-3.7keV	0.2-1.7keV	<200eV	-
(no E.C.)	5-8.3keV	3-5.5keV	1.5-3keV	0.75-1.5keV	0.4-1keV
PLAD	1.6-3kV	1.2-2keV	0.5-1.2kV	200-600V	100-300V

Control of junction leakage is a major concern but on non-PAI wafers excellent junction leakage can be achieved with low temperature annealing below 600°C but at the expense of Rs as reported by Tsuji et al and Kanemoto et al [31, 32]. With PAI techniques on the other hand, they were able to achieved low Rs but at the expense of degrading junction leakage. To improve junction leakage they both had to add an additional high temperature RTA step similar to that reported earlier by Talwar et al for laser melt annealing. Tsuji's anneal at 900°C pushed the junction an additional 20.0nm. Therefore to overcome the need for the additional higher temperature anneal and dopant diffusion an experiment to look at the relationship between dopant Xj and PAI EOR depth (Xj-EOR) was conducted as illustrated in figure 7 [33]. Since SPE Rs values with Si-PAI was about 20% lower than with Ge-PAI, Si-PAI SPE wafers were used in the study. However, for a 1E15/cm^2 dose, the amorphous interface with Ge-PAI is about 4x smoother compared to Si-PAI [34]. Figure 8 shows results of Rs-vs-junction leakage for both non Si-PAI and Si-PAI wafers. Low Rs values were achieved with PAI. When the junctions remained inside the Si-PAI material, poor junction leakage resulted and as the junction extended beyond the Si-PAI EOR damage leakage improved as illustrated in figure 9. Through further optimization of the PAI process by using other implant species to improve the amorphous interface roughness it is believed further improvement in leakage can be realized. However, the current level of junction leakage with SPE is within the acceptable range for low power CMOS as reported by Gossmann et al and labeled in figure 9 for high performance CMOS at <20nA/um and low power CMOS at <20pA/um [14].

Some companies are developing the disposable spacer process for 100nm technology node and with this technique, low temperature SPE formation for the SDE structure is an option being considered for ease of process integration [16]. Alternatively, elevated S/D structures can also be integrated with low temperature SDE formation for improved salicide contacting.

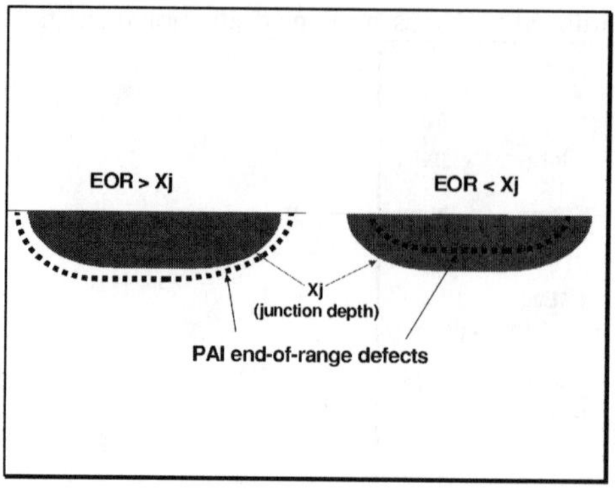

Figure 7. SPE leakage study for Xj-EOR sensitivity.

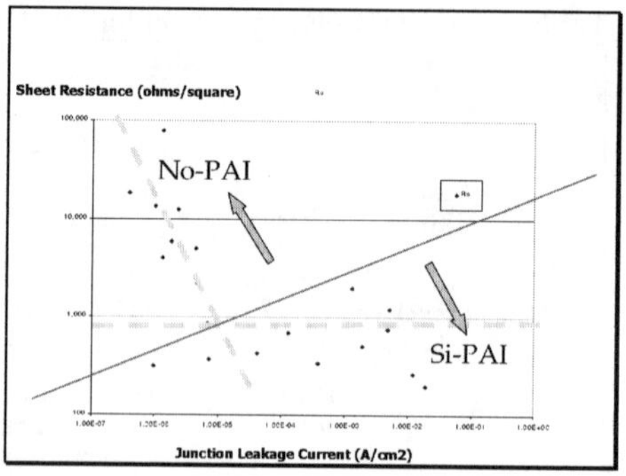

Figure 8. Influence of PAI on Rs-vs-junction leakage.

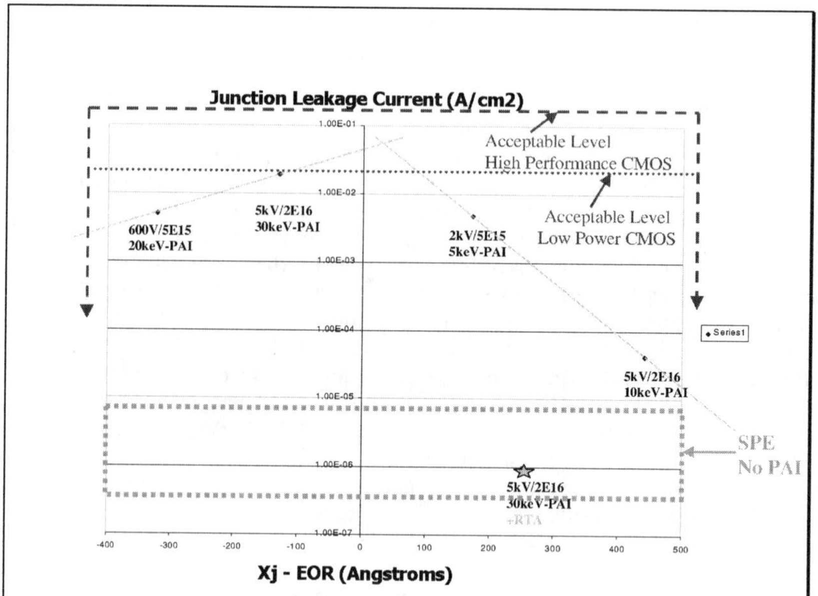

Figure 9. Sensitivity of Xj-EOR on junction leakage current.

SUMMARY

With oxide and oxynitride/polysilicon gate stack structures, high temperature (flash or laser melt) and low temperature annealing techniques can be used for implanted shallow junction formation in the conventional process flow or reverse flow disposable spacer technique for the

70nm technology node requirements and may be extended beyond with laser melt or SPE to 25nm node. With the planned insertion of high-k gates with metal electrodes for low power CMOS starting at 70nm technology node in 2006, new low temperature shallow junction processing techniques will be needed. Using various PAI techniques, shallow (8.0nm) & abrupt (<2.0nm/decade) junctions with low Rs (2.5E20/cm^3) and acceptable quality junction leakage (<20pA/um) have been achieved at 600°C satisfying the 2001 ITRS roadmap requirements for sub-25nm technology node but this would require boron implant energy below 100eV. The process integration constraints imposed with high-k gate stack structures requiring low temperature shallow junction formation makes SPE processing very attractive and the leading candidate for 70nm and beyond USJ formation for both classical and non-classical CMOS using bulk or SOI wafer technology.

ACKNOWLEDGEMENTS

The author a grateful to the support by Sue Felch, Zeiwi Fang and John Koo of VSEA for plasma ion implantation and SPE annealing. Also to Anthony Fiory of New Jersey Institute of Technology for the SPE junction leakage measurements and Hans Grossmann of Agere for the preparation of the wafers used in the SPE leakage study. Additional thanks goes to T. Matsuda and K. Sakamoto of NEC, K. Jones and C. Linford of Univ. of Florida for support in low temperature SPE studies and the many individuals in our industry investigating shallow junction formation techniques for their useful discussions and inputs.

REFERENCES

1. M. Niwa, short course presentation material at IEEE 2000 Symposium on VLSI Technology, June 2000.
2. C. Osburn, North Carolina State Univ., presentation material on replacement gate process at I-SEMATECH gate stack working group meeting June 2000.
3. H. S. Phillip Wong, short course presentation material at IEEE IEDM 1999 on sub-100nm CMOS showing TI's replacement gate results, Dec. 1999.
4. 2001 International Technology Roadmap for Semiconductors (www.public.itrs.net).
5. T. Ghani et al., VLSI Symposium 2000, section 18.1, p.174, June 2000.
6. C. Osburn et al, Jour. of Vac. Science and Technology, vol.18, Jan-Feb. 2000, p.338.
7. S. Narasimha et al, IEEE, IEDM-2001, section 29.2, p.625, Dec. 2001.
8. J. Borland, USJ presentation material at the 4th National Implant Users Group Meeting, Austin, Texas, Oct. 18, 2001.
9. S. Shishiguchi et al., the ECS PV99-10, ECS Conf., May 1999, p.105.
10. C. Osburn, NCSU private communications.
11. B. Murto, presentation material at I-SEMATECH source drain meeting March 2000.
12. J. Borland, Semiconductor International, April 2001, p.70.
13. S. Kim et al., IEEE, IEDM 2000, section 31.3, p.723, Dec. 2000.
14. H. Gossmann et al., Material Research Society, vol. 610, Spring MRS Meeting April 2000, p. B1.2.1..
15. VSEA internal results for 500eV implantation.
16. J. Borland, private communications with customers on 70nm & 100nm targets.

17. K.Bourdelle et. al, IIT-2000, Sept. 2000, p.25, Alpbach, Austria.
18. A. Jain of TI, discussions at the I-SEMATECH source drain engineering working group meeting Nov.8, 2001.
19. D. Downey et al, the Electrochemical Society, Spring Meeting May 1999, PV 99-10, p.151.
20. T. Matsuda et al, extended abstract of the International Workshop on Junction Technology, section 2-3, Dec. 2000, p.29, Chiba, Japan.
21. D. Jacobson et al, IIT-2000, Sept. 2000, p.300, Alpbach, Austria.
22. J. Gelpy of Vortek, presentation material at the I-SEMATECH source drain engineering working group meeting Nov.8, 2001.
23. S.Felch et al, joint VSEA/Verdant patent filing on sub-melt laser annealing, April 2000.
24. T.Ito et al, extended abstract of the 2001 SSDM conference, section A-5-3, p.182, Sept. 2001.
25. Y. Takamura et al., Material Research Society, vol. 669, Spring MRS Meeting April 2001, p. J7.3.1.
26. S. Talwar, et al., presentation material on laser melt annealing at the Spring MRS 2001 meeting, April 2001.
27. C. Osburn et al, IIT-96, p.607, June 1996, Austin, Texas.
28. D. Jacobson, private communications.
29. J. Borland & C. Galewski, IIT-98, p.1211, June 1998, Kyoto, Japan.
30. Y. Ponmanev et al., IEEE 2001 Symposium on VLSI Technology, section 4A-1, p33, June 2001.
31. K. Tsuji, et al., IEEE 1999 Symposium on VLSI Technology, section 2-1, June 1999.
32. K. Kanemoto, et al., extended abstract of the 2000 SSDM conference, section A-7-4, Aug. 2000, p. 406.
33. J. Borland et al., new SPE patent filing Nov. 2001.
34. A. Al-Bayati, et al, IIT-2000, Sept. 2000, p.54, Alpbach, Austria.

Mat. Res. Soc. Symp. Proc. Vol. 717 © 2002 Materials Research Society

Electromagnetic Induction Heating for the 70 nm Node

Keith Thompson, John H. Booske, R.F. Cooper, Y.B. Gianchandani[1]
University of Wisconsin, Madison 53706
[1] University of Michigan, Ann Arbor

ABSTRACT

Electromagnetic heating provides a novel alternative to "illumination-based" rapid thermal processing techniques. Exposure to radiation, in the RF and microwave frequency regimes, rapidly heats silicon (~125°C/sec) to temperatures in excess of 1000°C without the use of a susceptor. These ramp rates make this technology suitable for the activation of shallow implanted dopants, and satisfaction of the 100 nm technology node has been achieved. Furthermore, the presence of high frequency electric fields creates ponderomotive forces that may alter the kinetics of dopant activation and diffusion. These additional driving forces could, once fully understood, lead to an enhanced activation mechanism that activates sufficient dopants to satisfy the 70 nm technology node at temperatures less than 1000°C.

INTRODUCTION

The 70 nm CMOS technology node requires source/drain regions as shallow as 25 nm with sheet resistances of 500 ohms/square. Optimal operation of these devices occurs when this profile is as "box-shaped" as possible, preferably with a surface concentration greater than $10^{20}/cm^3$ and a profile slope greater than 5 nm/decade. [1] Obtaining these junctions requires sophisticated rapid thermal activation techniques that restrict the thermal budget, limiting the diffusion of the implanted species, while activating a significant percentage of the dopants. "Conventional" spike anneal technologies (ramp rates of 425°C/sec in a low ppm oxygen ambient), currently satisfy the 100 nm technology node [2], but may have reached their limit. Techniques that can achieve higher levels of activation, relative to conventional techniques, need to be investigated if the 70 nm technology node is to be realized.

A novel technique, electromagnetic induction heating (EMIH) [3], may provide a solution to the 70 nm node. EMIH utilizes RF and microwave radiation (1 MHz – 10 GHz) to rapidly (125°C/sec) heat silicon to temperatures greater than 1000°C. In addition to heating the silicon, the presence of these high frequency electric fields drives additional ponderomotive forces within the silicon lattice [4] that may enhance dopant activation without causing increased diffusion.

EXPERIMENTAL APPARATUS

The numerous and complex requirements for rapidly heating silicon with microwave or radio frequency radiation are outlined in detail in a previous publication [5]. For all of the experiments contained within this paper, the silicon was heated using the resonant cavity shown in Fig. 1. The diameter of the cavity was 17 cm and the height was left adjustable to tune in specific modes. A radial tuning stub helped minimize reflected power and up to 3000 watts could be supplied from a fixed frequency (2450 MHz) microwave source. Although several resonant modes were available, the dominant TM111 and TM011 modes were primarily used. The magnetic field patterns of the mode determined the current distribution through the wafer, which was supported by a hollow quartz cylinder positioned 1 mm above the cavity bottom to ensure that the wafer was located in a magnetic field maximum.

EXPERIMENTAL RESULTS

Heating rates of 125°C/sec up to and beyond 1000°C were achieved. A typical temperature transient is shown in Fig. 2. A semilog plot of steady state temperature and power is shown in Fig. 3. As expected, the steady state temperature is dependent on the fourth root of the input power, indicating a loss mechanism dominated by radiation. This excellent fit between the expected loss mechanism and the input power indicates significant power absorption efficiency throughout the temperature range.

The most important issue with any rapid heating technique is the overall temperature uniformity. In electromagnetic heating, the magnetic flux pattern, dictated by the specific mode, determines the heating pattern across the wafer surface. As a consequence, a non-uniform mode pattern directly leads to a non-uniform heating pattern. This dependence can be seen in Fig. 4, which shows a picture of the

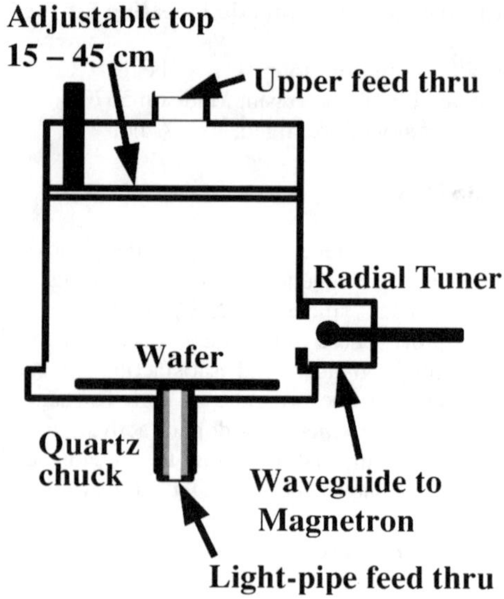

Fig. 1. Resonant cavity used for microwave (2.45 GHz) heating. The silicon wafer was placed at the bottom of the cavity, where the transverse components of the magnetic fields and the normal component of the electric field were strongest.

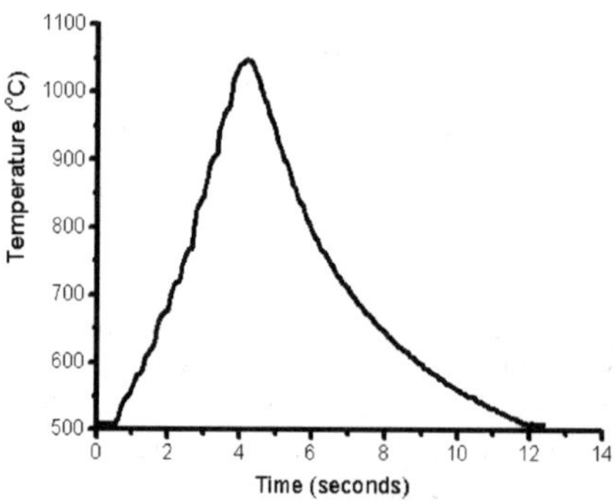

Fig. 2. A typical temperature transient for EMIH heating.

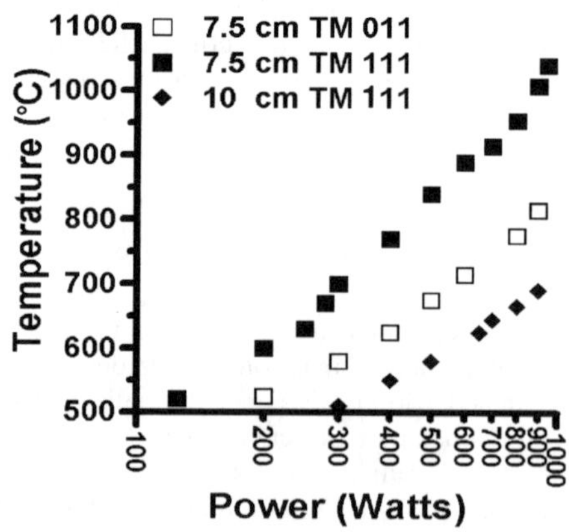

Fig. 3. Semilog plot of temperature and power.

wafer being heated at 1000 watts in the TM111 mode. The temperature at different points in the wafer, Fig. 5, was deduced by monitoring solid-state reactions as detailed in a previous publication [3]. Rotating the wafer can average out much of this mode dependence. An improvement in cross-wafer uniformity from 12% (no rotation) to 3.6% (rotating at 120 rpm) was achieved and these results are also shown in Fig. 5.

While spinning the wafer has achieved an azimuthally symmetric heating profile, a center hot radial temperature gradient is still present. It is expected that this can be removed through the introduction of a second resonant mode. The heating pattern of the TM011 mode, Fig. 4, is edge hot, center cool; exactly opposite to that seen when spinning the wafer in the TM111 mode. It is speculated that the simultaneous application of both of these modes would result in a very uniform heating profile across the entire wafer. Furthermore, adjusting the intensity of one mode with respect to the other would allow the uniformity profile to be optimized.

The main drawback to this solution is the design complexity. A second power source, whose frequency was such that both modes were simultaneously in resonance with the cavity, would be required. This would necessitate a complex matching scheme that adjusted both the height and effective radius of the cavity on a time scale fast enough to guarantee dual resonance during the critical temperature range of the spike anneal process. While constructing such a system is plausible, the complexity makes it somewhat less appealing for commercial use.

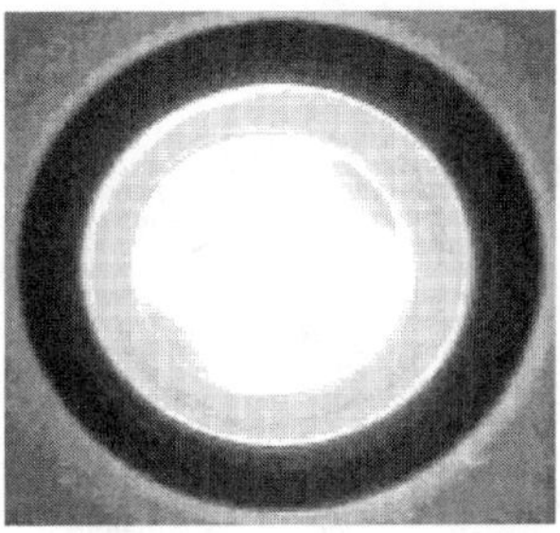

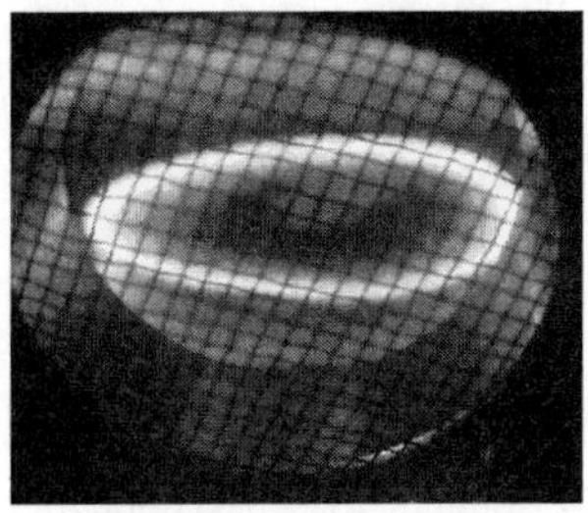
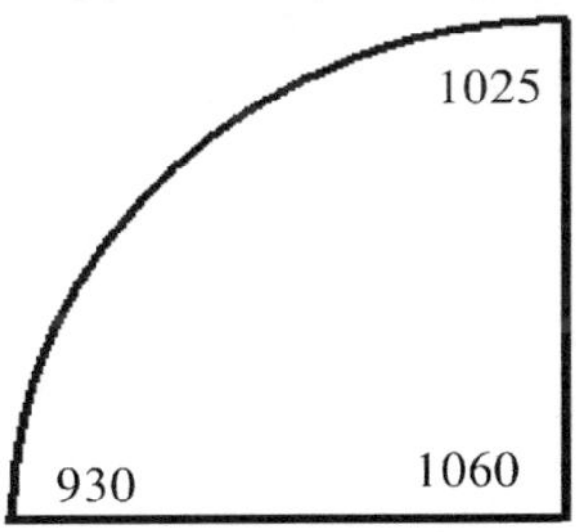

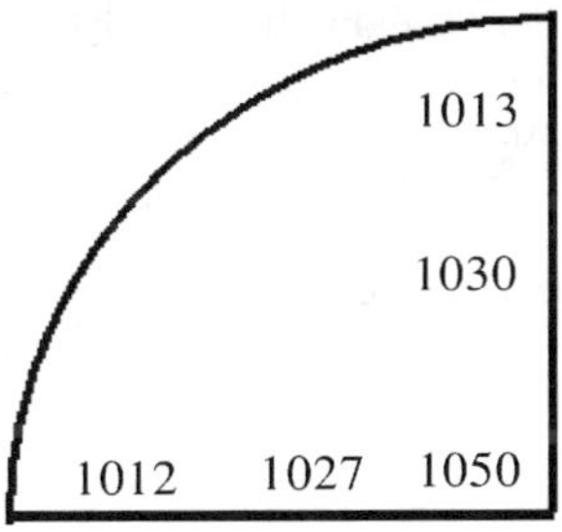

Fig. 4. Pictures of actual heating patterns for the TM111 mode (top left) TM111 mode during rotation at 120 rpm (top middle) and the TM011 mode (top right).

Fig. 5. The temperature uniformity was evaluated qualitatively through solid-state reaction mechanisms [3]. The temperature non-uniformity falls from 12% to 3.6% when the wafer is rotated. Further improvement may be possible through the simultaneous use of the TM111 and TM011 modes, although it would require a significant increase in equipment complexity.

An alternative solution to the uniformity issue would be the introduction of a fully multi-moded cavity. Such a system would involve a single power source launching hundreds of unique modes into the cavity at the same time. Because no specific mode would be resonant, the mode patterns would average out across the wafer, resulting in an extremely uniform heating pattern. To achieve this high number of modes, the cavity dimensions would need to be much larger than the wavelength of the frequency employed, a requirement that fits well with the current scale-up to 300 mm wafers. For example at 2450 MHz (wavelength of 125 mm) a system 500 mm in length would be 4 times the wavelength of the radiation. Moving to a higher frequency or extending the length of the cavity would continue to increase the number of modes generated within the cavity. The main drawback to this technique is the loss in power efficiency. Because none of the modes are resonant, the stored circulating power density within the chamber will be lower than the single mode resonant case. Recent experiments [6] indicate that the feed power would therefore need to be as much as 30 times higher than the single mode resonant approach.

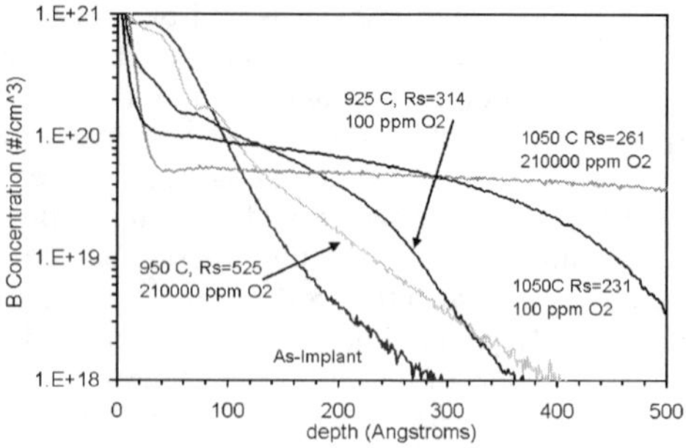

Fig. 6. 4500 eV BF_2 anneal results

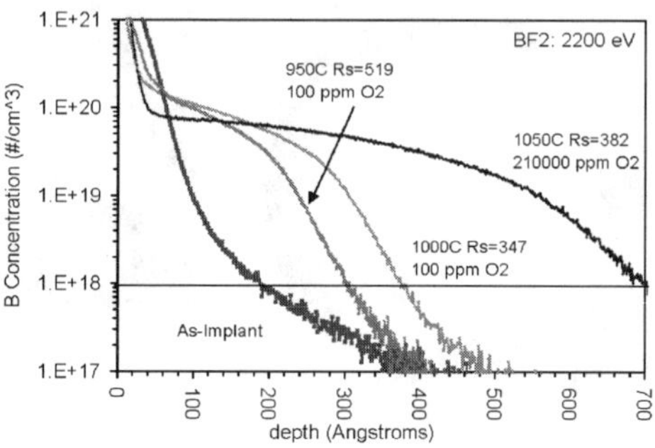

Fig. 7. 2200 eV BF_2 anneal results

USJ RESULTS

The effectiveness of microwave heating for ultra-shallow junction formation was evaluated by spike annealing several shallow implanted samples. The implant conditions of the samples shown were BF_2 at 4500 eV, Fig. 6, BF_2 at 2200 eV, Fig. 7, and BF_3 at 200 eV, Fig. 8. The BF_2 species were beam-line implanted, and the BF_3 species were implanted by plasma doping. All samples had a dose of $10^{15}/cm^2$. It has been shown for lamp based RTP [2,7] that reducing the oxygen partial pressure of the ambient during the anneal process limits the diffusion of boron with an optimal anneal condition occurring at 100 ppm oxygen in a nitrogen purge. As a consequence, the microwave anneal experiments were split between an uncontrolled ambient, ~ 21% O_2, and an ambient that was purged with N_2 to a controlled oxygen concentration of 100 ppm.

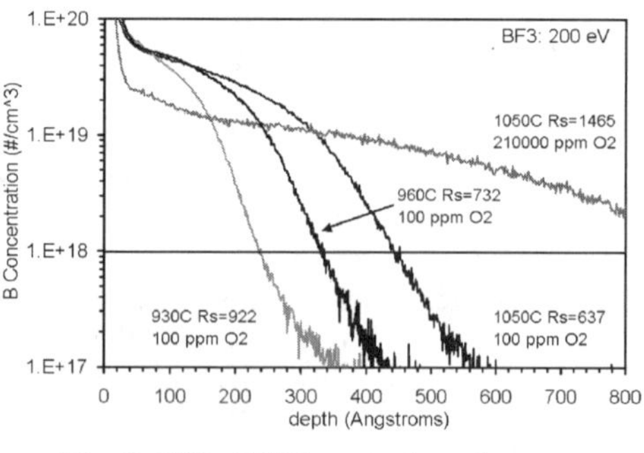

Fig. 8. 200 eV BF_2 anneal results

The 4500 eV anneals performed in a 21% O_2 ambient show significant diffusion at 1050°C. The 100 ppm anneal at 1050°C showed less diffusion, as expected, with an improved Rs. At 950°C, only moderate diffusion is seen, but instead of becoming "box-shaped" the profile tails out. Annealing in a controlled O_2 concentration, at a slightly lower temperature, the profile becomes "box-shaped" and the Rs drops considerably.

These differences are directly attributed to the oxidation of the silicon during the anneal process. The largest fraction of the implanted dose lies in the extreme near surface, top 5 nm, of the silicon, most of which is well above the solid solubility limit for boron in silicon. For superior junctions to form, this dose must diffuse into regions where the boron concentration is below the solubility limit so that it can become substitutional. When annealing in an oxidizing ambient, most of this dose is trapped in the oxide layer formed during the anneal. However, holding the O_2 concentration at 100 ppm limits the oxidation and frees the boron to diffuse into the silicon. If the thermal budget is adequately restricted, the boron will diffuse from this super-saturated surface region into the next 20 nm of silicon without being driven deep into the wafer. This concentration enhanced diffusion accounts for the formation of the "box-like" profile and much lower Rs achieved during annealing in a 100 ppm O_2 ambient. Furthermore, the removal of oxidation enhanced diffusion effects [7] reduces the tailing seen at the end of the profile, further restricting the movement of the junction.

The 2200 eV anneals, Fig. 7, also show considerable diffusion when performed in an uncontrolled ambient. Controlling the O_2 partial pressure and reducing the temperature again limited the diffusion and improved the Rs. For example, the 950°C anneal had an Rs of 519 ohms/square and a junction depth of 30 nm, placing this data point well within the technology node requirements for 100 nm devices.

The effect of O_2 concentration is illustrated yet again during the 200 eV BF_3 anneals, Fig. 8. At 1050°C, the junction depth and Rs are reduced by 50 nm and 828 ohms/square respectively. Anneals performed at subsequently lower temperatures result in similar "box-like" profiles, slopes of ~ 6 nm/decade, with consistently shallower junction depths and higher sheet resistances. These points also form a curve that satisfies the 100 nm technology node as defined by SEMATECH [1].

CONCLUSION

These results clearly indicate that electromagnetic annealing, a novel RTP technique for ultra shallow junction formation, satisfies the 100 nm technology node. The optimal anneal temperature of 950°C is lower than the 1070°C reported by SEMATECH [8]. While the anneals reported here were performed at an oxygen concentration of 100 ppm, the optimal process point for lamp-based RTP [2], the optimal oxygen concentration may be somewhat different for electromagnetic annealing. Adjusting this process parameter may further improve the USJ results by lowering the process temperature, restricting diffusion, or enhancing the activation. The heating rate used in these experiments was 125°C/sec. Increasing this rate to above 500°C/sec should further restrict junction diffusion. It is hoped that improvements in these areas, as well as future research and development of this technique, will provide the necessary technology to satisfy the 70 nm technology node at temperatures that do not exceed 900°C. Finally, the temperature uniformity during the heating process has been considerably improved over previously reported results [3], and alternative cavity designs or multi-mode techniques should allow for further reductions in temperature non-uniformity.

ACKNOWLEDGMENTS

The authors would like to acknowledge Dan Downey and Edwin Arevalo of Varian Semiconductor Equipment Associates for their generous help with materials and analysis. We would also like to thank Robin Tichy at SEMATECH and Steve McCoy at Vortek for helping us run a suite of comparison experiments. Finally, funding for this research was supplied by the Wisconsin Alumni Research Foundation.

REFERENCES

1. www.sematech.org
2. W. Lerch, B. Bayha, D.F. Downey, E.A. Arevalo, "State of the art techniques for Ultra-Shallow junction Formation," in Rapid Thermal and Other Short-Time Processing Technologies II, p. 321 of Proceedings of the 199th meeting of the Electrochemical Society, 25 March 2001, Washington DC.
3. K. Thompson, J. H. Booske, R.F. Cooper, Y.B. Gianchandani, D. F. Downey, "RF and Microwave Annealing for Ultra Shallow Junction Formation," in Rapid Thermal and Other Short-Time Processing Technologies II, p. 121 of Proceedings of the 199th Meeting of the Electrochemical Society, 25 March 2001, Washington DC.
4. J.H. Booske, R.F. Cooper, S.A. Freeman, K. I. Rybakov, V.E. Semenov, "Microwave ponderomotive forces in solid-state ionic plasmas," Phys. Rev. B, vol. 55 p. 3559-3567. (1997)
5. K. Thompson, Y.B. Gianchandani, J.H. Booske, R.F. Cooper, "Direct Si-Si Bonding by Electromagnetic Induction Heating," Journal of Micro-Electro-Mechanical Systems. Accepted for publication. (April 2002)
6. K. Thompson, J.H. Booske, Y.B. Gianchandani, R.F. Cooper, Y. Bykov, A. Eremeev, I. Plotnikov, "Electromagnetic Induction Heating for Cold Wall Rapid Thermal Processing," in Proceedings of the 9th International Conference on RTP, IEEE, September 2001, Anchorage Alaska.
7. D.F. Downey, S. W. Falk, S.D. Marcus, Proc. of 7TH International Conference on Advanced Thermal Processing of Semiconductors. pp. 229-236 (1999)
8 R.S. Tichy, K. Elliott, S. McCoy, D.C. Sing, "Annealing of Ultra-Shallow Implanted Junctions Using Arc-Lamp Technology: Achieving the 90 nm Node," 9th International Conference on Advanced Thermal Processing of Semiconductors, IEEE RTP 2001, p. 82, Anchorage, Alaska. (25-29 Sept. 2001)

Mat. Res. Soc. Symp. Proc. Vol. 717 © 2002 Materials Research Society C1.4

Study of the Effects of a Two-Step Anneal on the End of Range Defects in Silicon

Renata A. Camillo-Castillo[a], Kevin. S. Jones[a], Mark E. Law[b], Leonard M. Rubin[c]
[a]Department of Materials Science and Engineering, University of Florida, Gainesville, FL
[b]Department of Electrical Engineering, University of Florida, Gainesville, FL
[c]Axcelis Techologies, Beverly , MA

ABSTRACT

Transient enhanced diffusion (TED) is a challenge that the semi-conductor industry has been faced with for more than two decades. Numerous investigations have been conducted to better understand the mechanisms that govern this phenomenon, so that scale down can be acheived. {311} type defects and dislocation loops are known interstitial sources that drive TED and dopants such as B utilize these interstitials to diffuse throughout the Si lattice. It has been reported that a two-step anneal on Ge preamorphized Si with ultra-low energy B implants has resulted in shallower junction depths. This study examines whether the pre-anneal step has a measurable effect on the end of range defects. Si wafers were preamorphized with Ge at 10, 12, 15, 20 and 30keV at a dose of $1x10^{15}cm^{-2}$ and subsequently implanted with $1x10^{15}cm^{-2}$ 1keV B. Furnace anneals were performed at 450, 550, 650 and 750°C; the samples were then subjected to a spike RTA at 950°C. The implant damage was analyzed using Quantitative Transmission Electron Microscopy (QTEM). At the low energy Ge preamorphization, little damage is observed. However at the higher energies the microstructure is populated with extended defects. The defects evolve into elongated loops as the preanneal temperature increases. Both the extended defect density and the trapped interstitial concentration peak at a preanneal temperature of 550°C, suggesting that this may be an optimal condition for trapping interstitials.

INTRODUCTION

Post implantation thermal annealing is commonly used to repair the lattice damage from the ion implantation process as well as to electrically activate the implanted dopant species. During these thermal anneals B diffusion has been observed[1] to be strongly enhanced; however the magnitude of this enhanced diffusion decreases with time and is known as transient enhanced diffusion (TED). TED arises from the coupling of the substitutional dopant with excess Si interstitials from the implantation process to form a mobile complex[2]. The increased interstitial concentration in the ion-implanted region shifts the local equilibrium between substitutional and mobile dopant[2] resulting in the observed enhanced dopant diffusivity.

Surface preamorphization is widely used prior to dopant implantation to achieve ultra shallow junction depths. Ge has gained much acceptance as an amorphization species because it has the advantage of inducing greater disorder in the crystal structure at lower doses than Si, owing largely to its bigger atomic structure. Complete removal of damage induced by Ge amorphization is possible and has been achieved upon annealing[3,4]. However, low energy Ge preamorphization alone has not been very effective in preventing the diffusion of excess Si interstitials under certain conditions, thereby inducing TED upon dopant activation [5]. For amorphizing implants, dislocation loop formation has been attributed to the primary defects that reside in the silicon structure near the amorphous/crystalline interface[6] known as the End of Range (EOR) defects. The thermal evolution of the EOR defects has been extensively studied[7-9] and dislocation loops are known to evolve from the dissolution or unfaulting of {311} type defects. Deep Level Transient Spectroscopy (DLTS) analysis [10-12] has confirmed the existence of submicroscopic interstitial clusters (SMIC) which are believed to act as precursors for {311} defect formation[13,14]. EOR defects are known to act both as a source and a sink for excess Si interstitials and can therefore be a driving force for TED. The goal of this study is to determine the effect of a two step anneal on the evolution of the EOR defects for an amorphizing implant. The preanneal temperatures investigated are between 450 and 750°C and the second stage anneal is a spike RTA at 950°C.

EXPERIMENTAL DETAILS

200 mm <100> n-type Czochralski grown wafers were implanted using a GSD implanter. The wafers were preamorphized with Ge^+ at energies 10, 12, 15, 20 and 30keV respectively, at a dose of $1x10^{15}cm^{-2}$ in order to create a continuous amorphous layer. Subsequently the amorphized wafers were implanted with 1keV B^+ at a dose of $1x10^{15}cm^{-2}$. The tilt/twist angles for each implant were 5°/0°. Post implantation annealing was performed on the samples which comprised two stages. The samples were initially subjected to pre-anneals for 300 seconds at 450, 500, 550, 600, 650 and 750°C in a tube furnace in a dry N_2 ambient. Subsequently the samples underwent a Rapid Thermal Anneal (RTA) in which the temperature was ramped up to 650°C and maintained at that temperature for 10 seconds, they were then spiked at 950°C. The RTA was conducted in an AG Associate 210T Rapid Thermal Annealer. Plan-view Transmission Electron Microscopy (PTEM) was utilized to analyze the EOR defects and cross-sectional TEM (XTEM) was used to confirm the amorphous layer depth via a JEOL 200CX. Plan-view images were taken at g_{220} weak beam dark field[16,17] diffraction condition and cross-sectional view images were taken at g_{011} bright field diffraction condition. The amorphous layer depth was measured using a multi-wavelength ellipsometer at a fixed angle of 75°C. The defect data was analyzed using a quantification technique of Bharatan et. al[18].

RESULTS AND DISCUSSION

Ge preamorphizing implants in the low energy regime have been investigated to determine the effect that a two step anneal has on the defect evolution. The energies of the Ge implant are 10, 12, 15, 20 and 30keV. Figure 1 is a plot of the amorphous layer depth as a function of the preamorphization energy after the 300s preanneal. As the energy increases the amorphous layer depth is seen to increase and for a 30keV Ge implant the amorphous layer depth is approximately 492Å. At the lowest Ge implant energy of 10keV the amorphous layer depth is approximately 230Å. Complete regrowth of the amorphous layer is achieved for the 650 and 750°C preanneals, but at the lower temperatures the amorphous layer does not fully regrow. Application of the spike RTA at 950°C results in complete regrowth of the amorphous layer.

Recently studies have been conducted on the defect evolution of low energy Ge implants in Si. The results of these experiments indicate at 30keV implant energy and a dose of $1x10^{15}cm^{-2}$

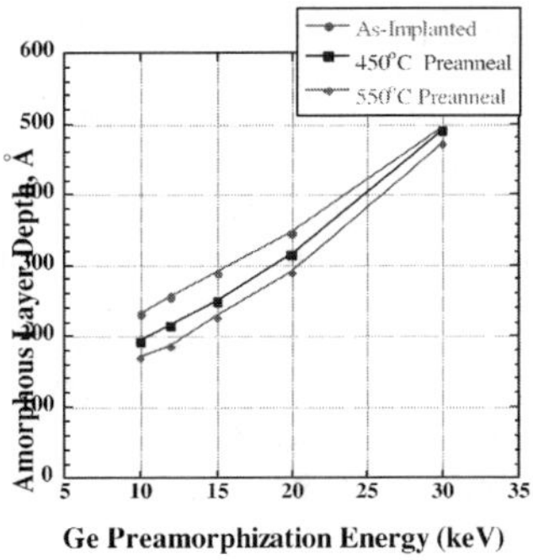

Figure 1. Amorphous layer depth versus Ge preamorphization energy indicating the regrowth after preanneals at 450 and 550°C for 300s.

the time constant for {311} dissolution at 750°C is 40 minutes[5]. In accordance with these kinetics, any {311}'s formed at the lower preanneal temperatures should undergo complete dissolution on application of a spike RTA at 950°C. TEM analysis indicates that {311} type defects are not present and dislocation loops are in fact the predominant defect for the entire range of energies and temperatures investigated. In the lower Ge preamorphization energy regime from 10 to 15keV the dislocation loops observed are small and the loop densities low. In particular at 10 and 12keV, the very low loop densities suggest that the TEM resolution limit has been exceeded. At 20 and 30keV Ge preamorphization energies however, larger dislocation loops are seen and the microstructure is populated with both small and elongated loops.

Figure 2 depicts the results of the two step anneal study of the EOR defect density for various energies of the preamorphization Ge implant. Visibly there are two distinct effects, the preanneal temperature and the Ge preamorphization energy. The former will now be discussed. If the

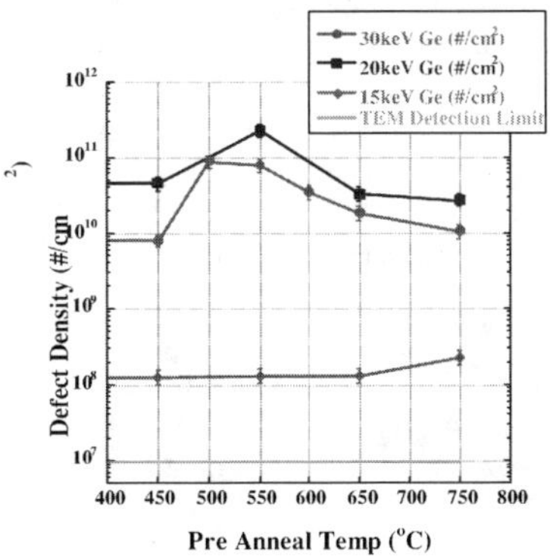

Figure 2. Defect Density versus preanneal temperature for samples which underwent 15, 20 and 30keV Ge preamorphization energies

formation of the SMIC is viewed as the homogenous nucleation of precipitates then the number and size of the defects are controlled by the super-saturation of the excess interstitials[19], consistent with Volmer and Weber's theory[15]. Using inverse modeling of the SMIC super-saturation derived from B TED, Cowern et.al[20] found that interstitial supersaturation as a function of time decreases with increasing temperature. Hence in the lower temperature regime there is a higher super-saturation of interstitials in the structure and therefore a larger nucleating flux. In accordance with these findings the defect density should be smaller at the higher temperatures. For preamorphization energies greater than 15keV, the defect density is seen to decrease for temperatures above 550°C. However, for temperatures between 450 and 550°C the density increases. Figure 3 demonstrates the evolution of the defects for 30keV Ge energy preamorphizations. As the temperature of the preanneal increases above 550°C the loops become larger and elongated. It is however clear that this does not hold true when the preanneal temperature increases from 450 to 550°C, as smaller dislocation loops are seen at the higher temperature. A similar trend is observed for the 20keV Ge preamorphization. The apparent peak in the defect density may be accounted for by the fact that there are two opposing effects occurring, the nucleation and dissolution of defects. The peak suggests that at the 550°C pre-anneal temperature optimum conditions may exist for trapping interstitials, as the number of defects and the number of trapped interstitials is a maximum. The TEM micrographs of the defects at 550°C indicate a high density of small loops. Similarly at 500°C the small loop density is high; however at 450°C the density of small loops is obviously less. The lower defect density at 450°C can be attributed to the rate of formation of the SMIC. As temperature is decreased the rate of formation is lower and the duration of the pre-anneal may have been insufficient for nucleation of the SMIC which will account for the lower loop density. We can also argue that there was sufficient time for nucleation; however the duration of time spent in the growth regime was inadequate to allow clusters large enough for TEM detection.

Comparison with the trapped interstitial concentration in Figure 4 indicates the trapped interstitial population unmistakably adheres to a similar trend to the defect density. At Ge preamorphization energies greater than 15keV, the trapped interstitial population is highest at 550 to 600°C. Simultaneously the defect density decreases at 600°C which indicates that the loops are in the coarsening regime and are undergoing conservative Oswald ripening, in which the larger loops grow at the expense of the smaller ones. Subsequently the trapped interstitial population decreases at 650°C where it appears to remain approximately constant for higher temperatures. This decrease in the trapped interstitial population observed at the higher temperatures may be accounted for by a decrease in the super-saturation of interstitials and hence a smaller nucleating flux. The effect of the preanneal is visible when the defect densities and the

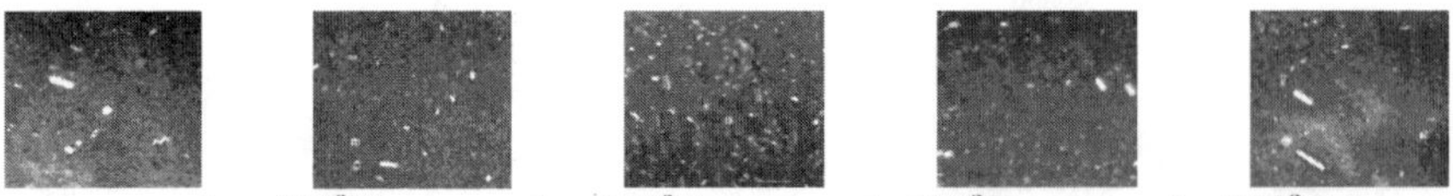

Figure 3: TEM pictures of the defect density for samples subjected to a 30keV Ge preamorphization and preanneals of 450, 550, 650 and 750°C and a spike 950°C RTA .

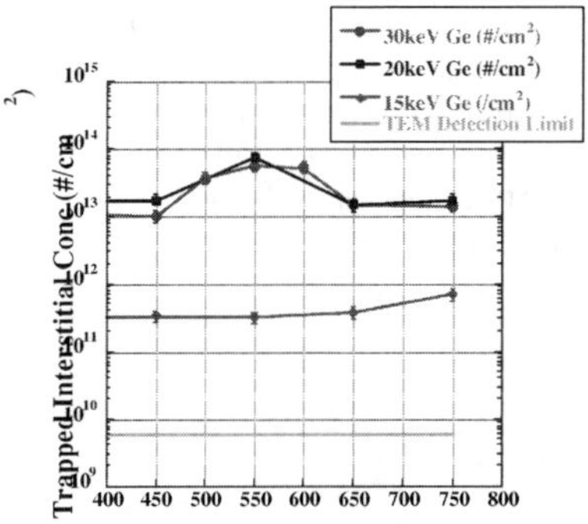

Figure 4. Trapped Interstitial Concentration versus preanneal temperature for 15, 20 and 30keV Ge preamorphization energy

trapped interstitial concentrations are compared to results obtained for a 950°C spike RTA only. The preanneal noticeably results in a greater defect density and trapped interstitial population for the temperature range 500 to 600°C. Within a 20% error the preanneal did not have a significant effect for the other temperatures investigated. It is worth noting that the defects observed for a 750°C preanneal are geometrically very similar to those for the spike RTA only.

The effect of the preamorphization energy is evident in the significant difference in the profiles observed at 20 and 30keV and that at 15keV Ge preamorphization energies. The profiles observed for the 20 and 30keV Ge preamorphization energies are remarkably similar as previously discussed. For amorphizing implants the number of interstitials in the tail region of the profile includes those recoils from the projected range in addition to the plus one concentration. As the energy of the amorphizing implant increases the number of recoils into the tail of the profile increases. Therefore we can expect that the number of trapped interstitials in the end of range to be greater. This is clearly not the case for the 20 and 30keV Ge preamorphized samples and may be indicative of a contribution from a second source of interstitials. The profiles observed at the lower preamorphization energy of 15keV indicate considerably lower defect densities and trapped interstitial concentrations. This could be due largely to the fact that there are less recoils at lower energies and hence less damage in the structure. The preanneal does not appear to have any effect on the defect density, which remains approximately constant as the preanneal temperature increases. The slight increase in the defect densities and trapped interstitial concentrations visible at 750°C may not be a true artifact of the increase in temperature and should be attributed to noise. The data illustrates that there may be an energy threshold at which the preanneal has a significant effect on the defect dynamics.

CONCLUSIONS

In conclusion, there are two distinct effects on the defect density and the trapped interstitial concentration - the preanneal temperature and the Ge preamorphization energy. The preanneal appears to have an effect for the temperature range 500 to 600°C, indicated by the larger defect densities and trapped interstitial concentrations observed over those of a spike RTA at 950°C only. Secondly there may be an optimal condition for trapping interstitials at which both the defect density and trapped interstitial concentration are maximum. This optimal condition occurs at 550°C. The observation of these preanneal effects appear to be dependent on the energy of the Ge preamorphization, and the results indicate that there may be an energy threshold above which the preanneal has an effect.

REFERENCES

1. E. Lampin, V. Senez, A. Claverie, J Appl. Phys. 1999, 85 (12): 8137
2. P.M. Fahey, P. B. Griffin, J.D. Plummer, Re. Mod. Phys. 1989, 61: 289
3. M. C. Ozturk, J.J. Wortman, C.M. Osburn, A.Ajmera, G.A. Rozgonyi, E. Frey, W.Chu, C. Lee, IEEE Transactions on Electron Devices. 1988, 35 (5): 659-667
4. A.C. Ajmera, G.A. Rozgonyi, Appl. Phys. Lett. 1986, 49(19): 1269-1271
5. A. F. Gutierrez, Master's Thesis, Materials Science and Engineering, University of Florida. 2001,16
6. R. Drosd and J. Washburn, J. Appl. Phys. 1982, 53(1) : 397-403
7. D. J. Eaglesham, P.A. Stolk, H.J. Grossman, J.M. Poate, Appl. Phy. Lett. 1994, 65 (18) :2305
8. C. Bonafos, D. Mathiot and A. Claverie, J. Appl. Phys. 1998, 83 : 3008
9. K.S. Jones, L.H. Zhang, V. Krishnamoorthy, M. Law, D.S. Simons, P. Chi, L. Rubin and R.G. Elliman, Appl. Phy. Lett. 1996, 68 : 2672
10. S. Libertino, S. Coffa, J.L. Benton, K. Halliburton, D.J. Eaglesham, Nuc. Ins. Met. Phys. Res. B, 1999, 148: 247
11. S. Coffa, S. Libertino, C. Spinella, Appl. Phys. Lett. 2000, 76(3): 321
12. J.L. Benton, K. Halliburton, S. Libertino, D.J. Eaglesham, J. Appl. Phys. 1998, 84(9): 4749
13. L.H. Zhang, K.S. Jones, P.H.Chi, D.S.Simmons, Appl. Phy. Lett.1995, 67: 2025
14. J.L. Benton, S. Libertino, D.J. Eaglesham, S. Coffa, Mat. Res. Soc. Symp. Proc., 1997, 469 : 193
15. M. Volmer and A. Weber, J. Phys. Chem. 1926, 119: 227
16. G. Thomas, M.J. Goringe,Transmission Electron Microscopy of Materials, John Wiley & Sons. 1979
17. M.H. Loretto, Electron Beam Analysis of Materials, Chapman & Hall, London. 1984
18. S. Bharatan, J. Desrouches, K.S. Jones, Materials and Process Characterization of Ion Implantation, Ion Beam Press. 1997. Chp.4: 222-243
19. C. Jasper, A. Hoover, K.S. Jones, Appl. Phy. Lett. 1999, 79(17): 2629
20. N.E.B. Cowern, G. Mannino, P.A. Stolk, F. Rooseboom, H.G.A. Huizing, J.G.M. van Berkum, F. Cristiano, A. Claverie and M. Jariaz, Phys. Rev. Lett. 1999, 82: 4460
21. A. Claverie, L.F. Giles, M. Omri, B. de Mauduit, G. Ben Assayag and D. Mathiot, MRS Proceedings 1998, J 1.1

Mat. Res. Soc. Symp. Proc. Vol. 717 © 2002 Materials Research Society C1.6

The Effect of Ge Content in MBE $Si_{(1-x)}$ $Ge_{(x)}$ on the Evolution of {311} Defects

Robert Crosby, Jackie Frazer, Dr. K. S. Jones, Dr. M. E. Law, A. Nylandsted Larsen[1], and J. Lundsgaard Hansen[1]
University of Florida, SWAMP Group Gainesville, Fl.
[1]Institute of Physics and Astronomy, University of Aarhus, DK-8000, Aarhus, Denmark

ABSTRACT

Molecular beam epitaxial $Si_{1-x}Ge_x$ layers of various Ge concentrations ranging from 0% to 50% were grown on top of a Si substrate. The wafers were then implanted with a 40 keV, 1 x $10^{14}cm^{-2}$ Si^+. To study the development of {311} defects, the samples were annealed at 750°C for times ranging from 0 to 20 minutes. TEM was utilized to observe both the formation and dissolution of the defects. The $Si_{1-x}Ge_x$ samples with ≤ 5% Ge exhibit {311} defect formation and dissolution; however, samples with the Ge content lying between 15% and 50% showed only dislocation loop formation. It is suggested that the decrease in bond strength with increasing Ge content is the reason for the lack of {311} defect formation.

INTRODUCTION

It has been discovered that SiGe alloys have a higher mobility (than that of Si), and its band gap can be engineered for certain applications [1]. Because of these features, $Si_{1-x}Ge_x$ alloys have been used in avalanche photodiode devices and heterojunction bipolar transistors [2-4]. The narrowing of the band gap is related to the germanium concentration [5] and the strain of the layers as well [6]. Moreover, the strain and layer critical thickness is dependent upon Ge content [5]. These thin $Si_{1-x}Ge_x$ films can be grown pseudomorphically (and are therefore strained) onto a Si substrate [7]. When the layer thickness grows larger than a concentration-dependent critical thickness, the structure will introduce dislocations with misfits dislocations parallel to the interface and threading defects propagating through the epitaxial layer to reach a relaxed state. To circumvent the issue of threading of dislocations in a layer above the pseudomorphic thickness, compositionally graded buffer layers have been used. This allows the formation of low defect density SiGe layers on Si with reasonable thicknesses [8]. These low defect density SiGe layers are ideal for studies of the effect of Ge content on the point defect evolution.

Annealing of ion implanted dopants into semiconductors can illustrate anomalous diffusion [9] such as transient enhanced diffusion (TED) resulting from excess interstitials injected into the damaged region [10]. During annealing, excess interstitials can migrate and form {311} defects for given conditions. The time-temperature profile for {311} defect formation ranges from hours at 670°C anneals to minutes at 815°C anneals [10]. These rod-like defects visible by transmission electron microscopy reside on the {311} plane as a single monolayer of hexagonal Si (or Ge) and grow in the <110> direction [11]. {311} defects extend uniaxially allowing for the incorporation of self-interstitials into the lattice while minimizing the number of dangling bonds. Upon annealing these {311} defects will experience ripening and dissolution, which ultimately contributes to TED [10]. It is of interest to investigate {311} defect evolution in implanted SiGe alloys to determine how Ge content affects the {311} defect formation and dissolution.

EXPERIMENTAL DESIGN

$Si_{1-x}Ge_x$ wafers were grown by molecular beam epitaxy. The Ge content and epitaxial growth methods were varied as follows: 0%, 2%, 5%, 15%, 35%, and 50%. The pure Si sample is grown by the "bottleneck" method to getter contaminants from the epitaxially grown top layer. The other samples (2% to 50% Ge) are grown via a compositionally graded buffer layer, in which the Ge is incorporated into the sample at an increasing rate of10% Ge/micron at high growth temperatures around 750°C to 800°C. The samples with Ge are also lightly doped (5×10^{15} cm^{-3}) with Sb. The top layer of constant composition is always 3 to 4 microns in thickness. The samples are subjected to a 40 keV, 1×10^{14} cm^{-2} Si implant and then annealed in a furnace tube under a N_2 ambient for a total of 20 minutes at 750°C in 5 minute intervals. These samples were then lapped and etched to attain a plan-view transmission electron microscopy (PTEM) sample. Analysis is conducted via plan-view TEM using a JEOL 200CX microscope, and the defects are imaged under weak beam, dark-field conditions at g_{220}. Loop density and {311} defect density is quantified to attain a correlation between defect growth and Ge content.

RESULTS

The evolution of the defects upon annealing at 750°C is shown in figures 1 and 2 for two different Ge concentrations. The resulting progression of defect growth is presented below for both a representative of the low Ge content (0-5%) samples and a representative of the high Ge content (≥ 15%) samples.

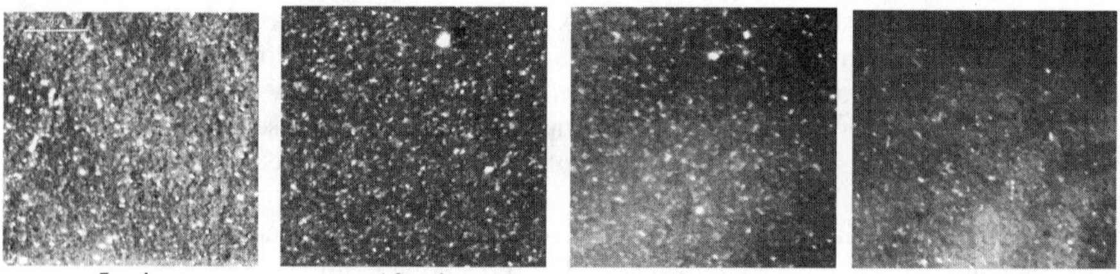

5 min 10 min 15 min 20 min

Figure 1. PTEM images of $Si_{.98}Ge_{.02}$ samples after a 40 keV, $1 \times 10^{14}cm^{-2}$ Si^+ implant annealed at 750°C for 5, 10, 15, and 20 minutes.

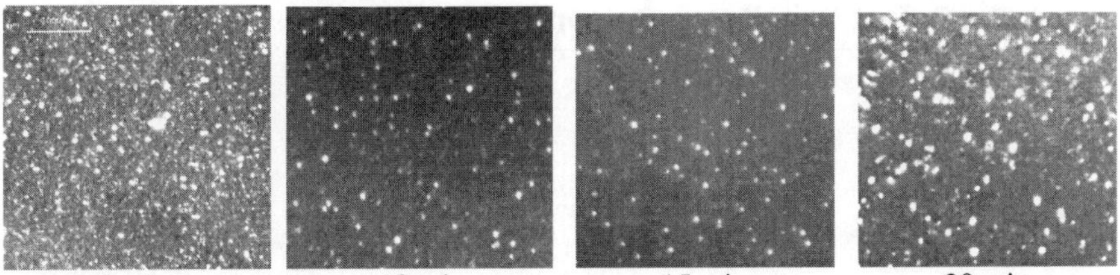

5 min 10 min 15 min 20 min

Figure 2. PTEM images of $Si_{.65}Ge_{.35}$ samples after a 40 keV, 1 x $10^{14}cm^{-2}$ Si^+ implant annealed at 750°C for 5, 10, 15, and 20 minutes.

Figure 1 depicts the defect evolution for a 2% Ge alloy. It is evident that at 5 and 10 minutes into the anneal, there are small clusters. However, at 15 minutes {311} defects are observed as well as small clusters. By 20 minutes, more {311} defects are detected.

In figure 2, we see a similar behavior at 5 minutes with cluster formation for the high Ge content samples. The early clusters are slightly larger in size. As time increases, these clusters begin to form dislocation loops. Small loop formation is visible at the 10 minute and 15 minute stage. After 20 minutes, governed by Ostwald ripening, the smaller loops have been engulfed by larger, more stable loops.

Those samples with ≤ 5% Ge all showed similar defect evolution in particular early {311} defect evolution. However, the defect evolution for the ≥ 15% Ge content samples differed in that there was no apparent {311} defect formation and only dislocation loop formation was observed. As a result, the defects were quantified by their respective Ge content subgroup.

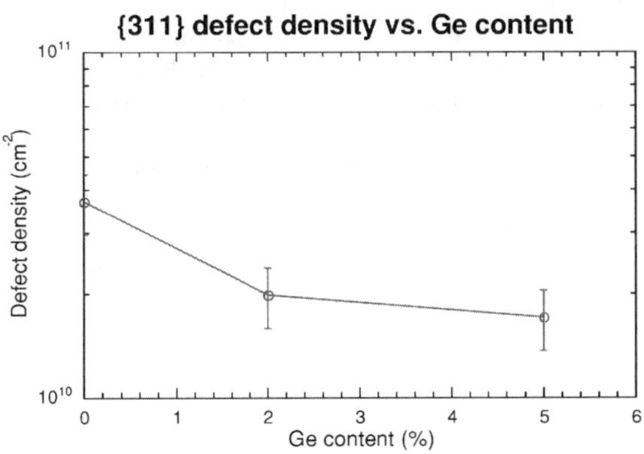

Figure 3. {311} defect density as a function of Ge content (0%, 2%, and 5%) after a 40keV, 1 x $10^{14}cm^{-2}$ Si^+ implant and a 20 minute, 750°C anneal.

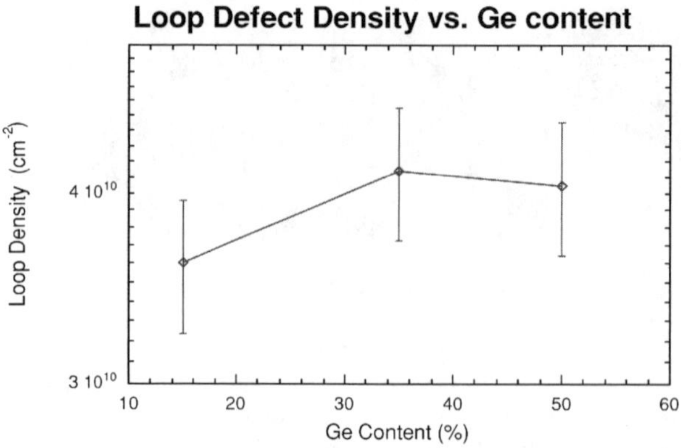

Figure 4. Dislocation loop density as a function of Ge content (15%, 35%, and 50%) after a 40keV, 1 x $10^{14}cm^{-2}$ Si^+ implant and a 20 minute, 750°C anneal.

Figures 3 and 4 illustrate the defect density as a function of Ge concentration for the low Ge concentration (≤ 5% Ge) samples and the high Ge concentration (≥ 15% Ge) samples. Figure 3 indicates that for the low Ge samples, the {311} defect density decreases with rising Ge content. On the other hand, Figure 4 displays that the loop density increases with increasing Ge content for the high Ge content samples. Evidently, Ge incorporation favors dislocation loop formation and over the formation of {311} defects.

DISCUSSION

From previous experiments from T. Y. Tan and many others, it has been found that {311} defects form readily in Si [11]. In our experiments, after 20 minutes of annealing at 750°C, {311} defects are present only in the low Ge content alloys. Ge is known to have a lower melting point than Si. This suggests that the covalent bond strength of Ge is less than that of Si. As one increases the Ge content of the alloy, the covalent bond strength of the alloy decreases. Since it is the high energy associated with the dangling bonds in Si that is believed to stabilize {311} defects [11] in favor of dislocation loops, it is reasonable to believe that as the bond strength decreases, the driving force for forming {311} defects also decreases. Thus, loop formation is favored over {311} defect formation with increasing Ge content.

ACKNOWLEDGEMNTS

The authors would like to acknowledge researchers at University of Aarhus for the processing of the samples and knowledgeable insight. We would also like to thank T. Saaverda, E. Kurilyw, and R. Camillo-Castillo for help with analysis. Funding this research was provided by the SRC.

REFERENCES

[1] G. L. Patton, J.H. Comfort, B.S. Meyerso, E.F. Crabbe, G.J. Scilla, E. De Fresart, J.C. Stork, J. Y.-C. Sun, D.L. Harame, and J.N. Burghartz, IEEE ED Lett. **11**,171, 1990.

[2] T. Tatsumi, H. Hirayama, and N. Aizaki, Appl. Phys. Lett. **52**, 895, 1988.

[3] G.L. Patton, S.S. Iyer, S. L. Delage, S. Tiwari, and J. M. C. Stork, IEEE Electron Device Lett. **9**,165, 1988.

[4] H.Temkin, A. Antreasyan, N. A. Olsson, T. P. Pearsall, and J. C. Bean, Appl. Phys. Lett. **49**, 809, 1986.

[5] R. People, IEEE J. Quantum Electron **22**, 1696, 1986.

[6] G. Abstreiter, H. Brugger, T. Wolf, H. Jorke, and H.-J. Herzog, Phys. Rev. Lett. **54**, 2441, 1985.

[7] E.A. Fitzgerald, Mater. Sci. Rep. **7**, 87, 1991.

[8] F. K. LeGoues, B. S. Meyerson, J. F. Morar, and P.D. Kirchner, J. Appl. Phys. **71**, 4230, 1992.

[9] N. E. B. Cowern, K. T. F. Jannsen, and H. F. F. Jos. J. Appl. Phys. **68**, 6191, 1990.

[10] D. J. Eagelsham, P. A. Stolk, H.-J. Gossmann, and J. M. Poate, Appl. Phys. Lett., **65**, 2305, 1994.

[11] T. Y. Tan, Philosophical Magazine A. **44**, 101, 1981.

Mat. Res. Soc. Symp. Proc. Vol. 717 © 2002 Materials Research Society C1.8

Single-Shot Excimer Laser Annealing and In Process Ellipsometry Analysis for Ultra Shallow Junctions

T. Noguchi[1], G. Kerrien[1], T. Sarnet[1], D. Débarre[1], J. Boulmer[1],
D. Zahorski[2], M. Hernandez[2], C. Defranoux[2], C. Laviron[3], M.-N. Semeria[3],
[1]Institut d'Electronique Fondamentale (IEF), UMR CNRS 8622,
Université Paris-Sud, Bât. 220, 91405 ORSAY Cedex, France
[2]SOPRA SA, 26, rue Pierre Joigneaux, 92270 BOIS-COLOMBES, France
[3]CEA-DRT/LETI/DTS, 17 Avenue des Martyrs 38054 GRENOBLE CEDEX 9, France

ABSTRACT

Single-shot Excimer Laser Annealing (ELA) was performed onto Si surface that was previously B^+ implanted with or without Ge^+ pre-amorphization. As a result, p^+ type USJ (Ultra-Shallow Junction) has been formed. In process analysis, using Infrared Spectroscopic Ellipsometry (IR-SE) has been performed and compared with conventional 4-point probe method. Also, the corresponding crystallinity for the USJ of Si surface has been studied using Ultraviolet-Visible (UV-Vis.) Spectroscopic Ellipsometry. In the case of pre-amorphization by Ge^+ implantation, the laser energy density threshold required for melting the surface, and therefore for electrical activation, decreased drastically because of the difference in the thermodynamic properties of the amorphized Si. Estimation of the junction depth shows a shallower junction when using UV-SE, as compared to IR-SE. This can be explained by the fact that, in the UV range, the crystallinity of the top layer is predominant while IR-SE is more sensitive to dopant activation. This efficient single-shot ELA is a candidate for the USJ formation for sub-0.1μm CMOS transistors. The effective method for investigating the activation state related to the crystallinity by using UV-SE and IR-SE is expected to apply as a non-contact analytical tool for USJ formation.

INTRODUCTION

The sub-0.1 μm transistor has been extensively studied, because it will play a leading role in ULSIs [1, 2]. In order to activate dopants in the silicon layer, Rapid Thermal Annealing (RTA) and Spike RTA are conventional techniques which usually follow ion implantation [3]. These techniques do not meet the International Technology Roadmap for Semiconductors (ITRS) specifications for sub-0.1 μm CMOS technology, which include ultra-shallow Source/Drain extensions (down to 10 nm), very doped (up to 10^{21} at/cm^3) and highly activated (sheet resistance down to 100 Ω/□) layers with box-like profiles. Other approaches like Gas Immersion Laser Doping (GILD) using a direct doping of the surface with excimer laser and precursor gas have also been reported [4, 5, 6, 7]. As far as GILD is concerned, the present authors (from IEF, SOPRA SA and CEA/LETI) and others have shown that this technique can already meet such specifications [4, 5, 6, 7]. In practical application, single-shot Excimer Laser Annealing (ELA) after shallow ion implantation is a promising technique in place of conventional RTA or advanced RTA for realizing sub-0.1 μm CMOS FETs, with USJ of high drivability that can suppress the Short Channel Effect (SCE). Some results concerning the fabrication of a transistor using ELA have been reported [8, 9, 10]. However, for Si LSI process application, USJ requires very uniform and controlled resistance and depth, which can be achieved only with a laser set-up having a good uniformity, repeatability and reproducibility. Also, a high crystallinity of the USJ is required in order to eliminate the p-n junction leakage, related to the efficient and uniform activation of impurity carriers after ELA [11, 12]. In order to control the formation by laser of an optimised USJ with a good efficiency and uniformity, a practical in-process analysis using a non-contact method is required in place of -or as well as - conventional destructive or contact analysis tool such as 4-point probe, Secondary Ion Mass Spectroscopy (SIMS) or Transmission Electron Microscopy (TEM). We have, therefore, used Spectroscopic Ellipsometry to characterize the USJ formation, and compared the results to conventional methods.

EXPERIMENTAL RESULTS

Description

Two different types of implanted Si wafers have been studied. They have been implanted at the CEA/LETI in Grenoble, France, with boron (B^+) at a dose of $2x10^{15}$ atoms/cm^2, at an energy of 3 keV, with and without pre-amorphization with Ge^+ ($1x10^{15}$ atoms/cm^2 at 10 keV). When using Ge^+ pre-amorphization, the Si surface is completely amorphous despite the low energy and low dose used for the implantation.

The surface was annealed at IEF using a single pulse of an excimer laser (XeCl, 308 nm wavelength, 30 ns pulse duration). The laser beam is normal to the surface and is made uniform with a beam homogenizer (inhomogeneity $\leq \pm$ 5%). Laser fluence on the sample surface may be adjusted with a variable attenuator in the range of 0.3 to 1 J/cm^2. Beam homogeneity and optical alignment are controlled with two CCD cameras optically conjugated with the Si surface.

Electrical activity of the laser annealed surface has been analyzed using conventional 4-point probe method. Crystallinity has been first analyzed by observing the Backscattering Kikuchi Diffraction (BKD) in a Scanning Electron Microscope (SEM).

These results have been compared with non-contact measurements from two SOPRA Spectroscopic Ellipsometers (SE): one in the UV-Visible range (from 0.19 to 0.85 µm) and the other in IR region (from 2 to 18 µm). We used a commercial software (Winelli®) for analysis and simulation of the Spectroscopic Ellipsometry data. Therefore, the variation of the refractive index (n) and the extinction coefficient (k), due to the laser annealing, has been analyzed and fitted, using the Drude Law to describe the free carrier absorption effects [13]. From these measurements we could evaluate the thickness, the concentration and therefore the resistivity, and also the fraction of crystallinity thanks to the different reflectance between c-Si and a-Si at 4.4 eV (inter-band absorption) [14].

Results and Discussion

Table I shows the resistivity of p^+ and n^+ Si wafers, used as references, measured by 4-point probe and IR SE. The results show almost similar values with both techniques and are in good agreement with the wafer specifications, which comfort us about the use of ellipsometry for resistivity measurements.

Table I. Comparison between resistivity measured by IR-SE and 4-point probe.

Material	Data Sheet Specifications	4-point probe	IR-SE
p-type (B)	0.05-0.01 ohm.cm	0.011 ohm.cm	0.009 ohm.cm
n-type (As)	<0.04 ohm.cm	0.002 ohm.cm	0.003 ohm.cm

Simulated (TRIM software) and measured (SIMS) boron profiles after implantation at 3 keV are shown in Figure 1. In the case of Ge^+ pre-amorphization the boron profile should be shallower (suppression of the tail) because of the limited channelling effect through the substrate. The extended tail depth might be close to 40nm, slightly deeper than the damaged depth that was estimated to less than 20nm after Ge^+ pre-amorphization.

The measured sheet resistance by 4-point probe are shown in Figure 2, with and without Ge^+ pre-amorphization, as a function of the laser energy density. We used a sheet resistance correction due to the substrate resistance. The energy density required for effective activation drastically decreases for the case of Ge^+ pre-amorphization. This effect is generally considered to be due to the change in the thermodynamic properties of the material, including the lowering of the melting point for amorphous Si [10, 14, 15]. This allows the use of a larger annealed area for the same laser energy [14, 16] which is favourable in the case of future lumped shot ELA applications.

These results have been compared to the sheet resistance measured by Infrared Spectroscopic Ellipsometry (IR-SE), and are reported in the same figure (see Figure 2). The resistivity given by IR-SE agrees with the 4-point probe in the energy density range that was analyzed, however more data are still needed to confirm this tendency.

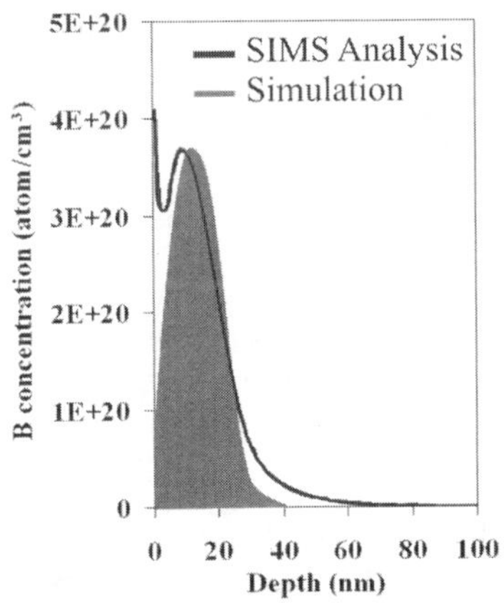

Figure 1. Simulated (TRIM) and measured (SIMS) boron profiles after implantation at 3 keV

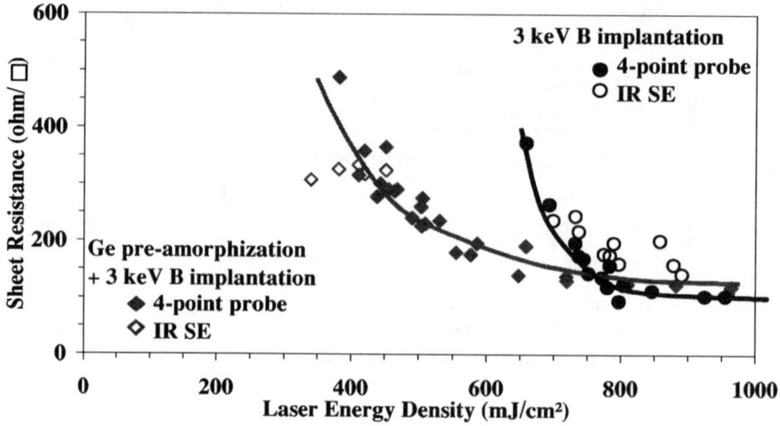

Figure 2. Sheet Resistance measured by 4-point probe and IR-SE, with and without Ge^+ pre-amorphization.

Related activated depth compared with the re-crystallized depth deduced from UV-SE is shown in Fig. 3a. In this range of laser energy density, the surface was partially crystallized and the estimation of the junction depth shows a shallower junction when using UV-SE, as compared to IR-SE. This can be explained by the present model parameters: in the UV range, the crystallinity of the top layer is predominant while IR-SE is more sensitive to activated dopants. However, more accurate study should be carried out in order to improve this model and to compare these results to the depth analysed by other destructive methods (such as SIMS, TEM etc…).

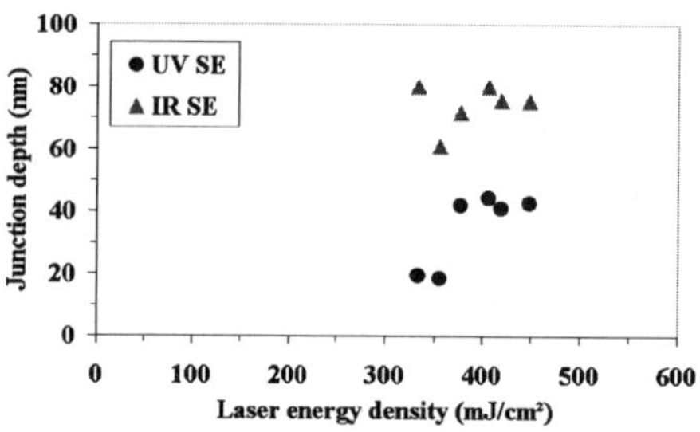

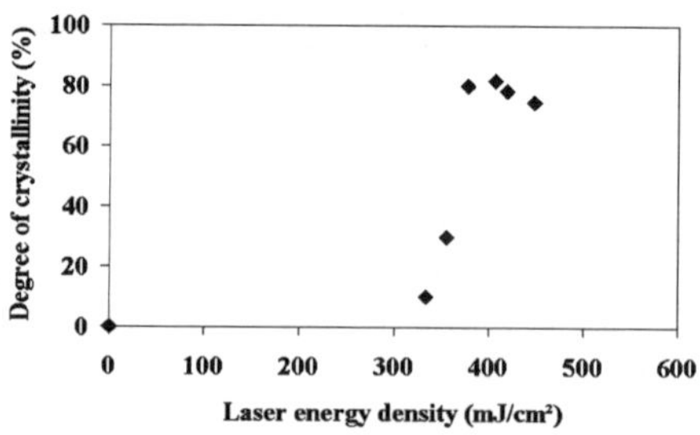

Figure 3a and 3b:
3a:Comparison of effective junction depth measurements by UV-SE and IR-SE
3b: Degree of Crystallinity of Ge^+ pre-amorphized doped Si for different energy densities.

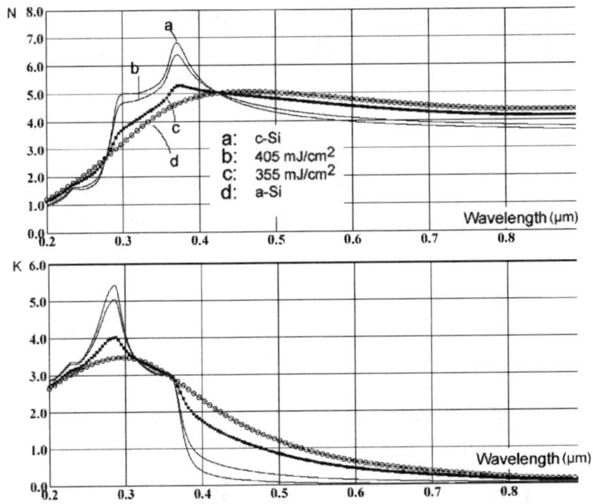

Figure 4a and 4b. Refractive Index (n) and Extinction Coefficient (k) for different Si wafers: crystalline Si (c-Si), amorphous Si (a-Si) and laser treated (406 and 355 mJ/cm^2).

An incorporation of little amount of Ge might affect slightly the surface crystallinity after ELA. Below 500mJ/cm^2, it is speculated that only the damaged layer was melted. But, the crystallinity is fairly improved for the ELA condition at about 400mJ/cm^2 depending on the increase in the re-crystallized depth as shown in Fig. 3b and Fig.4a,b. Corresponding diffraction patterns have been observed using BKD. If the edge of USJ depending on the activated depth is deeper than the re-crystallized depth as shown in Fig. 3a, the leakage for junction can be eliminated using laser energy density between 400 and 500mJ/cm^2 in the transistor fabrication process. Although more detailed results should be obtained by using substrate with higher resistivity or by changing damaged or implanted depth, which should lead to the improvement of the efficiency of the ELA for USJ formation in practical Si process.

CONCLUSION

Ultra-Shallow Junction (USJ) has been formed by single-shot Excimer Laser Annealing (ELA) after shallow ion implantation with boron. The laser energy density threshold for the activation is reduced when using Ge^+ pre-amorphization, because of the difference in the thermodynamic properties of the material. In the case of Ge^++ B^+ implantation, the activated junction and re-crystallized junction depths were deduced by using a non-destructive spectroscopic ellipsometry analysis both in the UV-Visible and IR range. Using these ellipsometry techniques, the activation depth (deduced from IR-SE) seems to be deeper than the re-crystallization depth (deduced from UV-SE).

Reducing the laser energy density, thanks to the Ge^+ pre-amorphization, has improved the process and has allowed USJ formation with single-shot ELA, which should be a candidate in Si LSI. UV-Vis. and IR Spectroscopic Ellipsometry analyses have been performed, first results are in fair agreement with conventional techniques, and show the potentiality of Ellipsometry, but additional work has still to be carried out in order to use SE as a non-destructive analysing tool in Si process for USJ.

ACKNOWLEDGEMENTS

The authors would like to thank Messrs. J.-L. and M. Stehle from SOPRA SA, Dr. S. Deleonibus from LETI for encouragement and Staff Members at IEF for technical support.

REFERENCES

1. G. Sai-Halaz, M. Wordeman, D. Kern, E. Ganin, S. Rishton, H. Ng. D. Zicherman, D. Moy, T. Chang and R. Dennard, Tech. Digest of IEDM, 397, (1987).
2. D. Lenoble, A. Halimaoui, O. Kermarrec, Y. Campidelli, D. Bensahel, J. Bounnovierr, O. Menut, E. Robiliart, E. Perrin, F. Arnaud, F. Boeuf, T. Stotniki, A. Grouilet, and G. Bignel, presented at International Workshop on Junction Technologies, Tokyo, Japan, p. 2-2-1, (2001).
3. W. S. Yoo, T. Fukada, T. Setokubo, K. Aizawa, J. Yamamoto and R. Komatsubara, presented at International Workshop on Junction Technologies, Tokyo, Japan, p. 4-4-4, (2001).
4. K-J. Kramer, S. Talwar, A. M. McCarthy and K.H. Weiner, IEEE Trans. Electron Dev. Lett. **17**, 4561 (1996).
5. D. Débarre, G. Kerrien, T. Noguchi and J. Boulmer, presented at IEICE transactions "*Fundamentals/Commun.,/Electron., Inf. & Syst.*" **8,** (2001) (accepted for publication).
6. G. Kerrien, J. Boulmer, D. Débarre, D. Bouchier, A. Grouillet, and D. Lenoble, Applied Surface Science, **186**, 45-51 (2002).
7. Jong Won Moon, Myung Geun Han and Seonghoon Lee, presented at the 2002 MRS Spring Meeting, San Francisco, CA, 2002 (to be published).
8. T. Noguchi, M. Kubota, H. Yamamoto, K. Matsumoto and M. Yamagishi, Tech. Report of IEICE, SDM2000-**45** (2000).
9. M.A. Crowder, P.G. Caray, P.M. Smith, R.S. Sposili, H.S.Cho and J.S. Im, IEEE Trans. Elect. Dev. Lett. **19,** 306 (1998).
10. C. Park, S.-D. Kim, Y. Wang, S. Talwar and J. C. S. Woo, Symp. on VLSI Tech. Dig., 69 (2001).
11. H. Tsukamoto, H. Yamamoto, T. Noguchi and T. Suzuki, Jpn. Appl. Phys. **32**, L659 (1992).
12. K. Goto, T. Yamamoto, T. Kubo, M. Kase, Y. Wang, T. Lin, S., Talwar and T. Sugii, Tech. Digest of IEDM, p. 931 (1999).
13. P. Boher, M. Bucchia, J. P. Piel, C. Defranoux, J. L. Stehle and C. Pickering in *Optical Metrology Roadmap for the Semiconductor, Optical, and Data Stotage Industries II*, edited by A. Duparre and B. Singh, (SPIE Proc. **4449**, 2001) pp 69-78.
14. T. Noguchi, H. Tsukamoto, T. Suzuki, and H. Masuya, Extended Abstracts of the Int. Conf. on SSDM, p. 620 (1991).
15. M. O Thompson, G. J. Galvin and J. W. Mayer, Phys. Rev. Lett. **27**, 375 (1975).
16. C. Laviron, M. N. Semeria, D. Zahorski, M. Stehle, M. Hernandez, J. Boulmer, D. Débarre, and G. Kerrien presented at International Workshop on Junction Technologies, Tokyo, Japan, p. 6-1-1, (2001).

Mat. Res. Soc. Symp. Proc. Vol. 717 © 2002 Materials Research Society C1.9

Effect of Laser Thermal Processing on Defect Evolution in Silicon

Erik Kuryliw, Kevin S. Jones, David Sing[1], Michael J. Rendon[1], and Somit Talwar[2]
SWAMP Center, University of Florida
Dept. of Materials Science and Engineering
P.O. Box 116130, Gainesville, FL 32611-6130
[1]Motorola/International Sematech, Austin TX
[2]Verdant Technologies, Santa Clara CA

ABSTRACT

Laser Thermal Processing (LTP) involves laser melting of an implantation induced pre-amorphized layer to form highly doped ultra shallow junctions in silicon. In theory, a large number of interstitials remain in the end of range (EOR) just below the laser-formed junction. There is also the possibility of quenching in point defects during the liquid phase epitaxial regrowth of the melt region. Since post processing anneals are inevitable, it is necessary to understand both the behavior of these interstitials and the nature of point defects in the recrystallized-melt region since they can directly affect deactivation and enhanced diffusion. In this study, an amorphizing 15 keV 1 x 10^{15}/cm^2 Si^+ implant was done followed by a 1 keV 1 x 10^{14}/cm^2 B^+ implant. The surface was then laser melted at energy densities between 0.74 and 0.9 J/cm^2 using a 308 nm excimer-laser. It was found that laser energy densities above 0.81 J/cm^2 melted past the amorphous-crystalline interface. Post-LTP furnace anneals were performed at 750 °C for 2 and 4 hours. Transmission electron microscopy was used to analyze the defect formation after LTP and following furnace anneals. Secondary ion mass spectrometry measured the initial and final boron profiles. It was observed that increasing the laser energy density led to increased dislocation loop formation and increased diffusion after the furnace anneal. A maximum loop density and diffusion was observed at the end of the process window, suggesting a correlation between the crystallization defects and the interstitial evolution.

INTRODUCTION

Ion implantation and annealing has long been the accepted method for introducing and activating dopants into silicon. During the implantation process point defects are generated, via nuclear collisions. Upon subsequent annealing to activate the dopant, these point defects, typically interstitial silicon atoms, promote enhanced dopant diffusion, agglomerate into extended defects, as well as participate in dopant clustering[1,2]. One technique that shows promise in circumventing conventional problems is laser thermal processing (LTP) [3]. Laser thermal processing begins by forming a pre-amorphized silicon layer via isoelectronic ion implantation. This layer is further implanted with dopant that is contained within the amorphous region. An excimer laser is then used to melt the pre-amorphized layer. The melt region is confined to the amorphous layer due to the reduced melting temperature of amorphous silicon compared to crystalline silicon. The junction depth forms at the amorphous-crystalline (a-c) interface for a range of energy densities termed the process window. For energy densities below the process window, polycrystalline growth is observed and typically identified in transmission electron diffraction. For energy densities above the upper edge of the process window, the melt front penetrates into the underlying crystal and is detectable by secondary ion mass spectroscopy

(SIMS). During the recrystallization process, dopant is distributed throughout the melt region into the silicon lattice at concentrations far above solid solubility. The result is a shallow, low-resistivity, highly abrupt, "box-like" dopant profile[3,4] that is capable of meeting ITRS specifications for the 35 nm technology node[5].

It has been previously reported that laser melting within the process window results in the formation of growth related defects often seen in solid phase epitaxy[6]. These defects include stacking faults, twins, and threading dislocations. Increasing the laser annealing energy density decreases the density of these defects. Once the process window is surpassed, these defects drop to below detectable limits. What is unclear, however, is the condition of point defects in the end-of-range (EOR) region beyond the amorphous-crystalline (a-c) interface after laser melting is performed. It would stand to reason that if the melt does not extend significantly into the crystal, a supersaturation of interstitials should exist. Since post processing at elevated temperatures is inevitable, it is necessary to understand the behavior of these point defects as they could lead to transient enhanced diffusion (TED), extended defects, and deactivation.

EXPERIMENTAL DETAILS

Czochralski-grown (100) Si wafers were implanted with 15 keV Si^+ to a dose of $1\times10^{15}/cm^2$ at room temperature. A second implant of 1 keV B^+ at a dose $1x10^{14}/cm^2$ was also performed. Only a native oxide served as an implant screen. This resulted in boron profile completely contained within a 320 Å thick continuous amorphous layer. The samples were then laser annealed using a 15 ns XeCl excimer laser pulse operating at wavelength of 308 nm. Energy densities for the laser anneal ranged between 740 and 900 mJ/cm^2. The samples were annealed in a furnace at 750 °C for 2 and 4 hours using a flowing N_2 ambient. Plan view transmission electron microscopy (PTEM) specimens were made by mechanical polishing and etching using $HF:HNO_3$ solution. PTEM micrographs were taken using $\pm g_{220}$ for weak beam dark field imaging. Secondary ion mass spectroscopy was used to determine initial and annealed boron concentration profiles.

RESULTS

LTP samples were imaged prior to anneal using PTEM to determine the process window. These images are shown in Figure 1. It has previously been found that growth related defects rapidly drop to undetectable levels once the process window is exceeded. Figure 1 shows that for samples laser annealed with energy densities above 811 mJ/cm^2, there are very few observable defects. Samples laser annealed at 742 mJ/cm^2 showed no evidence of poly-crystalline diffraction ring patterns, indicating that the melt had reached the a-c interface for these samples.

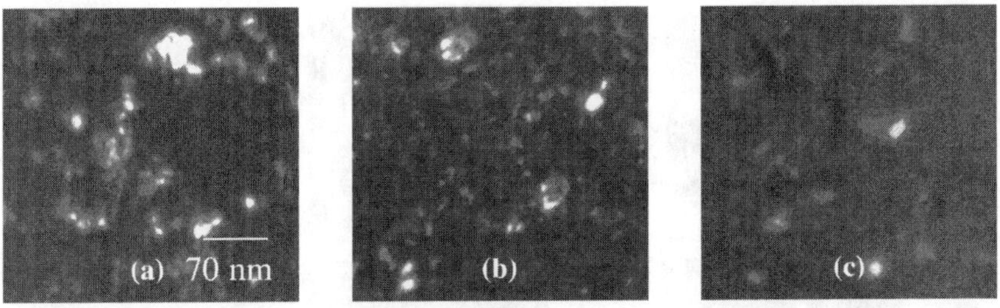

Figure 1. Plan view TEM of silicon following laser annealing at (a) 784, (b) 811, and (c) 842 mJ/cm^2. All samples showed a single crystal diffraction pattern. The 784 and 811 mJ/cm^2 samples show crystallization defects normally observed in the process window. The 842 mJ/cm^2 sample showed only a few crystal defects.

In order to understand what effect the laser energy density and the growth related defects have on the interstitial profile and extended defect evolution, samples were annealed at 750 ºC for 2 hours and imaged in PTEM, as shown in Figure 2. The non-LTP sample showed a loop and elongated loop morphology generally seen for this implant and anneal, however the laser annealed samples showed a dramatically different evolution. While loop formation was still observed, the size of the loops was much smaller than generated through conventional processing. It was also observed that increasing the laser annealing energy density in the process window resulted in an increase in the number of dislocation loops. As the energy density was increased through the end of the process window, a maximum number of loops was observed. For energy densities beyond the process window, increases in laser energy density resulted in drastically fewer loops. Dissolution of crystallization defects was observed between 2 and 4 hours at 750 ºC, for the 784 mJ/cm^2 sample, as shown in Figure 3. This dissolution coincided with the formation of additional loops of a number density similar to what was observed in the 811 mJ/cm^2 material for the two hour anneal.

Secondary ion mass spectroscopy was performed on the above samples to determine how the laser energy density affected the diffusion characteristics during a 750 ºC 2 hour anneal. It was seen that as the laser energy density was increased throughout the process window, the amount of diffusion during the anneal also increased, Figure 4. Once the end of the process window was reached a substantial increase in diffusivity was observed. As the process window was exceeded, this diffusivity remained high, though a slight decrease in the total junction motion was seen.

DISCUSSION

Comparison of TEM and SIMS results show that for variations of laser annealing energy densities that are within the process window, there are significant effects on both extended defect formation and enhanced diffusion of boron during subsequent thermal processing. For dislocation loop formation, an increase in loop density is observed as the laser energy density is increased, with a maximum at the end of the process window. However, there is the possibility that the crystallization defects are blocking ability to see the dislocation loops

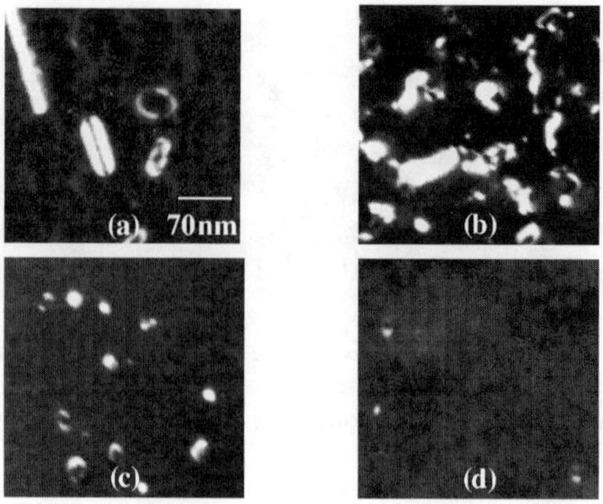

Figure 2. Plan view TEM showing the effect of laser energy density on the extended defect evolution after a 750 ºC 2 hour anneal on LTP silicon for samples receiving (a) no LTP, (b) 784, (c) 811, and (d) 842 mJ/cm^2. The laser melting is shown to reduce dislocation loop formation. As the laser energy density is increased, more loops are observed to form, with a maximum density of loops at the end of the process window. Once the process window is exceeded, loop formation dramatically decreases.

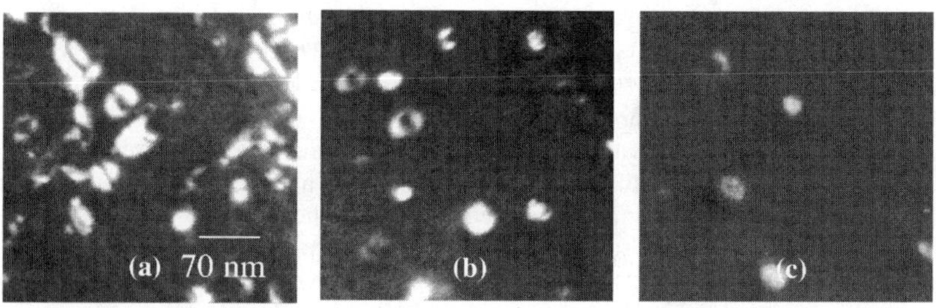

Figure 3. Plan view TEM showing the dissolution of crystallization defects and subsequent formation of dislocation loops after a 750 ºC 4 hour anneal for samples laser annealed at (a) 768, (b) 784 and (c) 811 mJ/cm^2. The highest energy density sample shows evidence of some loop dissolution.

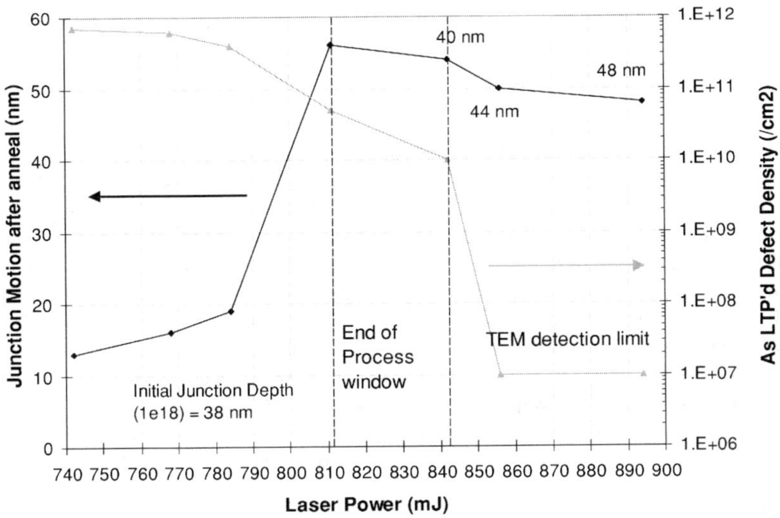

Figure 4. Plot showing the correlation between the initial crystallization defect density measured quantitatively through TEM and the junction motion observed after a 750 ºC 2 hour furnace anneal measured via SIMS at a concentration of $1x10^{18}/cm^3$. Also shown is the initial junction depth for each laser energy density.

in plan view. Further annealing reveals that samples near the end of process window show loop dissolution, while those in the process window show crystallization defect dissolution and loop formation. This supports the theory that the crystallization defects are impeding the formation of EOR dislocation loops. Once the process window is exceeded, dislocation loop formation abruptly ends. While there are little to no crystallization defects present, the laser has melted into the exponentially decaying end of range profile, thus reducing the number of interstitials available for loop formation.

A review of the enhanced diffusion shows a trend similar to that of the dislocation loop formation for laser energy densities in the process window. As the laser energy density is increased, an increase in enhanced diffusion is observed, with a maximum occurring at the end of the process window. However, unlike the loop evolution, once the process window is exceeded, a significant enhanced diffusion is still observed. Given that for the highest energy density the laser melted 10 nm into the exponentially decreasing EOR interstitial profile, it is unlikely that this could be the source of interstitials for such a large diffusion enhancement. Thus there is evidence to suggest that interstitials are quenched into the solid as solidification occurs.

Both extended defect evolution and the enhanced diffusion of boron are known to be mitigated by interstitial point defect populations. The similar trend of both within the process window with respect to the crystallization defects gives rise to the theory that these defects are trapping interstitials thus reducing both the loop formation and enhanced diffusion. The formation of loops following the dissolution of the crystallization defects further supports this theory.

CONCLUSIONS

In this work it was observed that changes in laser annealing energy density within the LTP process window had a direct impact on the density of crystallization defects viewed in PTEM images. In turn, these defects were shown to suppress the formation of dislocation loops and the enhanced diffusion of boron in subsequent thermal anneals. A maximum number of dislocation loops and enhanced boron diffusion was observed at the end of the process window where very few crystallization defects existed. Since both loop formation and enhanced diffusion of boron are governed by the presence of interstitials, there is strong evidence to suggest that crystallization defects, while present, trap interstitials. Once the laser energy density was increased beyond the process window, dislocation loop formation was no longer seen, but a large diffusion enhancement still existed despite the EOR interstitial profile having been dramatically truncated. This phenomenon supports previous findings that suggest the presence of interstitial point defects that were quenched into the epitaxially solidified region during the LTP process.

REFERENCES

1 S. J. Pennycook, J. Narayan, and O. W. Holland, "Point Defect Supersaturation and Enhanced Diffusion in Supersaturated Silicon Alloys," Electrochemical Society Extended Abstracts **84** (2), 672 (1984).

2 R. J. Culbertson and S. J. Pennycook, "Trapping of Interstitials During Ion Implantation in Silicon," presented at the MRS Proceedings, Beam-Solid Interactions and Phase Transformations, Boston, MA, 1986.

3 S. Talwar, G. Verma, and K. Weiner, "Ultra-shallow, abrupt, and highly-activated junctions by low-energy ion implantation and laser annealing," presented at the Ion Implant Technology, 1999.

4 R. Murto, K. Jones, M. Rendon *et al.*, "Activation and deactivation studies of laser thermal annealed boron, arsenic, phosphorus, and antimony ultra-shallow abrupt junctions," presented at the Ion Implant Technology, 2000.

5 Semiconductor Industry Association, *International Technology Roadmap for Semiconductors: 2001 Edition* (International SEMATECH, Austin, TX, 2001).

6 H. Banisukas, K.S. Jones, S. Falk *et al.*, "Defect reduction in laser thermal processing," presented at the Si Front-end Processing -Physics and Technology of Dopant-Defect Interactions II, San Francisco, CA, 2000.

Mat. Res. Soc. Symp. Proc. Vol. 717 © 2002 Materials Research Society C1.10

Electrical and Structural Characterization of Boron Implanted Silicon Following Laser Thermal Processing

K. A. Gable[1□], K. S. Jones[1], M. E. Law[2], L. S. Robertson[3], S. Talwar[4]
[1] Materials Science and Engineering, University of Florida, Gainesville, FL 32611
[2] Electrical and Computer Engineering, University of Florida, Gainesville, FL 32611
[3] Texas Instruments, Dallas, TX 75265
[4] Verdant Technologies, San Jose, CA 95134
□ kgabl@mse.ufl.edu

ABSTRACT

One alternative to conventional rapid thermal annealing (RTA) of implants for ultra-shallow junction formation is that of laser annealing. Laser thermal processing (LTP) incorporates an excimer pulsed laser capable of melting the near surface region of the silicon (Si) substrate. The melt depth is dependent upon the energy density supplied by the irradiation source and the melting temperature of the substrate surface. A process window associated with this technique is able to produce similar junction depths over a range of energy densities due to the melting temperature depression established with pre-amorphization of the substrate surface prior to dopant incorporation. The process window of germanium (Ge) pre-amorphized, boron (B) doped Si was investigated. 200 mm (100) n-type Si wafers were pre-amorphized via 18 keV Ge^+ implantation to $1x10^{15}/cm^2$ and subsequently implanted with 1 keV B^+ to doses of $1x10^{15}/cm^2$, $3x10^{15}/cm^2$, $6x10^{15}/cm^2$, and $9x10^{15}/cm^2$. The wafers were laser annealed from 0.50 J/cm^2 to 0.88 J/cm^2 using a 308 nm XeCl excimer irradiation source. Transmission electron microscopy (TEM) was used to determine the process window for each implant condition, and correlations between process window translation and impurity concentration were made. Four-point probe quantified dopant activation and subsequent deactivation upon post-LTP furnace annealing.

INTRODUCTION

Silicon (Si) technology is approaching a transition period for which novel front end processing techniques must be developed and implicitly understood in order to maintain the aggressive scaling trend outlined by the International Technology Roadmap for Semiconductors (ITRS).[1] Currently, ion implantation and rapid thermal annealing (RTA) are the methods by which dopants are introduced into the substrate and subsequently activated, respectively. While advances in RTA have extended its use, its isothermal nature introduces limits associated with the annealing parameters that cause anomalous diffusion[2] and electrical saturation[3] of the dopant during the thermal process. As a result, RTA is unable to produce junctions with the performance characteristics required for critical scale integration (CSI) and the need for investigating alternate annealing techniques becomes apparent.

One potential alternative to conventional thermal annealing is that of laser annealing. Laser thermal processing (LTP) incorporates an excimer pulsed laser capable of melting the near surface region of the crystalline Si (c-Si) substrate.[4] An issue with LTP is the fact that the melt depth displays a linear dependence with energy density, limiting a specific fluence to a particular melt depth. This is circumvented by pre-amorphization of the substrate surface prior to dopant incorporation, introducing a $225 \pm 50°C$ melting temperature depression associated with the amorphous Si (α-Si) phase transition.[5] In this respect, when the liquid front reaches the amorphous/crystalline (α/c) interface, the difference in melt thresholds disallows further melting. This results in a process window corresponding to the difference between the energy required to melt to the α/c interface and the energy required to melt the underlying c-Si. When the energy supplied to the system per unit time is less than that needed to sustain the Si melt, the severely undercooled liquid regrows by liquid phase epitaxy with regrowth velocities on the order of 3 m/s[6], which is sufficient for incorporating dopants on substitutional sites. This adiabatic annealing technique is beneficial in that the dopant diffusivities are on the order of 10^8 times greater in the liquid phase, and the segregation coefficient for most impurities approaches unity[7], allowing for the desired hyper-abrupt box like profile. By this method, LTP is able to reproducibly create junctions with depths defined by the α/c interface and electrical activation levels greater than those obtained by traditional means.

Although this technique is capable of producing junctions with performance characteristics greater than those obtained through conventional standards, implications of the implant conditions on the process window are not well defined and must be investigated in order to understand their effects. A number of analytical techniques are available for process window investigation. Dynamic secondary ion mass spectrometry (SIMS) quantifies impurity concentration as a function of depth and characterizes the process window by illustrating similar junction depths over a range of energies densities.[8] The Boxer Cross Carrier Illumination (CI) technique induces an excess carrier distribution as a function of depth, which influences the materials index of refraction allowing for quantifiable differences in sample reflectivity by which the process window may be defined.[9] Transmission electron microscopy (TEM) is capable of determining the process window by imaging characteristics indicative of the window bounds. The lower bound is defined as the transition between polycrystalline Si (p-Si) regrowth, due to solidification nucleating from a partially melted amorphous layer, and c-Si regrowth as determined by the presence of p-Si rings during selected area diffraction (SAD). The upper bound is defined as the energy density prior to that required to result in defect free c-Si regrowth. This latter technique was used to determine the process window for each implant condition of the study.

EXPERIMENTAL DESIGN

200 mm (100) n-type Czochralski (CZ) grown Si wafers were pre-amorphized via 18 keV germanium (Ge^+) implantation to $1x10^{15}/cm^2$ and subsequently implanted with 1 keV B^+ to doses of $1x10^{15}/cm^2$, $3x10^{15}/cm^2$, $6x10^{15}/cm^2$, and $9x10^{15}/cm^2$. The pre-amorphization implant was chosen to result in a continuous amorphous layer extending 300 Å from the substrate surface. The dopant implants were to result in B concentrations of $0.3x10^{21}/cm^3$, $1x10^{21}/cm^3$, $2x10^{21}/cm^3$, and $3x10^{21}/cm^3$, assuming a perfectly abrupt profile within the process window. The wafers were subsequently laser annealed with a 308 nm wavelength XeCl excimer irradiation source with a pulse duration of 18 ns from energy densities of 0.50 J/cm^2 to 0.88 J/cm^2, in increments of 0.02 J/cm^2, with exposure areas of 5 mm x 5 mm. Each exposed area was then diced into 4 mm x 4 mm samples in order to reduce the measurement error associated with edge rounding of the perimeter of the sample exposure area as a result of the non-uniform laser pulse cross-section.[10]

Cross-sectional transmission electron microscopy (XTEM) and variable angle spectroscopic ellipsometery (VASE) were used to determine the thickness of the amorphous layer. The XTEM sample was thinned by 5 kV argon (Ar^+) ion milling, with the plasma sources tilted 12° to the surface plane. All XTEM images were captured on a JEOL 200 CX TEM operating at 200 kV under a bright field (BF) imaging condition with the objective aperture centered over the transmitted beam. Plan-view TEM (PTEM) was used to investigate defect formation as a function of laser energy density. The samples surface and backside periphery were insulated from the 75% HNO_3/25% HF solution used to introduce an electron transparent edge surrounding a hole. The PTEM images were captured on a JEOL 200 CX TEM operating at 200 kV in **g**.3**g** centered weak beam dark field (WBDF) using a $\mathbf{g}_{220}$ two beam imaging condition. A Jandel four-point probe with spacings of 1.016 mm quantified dopant activation and subsequent deactivation upon post-LTP furnace annealing by sourcing 3.0724 mA, which was chosen to result in unity when multiplied by the sample geometric correction factor.

RESULTS & DISCUSSION

The 18 keV Ge^+ $1x10^{15}/cm^2$ implant generated a continuous amorphous layer extending 300 Å below the substrate surface, which was determined by XTEM and verified by VASE. Figure 1 shows PTEM images of the 1 keV B^+ $1x10^{15}/cm^2$ implant as a function of irradiation energy density with the corresponding diffraction pattern in the image inset. The transition from p-Si to c-Si regrowth can be determined from the evolution of the (111) p-Si ring surrounding the transmitted beam. The ring reduces in both intensity and continuity until no longer visible, representing c-Si regrowth and the initiation of the process window.

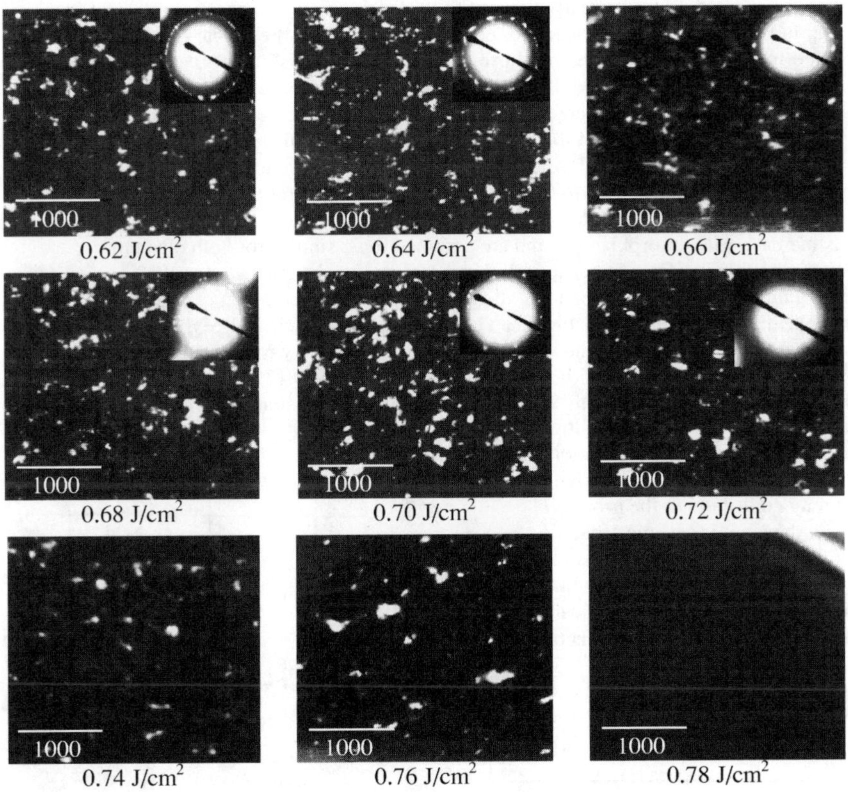

Figure 1: PTEM images showing the derfect evolution as a function of incident irradiation energy density

The beginning of the process window for this implant condition corresponds to an energy density of 0.72 J/cm^2. Energies below 0.72 J/cm^2 produce volumes of c-Si surrounded by regions of defects associated with the introduction of two liquid interfaces during regrowth, resulting in grain boundary formation. These grains increase in size with increased energy density.[11] The absence of an amorphous halo within the diffraction patterns is due to a phenomenon known as explosive recrystallization.[5] This recrystallization occurs due to partial melting of the amorphous layer and the latent heat associated with its solidification, which is sufficient to produce a self-sustaining melt front propagating away from the substrate surface. Although this event is capable of quenching before complete consumption of the α-Si[7], it terminated at the α/c interface resulting in p-Si formation. The upper bound of the process window is realized when comparing the images of 0.76 J/cm^2 and 0.78 J/cm^2. The latter resulted in no visible defect formation suggesting a melt front that reached below the α/c interface. This results in a process window of 0.72 J/cm^2 to 0.76 J/cm^2. The images of 0.72 J/cm^2 to 0.76 J/cm^2 show a decrease in defect density with increasing energy density, suggesting improved regrowth within the process window. The same structural characteristics and imaging techniques were used to determine the process window for all the implant conditions, and are given in Table 1. As can be seen by the data, the bounds decrease with increasing B concentration, while maintaining a consistent 0.04 J/cm^2 process window.

The process window was initially thought to be dependent upon the amorphous layer thickness and independent of dopant specie or concentration. While the specie dependence cannot be experimentally confirmed here, Murto et al. reported a possible process window translation dependence with implanted dopant, suggesting either a specie dependent absorption effect or a repeatability issue associated with the irradiation source.[8] Although Ge was used for the pre-amorphization of the Si substrate, the absorption characteristics of the isovalent species are similar for the wavelength under consideration, and any differences in the process window can be attributed to the differences in B concentration.[12] It should be noted that this absorption characteristic is based on crystalline material, and that absorption coefficients increase with amorphization and are assumed to be similar for both elements.[7]

Differences arising from increasing B concentration may be attributed to the introduction of an absorption phenomena or alloy induced melting temperature depression. The latter suggestion seems unreasonable when considering the Si rich side of the Si-B equilibrium phase diagram.[13] Since the temperature shift associated with the amorphous phase is overcome after 0.04 J/cm^2 within the process window, the 0.12 J/cm^2 difference between the 1×10^{15}/cm^2 and 9×10^{15}/cm^2 B^+ implants suggests a temperature difference much greater than that predicted by eutectic lowering, even when non-equilibrium considerations are taken into account.[14] Although an alloying component may contribute to the process window dependence, it is assumed that the majority of the difference is due to an enhanced absorption effect, and is to be confirmed.

Figure 2 shows the activation data represented as sheet resistance as a function

B^+ Implant Condition (/cm^2)	Lower Bound (J/cm^2)	Upper Bound (J/cm^2)
1×10^{15}	0.72	0.76
3×10^{15}	0.70	0.74
6×10^{15}	0.64	0.68
9×10^{15}	0.60	0.64

Table 1: Experimental upper and lower bounds of the process window

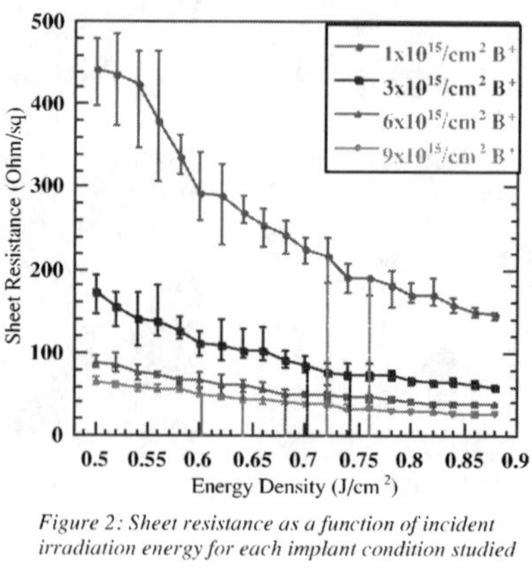

Figure 2: Sheet resistance as a function of incident irradiation energy for each implant condition studied

of energy density for the implant conditions studied. Each measurement consisted of an 18-sample set average and the error bars represent the measured error found within each population. The vertical lines on the graph represent the bounds of the process window for each implant condition. The measurement error decreased with increasing B concentration and energy density. The sheet resistance continually decreases with increasing energy density for each implant condition. This decrease in sheet resistance may result from the reduction in defect density or more effective solute trapping due to an increased regrowth velocity associated with increasing energy densities within the process window.[7, 15]

Isochronal deactivation anneals were performed in a conventional tube furnace under a nitrogen (N_2) ambient for 60 minutes from temperatures of 650°C to 850°C. The deactivation data reported for each implant condition was limited to the energy densities within the process window. Figure 3 shows the isochronal nature of the 1 keV B^+ $3x10^{15}/cm^2$ implant. The $3x10^{15}/cm^2$ implant was focused upon to allow for direct comparisons between similar data found in recently published articles.[8, 16] Each energy density shows an initial decrease in sheet resistance for the 650°C anneal, and is attributed to a decrease in defect density. The higher temperature anneals show increased sheet resistance due to B deactivation. This deactivation decreases with increasing energy density suggesting process window dependent deactivation kinetics.

Figure 4 shows the sheet resistance as a function of isochronal annealing temperature. The data is normalized to account for disparity in implanted B dose and the energy densities shown correspond to the middle of each individual process window. It can be seen that deactivation initiates at lower temperatures for higher B doses, showing that either B precipitation or inactive cluster formation occurs more readily with increasing concentration. Takamura et al. have shown activation and subsequent deactivation data for a number of conventional species[17], and reported stable B activation up to temperatures of 800°C. Figure 4 shows comparable data although it is apparent that this level of activation is dependent upon the implanted dose and deactivation anneal.

CONCLUSIONS

The structural and electrical characteristics of B implanted Si following LTP was investigated. TEM images show a B induced process window dependence that suggests a concentration dependent absorption phenomenon unexpected from initial reasoning. Activation data illustrated a continual decrease in sheet resistance with increasing energy density for each implant condition studied. The deactivation kinetics increased with dopant

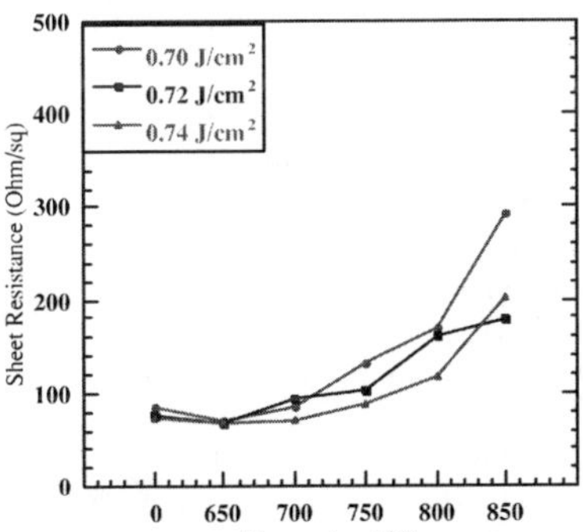

Figure 3: Sheet resistance as a function of isochronal annealing tenperature within the process window

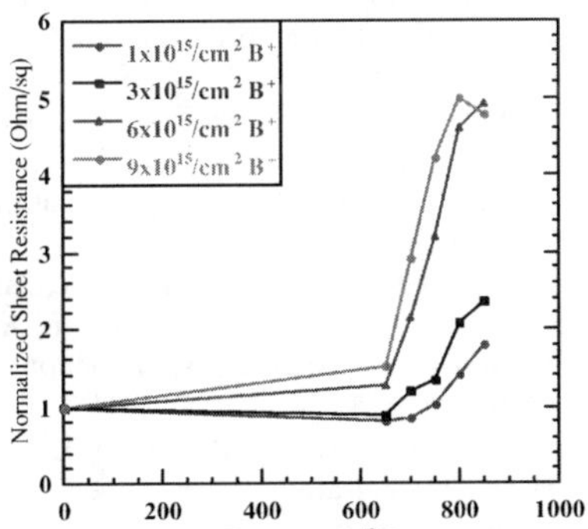

Figure 4: Sheet resistance as a function of isochronal annealing temperature for each implant condition studied

concentration suggesting a more efficient precipitation or inactive cluster formation process with increased implanted dose. The data show process window dependent activation and deactivation characteristics, which may be attributed to the reduction in defect density associated with increased energy densities within the process window.

REFERENCES

1) Semiconductor Industry Association, *International Technology Roadmap for Semiconductors: 1999 Edition*, Austin, TX: International SEMATECH, 1999.
2) D. J. Eaglesham, P. A. Stolk, H.-J. Gossmann, J. M. Poate, "Implantation and transient B diffusion in Si: The source of the interstitials," *Appl. Phys. Lett.* **65**, 2305 (1994).
3) F. A. Trumbore, Solid Solubility of Impurity Elements in Germanium and Silicon, *Bell Sys. Tech. J.* **39**, 205 (1960).
4) S. Talwar, S. Felch, D. Downey, Y. Wang, "Study of Laser Thermal Processing (LTP) to Meet Sub 130 nm Node Shallow Junction Requirements," *Proc. of the Conf. on Ion Implantation Technology*, (2000), p. 175.
5) M. Thompson, G. Galvin, J. Mayer, P. Peercy, J. Poate, D. Jacobson, A. Cullis, N. Chew, "Melting temperature and explosive crystallization of amorphous silicon during pulse laser irradiation," *Physical Review Letters* vol. 52, 1984, pp.2360-2363.
6) P. S. Peercy, M. O. Thompson, and J. Y. Tsao, "Dynamics of Rapid Solidification in Silicon," *Mat. Res. Soc. Symp. Proc.* vol 74, 1987.
7) J. Poate and J. Mayer, *Laser Annealing of Semiconductors*, Academic Press, New York, 1982.
8) R. Murto, K. Jones, M. Rendon, S. Talwar, "An investigation of Species Dependence in Germanium Pre-amorphized and Laser Thermal Annealed Ultra-Shallow Abrupt Junctions," *Proc. of the Conf. on Ion Implantation Technology* (2000), p. 182.
9) D. Sing, P. Borden, L. Bechtler, R. Murto, S. Talwar, "Boxer Cross Measurements of Laser Annealed Shallow Junctions," *Proc. of the Conf. on Ion Implantation Technology* (2000), p. 646.
10) A. G. Cullis, H. C. Webber, P. Bailey, "A Device for Laser Beam Diffusion and Homogenisation," *Journal of Physics E* (1979) **12**, 688.
11) W. F. Tseng, J. W. Mayer, S. U. Campisano, G. Foti, E. Rimini, "Grain Size Dependence in a Self-Implanted Silicon Layer on Laser Irradiation Energy Density," *Appl. Phys. Lett.* **32**(12), June 1978.
12) P. Griffin, J. D. Plummer, Private Communication.
13) http://cyberbuzz.gatech.edu/asm_tms/phase_diagrams/pd/b_si.jpg.
14) F. N. Rhines, *Phase Diagrams in Metallurgy*, McGraw-Hill, New York, 1956.
15) W. A. Tiller, K. A. Jackson, J. W. Rutter, B. Chalmers, "The Redistribution of Solute Atoms During the Solidification of Metals," *Acta. Metallurgica* vol. 1, July 1953.
16) R. Murto, K. Jones, M. Rendon, S. Talwar, "Activation and Deactivation Studies of Laser Thermal Annealing Boron, Arsenic, Phosphorus, and Antimony Ultra-Shallow Abrupt Junctions," *Proc. of the Conf. on Ion Implantation Technology* (2000), p. 155.
17) Y. Takamura, S. Jain, P. Griffin, J. D. Plummer, "A Study of the Deactivation of High Concentration, Laser Annealed Dopant Profiles in Silicon," *Mat. Res. Soc. Symp. Proc.* vol 669, 2001.

Mat. Res. Soc. Symp. Proc. Vol. 717 © 2002 Materials Research Society C1.11

Laser Thermal Processing of Alternate Dopants in Silicon

Mark H. Clark, Kevin S. Jones, Michael Rendon[1] and Kevin A. Gable, University of Florida, Department of Materials Science and Engineering, 525 NEB #33 Gainesville, FL 326-6130, U.S.A.
[1]Motorola Advanced Products R&D Lab, 3501 Ed Bluestein Blvd. MD-K10 Austin, TX 78721, U.S.A.

ABSTRACT

Traditionally, the choice of dopant has been limited to those species with the highest solid-solubility, however, Laser Thermal Processing (LTP) is not fundamentally limited by solid-solubility. Therefore, it is of interest to evaluate alternate dopants that have previously been excluded due to low solid-solubility. To this end, alternate dopants of ^{14}N, ^{121}Sb (n-type), ^{27}Al, ^{70}Ga, and ^{115}In (p-type) are compared to conventional dopants of As and B respectively, after LTP and post-LTP thermal processing. Dopants were implanted into <100> silicon wafers of opposite background doping type that had previously been amorphized to a depth of approximately 300 angstroms by a 15 keV $^{28}Si^+$ implant of $1x10^{15}/cm^2$ dose. An implant energy of 5 keV was sufficiently low to confine the implanted ions to the amorphous layer, with the exception of B, which required an energy of 2 keV. All species were implanted at doses of $1x10^{14}$, $5x10^{14}$ and $1.5x10^{15}/cm^2$. Samples were LTP utilizing a 308 nm, 18 ns laser pulse with a fluence of 0.680 J/cm^2. Post-LTP thermal processing of the samples consisted of a 900 °C rapid thermal anneal (RTA) in a nitrogen ambient for a duration ranging from spike to 300 seconds. Measurements of the sheet resistance, mobility and carrier concentration were taken after both LTP, and the post-LTP thermal processing. Experimental results show that LTP of alternate dopants increases the electrically active carrier concentration of Ga, Al and Sb above solid-solubility. Additionally, the amount of deactivation upon post-LTP thermal processing depends on the alternate dopant species.

INTRODUCTION

Variations of Laser Thermal Processing (LTP) have been studied to varying degrees for over thirty years. More recently, LTP has received renewed interest for the unique advantages it offers over conventional ion implantation coupled with rapid thermal annealing (RTA) for fabricating highly activated, ultra-shallow junctions necessary for future technology nodes. Traditionally, electrical activation has been limited by the solid-solubility of the dopant species in the silicon lattice. Therefore, in order to maximize the electrically active concentration, dopant species are chosen with the highest solid-solubility. In contrast, due to the rapid rate of the liquid-phase-epitaxy, LTP makes it possible to incorporate electrically active dopants in excess of solid-solubility, thus resulting in a low sheet resistance [1]. Consequently, LTP is not fundamentally limited by solid-solubility and thus allows consideration of unconventional/alternate dopants that have been historically excluded due to low solid-solubility. By utilizing LTP, unconventional dopants may be electrically activated to sufficient concentrations to provide a viable alternative to conventional dopants. Additionally, due to the meta-stable nature of the dopants activated by LTP, post-LTP thermal cycles (e.g., deep source drain anneal or second silicide anneal) can result in significant deactivation. The amount of deactivation that occurs during post-LTP thermal cycles depends on the dopant species [2], therefore, alternate dopants could potentially offer favorable deactivation kinetics (i.e., less deactivation) over conventional dopants. The two-fold objective of this study is to first, determine if alternate dopants can be electrically activated to significantly high concentrations by LTP and second, to determine if alternate dopants offer favorable deactivation kinetics after post-LTP thermal cycles, yielding a sufficiently low sheet resistance to provide a viable alternative to conventional dopants.

EXPERIMENTAL DETAILS

In order to study the effect of LTP and post-LTP thermal cycles on both p- and n-type alternate dopants in silicon the test matrix consists of two series of wafers. Both wafer series are silicon <100> wafers, differing in background doping type (e.g., n-type background for p-type dopants and vice versa). Processing of the wafers begins with a 15 keV, $1x10^{15}/cm^2$ preamorphizing Si^+ implant, which results in an amorphous crystalline depth of approximately 300 angstroms for all wafers in each series. After the preamorphizing implant the n-type wafer series was implanted with p-type dopants and the p-type wafer series was implanted with n-type dopants. The p-type alternate dopants consisted of aluminum, gallium and indium, with boron as the control. The n-type alternate dopants consisted of nitrogen and antimony, with arsenic as the control. Phosphorous was excluded from this study due to the previously published work performed by P. Griffin et. al., [3]. Implant energies were chosen to confine the dopant ion distribution to the amorphous region. For all species with the exception of boron (required a low energy 2 keV implant) and nitrogen (implanted as a diatomic species, resulting a 2.5 keV implant) all implants were performed at 5 keV. All dopant species were implanted at three different doses of $1x10^{14}$, $5x10^{14}$ and $1.5x10^{15}/cm^2$. LTP was performed on a 308 nm XeCl laser with a pulse length of approximately 18 ns. Exposures were 5 mm squares. In order to realize the process window for all dopant species and doses [4] laser fluences were varied from 0.680-0.800 J/cm^2. Upon completion of the LTP, all samples were diced into 4 mm square samples (to remove roll-off associated with edges) and subjected to a post-LTP thermal cycle consisting of a rapid thermal anneal (RTA) at 900 °C for spike to 300 second durations. Sheet Resistance measurements were made using a Four-Point-Probe for all samples both after LTP and post-LTP thermal cycles. Additionally, the mobility and carrier concentration was measured for each sample by Hall Effect.

DISCUSSION

Based on preliminary Plan view Transmission Electron Microscopy (PTEM) analysis, an optimal laser fluence of 0.680 J/cm^2 was chosen for analysis, because it largely achieves the process window for all species and doses. Figure 1 shows the sheet resistance for all dopant species and doses after LTP. As would be expected the sheet resistance for both n- and p-type dopant decreases with increasing dose. However, the most salient feature of Figure 1, is that the two lower doses ($1x10^{14}$ and $5x10^{14}/cm^2$) for both aluminum and indium and all doses of nitrogen do not activate sufficiently to achieve a sheet resistance that deviates from the bulk wafer, consequently they are not shown. Therefore, due to the lack of data for the lower doses of aluminum and indium the remainder of the study will focus on the highest dose of $1.5x10^{15}/cm^2$ for all dopant species in order to provide an equal comparison (for a 300 angstrom junction depth this is equivalent to a $5x10^{20}/cm^3$ concentration). Figure 1 also shows that for the p-type dopants the conventional dopant of boron has the lowest sheet resistance followed distantly by indium, gallium and aluminum. For the n-type dopants the conventional dopant of arsenic has the lowest sheet resistance followed closely by antimony.

To provide a greater insight into the electrical properties of the dopants after LTP, Figure 2 shows the carrier concentration as a function of dopant for a dose of $1.5x10^{15}/cm^2$. It is clear from figure 2 that not all of the dopants activate the same amount despite identical doses and LTP. In fact, it is shown that for the n-type dopants, the conventional dopant of arsenic exhibits the lowest sheet resistance followed closely by antimony, while for the p-type dopants, boron exhibits the highest carrier concentration followed by gallium, aluminum and indium. The carrier concentration data for both type dopants is in good agreement with the sheet resistance data (i.e., increase in carrier concentration results in decrease sheet resistance) with the exception of indium, which in comparisons to the other p-type alternate dopants exhibits a low carrier concentration despite a low sheet resistance. The discrepancy for the indium data is explained by Figure 3, which shows that the lack of carrier concentration for indium is compensated for by a high mobility.

Based on the electrical data presented in Figures 1-3, for the p-type dopants, boron has the most favorable electrical properties followed distantly by indium, gallium and aluminum. In the case of the n-type dopants arsenic is electrically favorable to antimony by a narrow margin, while nitrogen exhibits no electrical properties deviating from the bulk wafer. Over all for the electrical properties of both the n- and p-type alternate dopants are inferior to the conventional/control dopants, however, in agreement with the original hypothesis, several of the alternate dopants do activate in excess of their respective solid solubility limit due to LTP as shown in Table 1.

(a) (b)

Figure 1. Sheet resistance measured by Four Point Probe after LTP for (a) p-type and (b) n-type dopants.

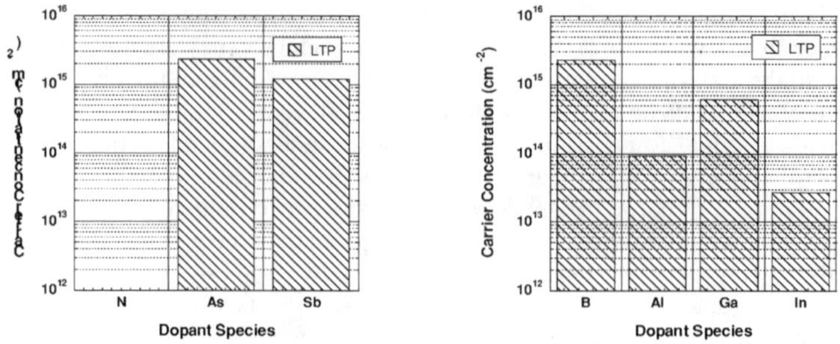

(a) (b)

Figure 2. Active carrier concentration measured by Hall Effect after LTP for (a) p-type and (b) n-type dopants of $1.5x10^{15}/cm^2$ dose.

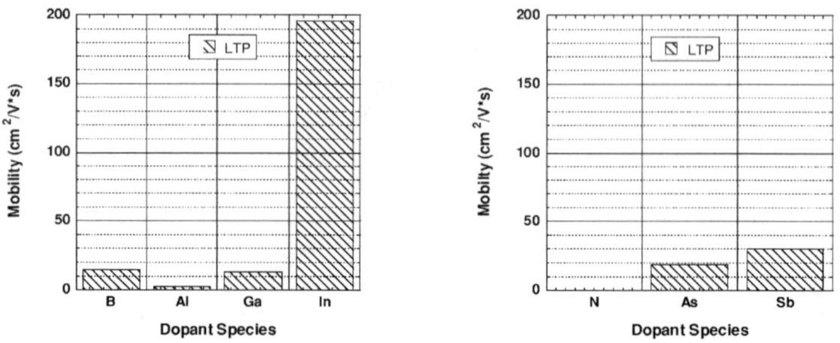

(a) (b)

Figure 3. Carrier mobility measured by Hall Effect after LTP for (a) p-type and (b) n-type dopants of $1.5x10^{15}/cm^2$ dose.

Dopant	Max Solid-Solubility (atoms/cm^3)	Xj (angstroms)	Active Carrier Conc. (atoms/cm^3)
Aluminum (Al)	**$2.1x10^{19}$**	**350**	**$2.7x10^{19}$**
Gallium (Ga)	**$4.0x10^{19}$**	**350**	**$1.8x10^{20}$**
Indium (In)	**$1.0x10^{19}$**	**350**	**$7.8x10^{18}$**
Antimony (Sb)	**$6.4x10^{19}$**	**350**	**$3.15x10^{20}$**

Table 1. Solid-solubility for alternate dopants and carrier concentration after LTP.

It is apparent from the electrical data presented thus far, that alternate dopants are not a viable alternative to conventional dopants after LTP. Despite this initial finding, alternate dopants may yet offer promise due to favorable deactivation kinetics upon post-LTP thermal processing. To observe the species dependence on deactivation, Figure 4 compares the carrier concentration achieved by LTP both before and after a 900 °C/300 second RTA in a nitrogen ambient. After the thermal cycle all dopants species experience a loss in carrier concentration, however, not all dopant species experience a proportional loss in carrier concentration. In fact, it is shown that for the p-type dopants, aluminum experiences a considerably larger deactivation than gallium or indium, while in contrast, boron exhibits only minor deactivation. For the n-type dopants the post-LTP thermal cycle results in a marginally greater deactivation for antimony than for arsenic. A further insight is gained from Figure 5, which plots the mobility as a function of dopant species, both before and after the post-LTP thermal cycle. With the exception of indium all of the dopants have an increase in mobility after the post-LTP thermal cycle. Unfortunately, as shown in Figure 6, the increase in mobility for the alternate dopants is insufficient to compensate for the lower carrier concentrations, thus resulting in a comparatively higher sheet resistance even after the post-LTP thermal cycle. Consequently, alternate dopants do not offer favorable deactivation kinetics after the post-LTP thermal cycle. In fact, based on the electrical data after post-LTP thermal processing, alternate dopants are not a viable alternative to conventional dopants when low sheet resistance is desired.

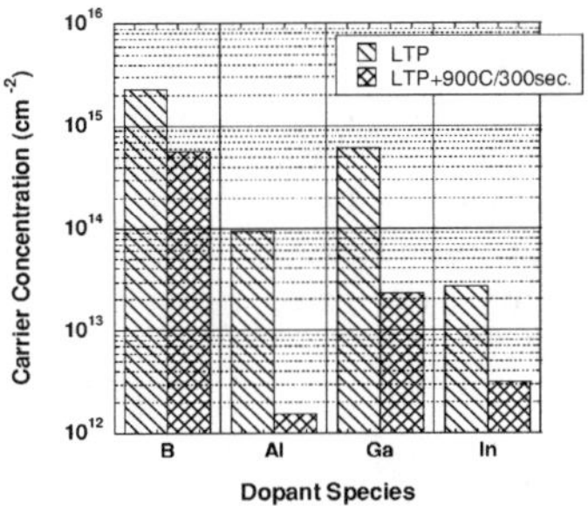

(a)

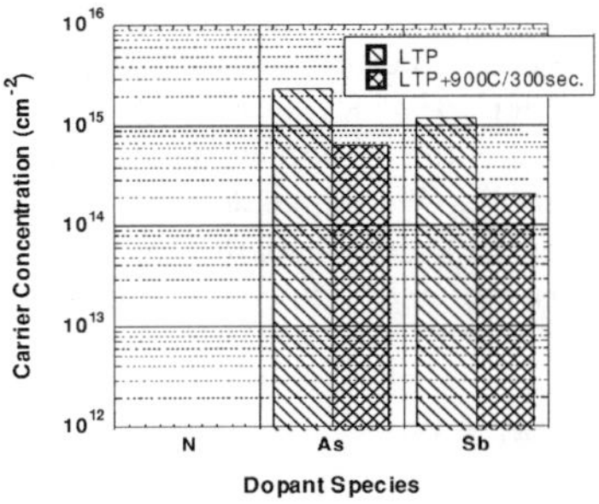

(b)

Figure 4. Carrier concentration resulting from LTP and LTP + 900 °C/300 sec. RTA for (a) p-type and (b) n-type dopants of $1.5x10^{15}/cm^2$ dose.

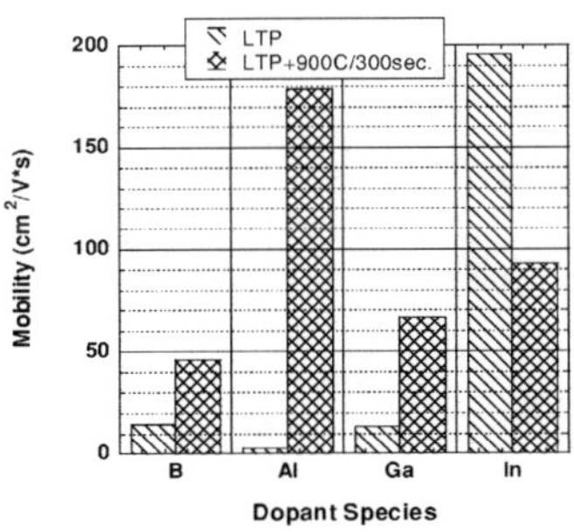

(a)

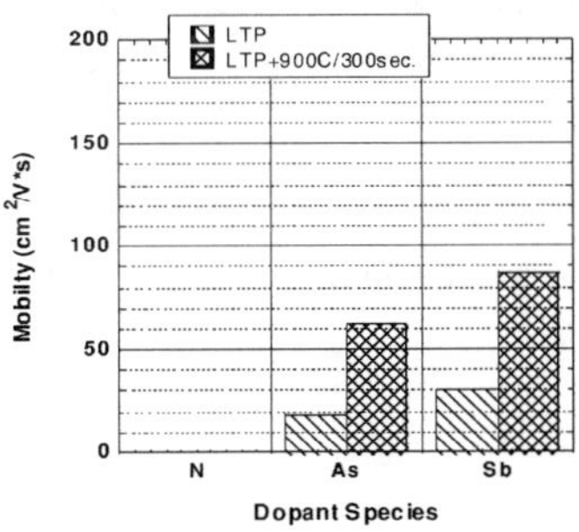

(b)

Figure 5. Carrier mobility resulting from LTP and LTP + 900 °C/300 sec. RTA for (a) p-type and (b) n-type dopants of $1.5x10^{15}/cm^2$ dose.

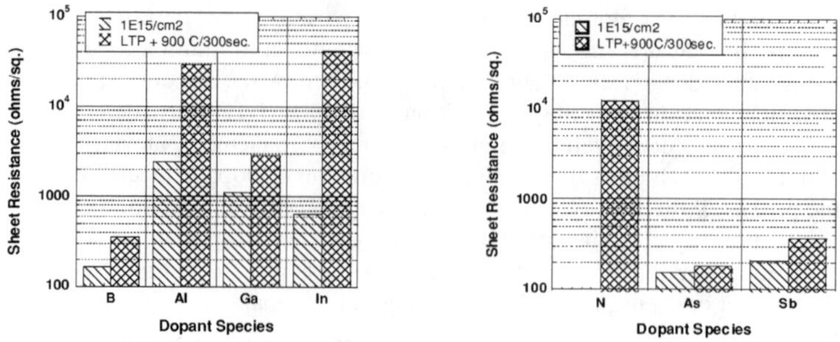

(a) (b)

Figure 6. Sheet resistance resulting from LTP and LTP + 900 °C/300 sec. RTA for (a) p-type and (b) n-type dopants of $1.5x10^{15}/cm^2$ dose.

CONCLUSION

Alternate dopants can be activated above solid-solubility by LTP, however, greater activation and low sheet resistance is realized for the conventional dopants. Additionally, while deactivation is effected by dopant species, alternate dopants do not offer favorable deactivation over that of conventional dopants after a 900 °C/300 second RTA. In summary, based on the electrical data, the conventional dopants of boron and arsenic out perform their respective p- and n-type alternate dopant counter parts investigated in this study.

AKKNOWLEDGEMENTS

This research was supported by SEMATECH.

REFERENCES

1 S. Talwar, G. Verma, and K. H. Weiner, IEEE, 1171-1174 (1999).

2 S. Talwar, G. Verma, K. Weiner, and C. Gelatos, in *Laser thermal processing for shallow junction and silicide formation*, Santa Clara, California, 1998, p. 74-81.

3 P. Griffin and J. D. Plummer, "Reduced MOSFET Series Resistance in Devices Below 70 nm," Report No. 661.001-2 (2000).

4 R. Murto, K. Jones, M. Rendon, and S. Talwar, IEEE, 182-185 (2000).

Mat. Res. Soc. Symp. Proc. Vol. 717 © 2002 Materials Research Society C1.12

The Influence of Low Temperature Pre-Annealing on the Defect Removal and the Reduction of Junction Depth in Excimer Laser Annealing

Sungkweon Baek, Taesung Jang, and Hyunsang Hwang
Department of Materials Science and Engineering,
Kwangju Institute of Science and Technology,
#1, Oryongdong, Buk-gu, Gwangju, 500-712, KOREA

ABSTRACT

The influence of low temperature pre-annealing on p^+/n ultra-shallow junction was investigated. An ultra-shallow junction was formed by means of B_2H_6 plasma doping at an energy of 500V. The activation was performed by excimer laser annealing. To study the low temperature annealing prior to laser annealing, furnace annealing at 300°C~500°C for 5min was performed. Compared with control samples with no pre-annealing, the low temperature pre-annealing significantly improves junction characteristics, resulting in a reduction of junction depth and a lower leakage current density. A cross-sectional transmission electron microscopy analysis confirmed the lower defect density, which explains the lower leakage current. By optimizing the process conditions, excellent electrical characteristics of the p^+/n ultra-shallow junction such as a junction depth of 28nm and a sheet resistance of 250Ω/sq. can be obtained.

INTRODUCTION

The scaling of a metal-oxide-semiconductor field-effect-transistor (MOSFET) device channel length for high-speed application requires an ultra-shallow junction depth in order to suppress the short channel effect and to obtain high current drivability. Although a shallow junction can be formed by a rapid thermal annealing (RTA) process, it is difficult to form ultra-shallow junctions due to transient enhanced diffusion (TED)[1]. The excimer laser annealing was chosen as a candidate for the rapid thermal annealing. Compared with a RTA, excimer laser annealing has advantages, which include uniform distribution and full activation.[2,3] However, since the laser annealing was performed for a very short time, all defects generated by implantation cannot be removed. To solve this problem, the pre-annealing is needed. H. Tsukamoto et al. reported on the effect of annealing at 600°C for 1h prior to the laser annealing of a shallow junction prepared by BF_2 implantation at 15keV. An improvement in crystallinity and a reduction in the generation-recombination centers due to annealing at 600°C for 1h were reported.[4] However they concluded that annealing at 600°C for 1h prior to laser annealing increases the junction depth.

In this paper, we report on an investigation of the structural and electrical characteristics of ultra-shallow p^+/n junctions formed by plasma doping at 500V and low temperature pre-annealing at 300°C~500°C for 5min followed by excimer laser annealing. The structural and electrical characteristics of ultra-shallow junction were improved through the low-temperature annealing prior to laser annealing. In addition, a significant reduction in junction depth was observed as the result of pre-annealing at low temperature.

EXPERIMENTAL DETAILS

An n-type (100) oriented, silicon wafers were used as a substrate material. Plasma doping was performed using B_2H_6 (3%) gas diluted with H_2 gas. The wafers were placed in the plasma and a negative DC bias was applied to the substrate. The plasma doping energy and dose were 500V and $1\times10^{16}cm^2$, respectively. After the plasma doping, laser annealing was performed with a homogenized KrF excimer laser. The 248nm excimer laser has a pulse duration of 25ns at a repetition rate of 1Hz. To study the pre-annealing effect, some samples were annealed in a furnace at 300°C~500°C for 5min in a nitrogen ambient. The sheet resistance and active doping concentration were measured by means of a Hall measurement system. The dopant profiles were analyzed by secondary ion mass spectrometry (SIMS). To investigate structure defects, cross-sectional transmission electron microscopy (XTEM) analysis was performed. In addition, a p^+/n junction diode was fabricated and the reverse bias junction leakage current was investigated.

RESULTS AND DISCUSSIONS

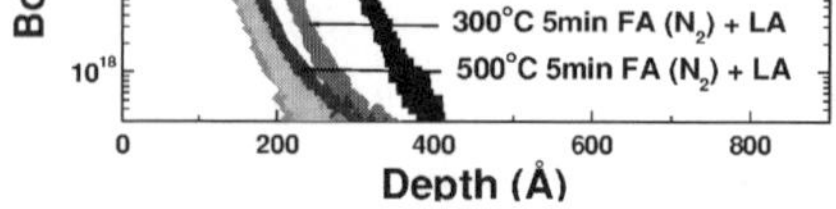

Figure 1. The boron SIMS depth profile of laser-annealed samples. With increasing the pre-annealing temperature, a reduction of junction depth was observed.

Figure 1 shows the SIMS boron depth profiles of laser annealed and pre-annealed samples. The boron depth profile of an as-implanted sample is exactly the same as that for a sample annealed at low temperature without additional laser annealing. In other words, a negligible boron redistribution was observed during the low temperature pre-annealing. We also investigated the boron depth profile of a laser-annealed sample using various pre-annealing conditions. The laser energy fluence was $500mJ/cm^2$ and a single pulse was used. The laser pulse at $500mJ/cm^2$ cannot melt the silicon.

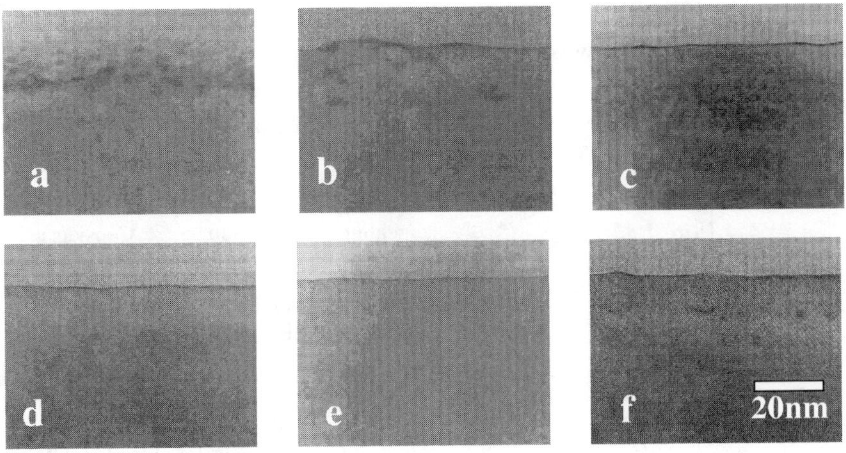

Figure 2. The XTEM images of the B2H6 plasma-doped and laser annealed samples. (a) as-implant, (b) excimer laser annealed sample without pre-annealing, (c) laser annealed sample with pre-annealing at 300°C, (d) 400°C, (e) 500°C for 5min, (f) 350°C 30min post-annealing.

Compared with the control sample without pre-annealing, the sample with an additional low temperature annealing and standard laser annealing causes a significant reduction in the boron depth profile. Considering the substrate doping concentration of 1×10^{18}atoms/cm^3, the junction depth of the control sample was determined to be 37nm. In contrast the junction depth was 28nm and 22nm for samples that had been annealed at 300°C and 500°C, respectively.

Using cross-sectional transmission electron microscopy (XTEM), we investigated structural defects as shown in Figure 2. An as-implanted sample has a 20nm-thick defect-rich layer (Fig. 2(a)). When laser annealing is performed without pre-annealing, the formation of dislocation loops was observed (Fig. 2(b)). However, as shown in Fig. 2(c), the sample that had been pre-annealed at 300°C shows a defect-free structure. This defect-free structure can also be obtained at higher pre-annealing temperature (Fig. 2(d),(e)). Figure 2(f) shows the XTEM image of laser-annealed sample followed by post-annealing at 350°C for 30min without pre-annealing. This indicates that the defects cannot be removed completely with low temperature post-annealing. Therefore, the defect-free structure can be obtained with low temperature annealing prior to excimer laser annealing. Although the exact mechanism of the shallow junction depth of the pre-annealed sample is not clear, the reduction in defect density which was confirmed by TEM analysis provides a satisfactory explanation of the reduction in junction depth.

Figure 3 shows the sheet resistance and active carrier concentration as a function of the pre-annealing temperature. For comparison, the sheet resistance of a control sample is also included.

A pre-annealing was performed in an N_2 ambient for 5min at various temperatures. After furnace pre-annealing, laser annealing was carried out in vacuum. The energy fluence was 500mJ/cm^2 and a single pulse was used. Compared with the control sample, the sheet resistances of the pre-annealed samples increased and carrier concentration decreased with increasing annealing temperature. Considering the technology roadmap, the sheet resistance of the pre-annealed sample is acceptable for use in sub-100nm MOSFET devices.

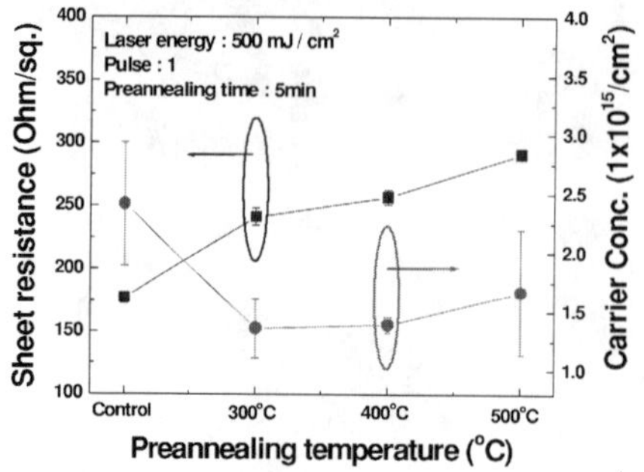

Figure 3. Sheet resistance and carrier concentration as a function of pre-annealing temperature, as measured by Hall measurement system.

To investigate the effect of pre-annealing on junction leakage characteristics, a p$^+$/n junction diode was fabricated. The excimer laser energy fluence was 500mJ/cm^2 and a single pulse was used. As shown in Figure 4(a), the reverse bias leakage current of the pre-annealed sample was lower than that of the control samples. Figure 4(b) shows the cumulative probability of a reverse bias leakage current, as measured at –2V. Compared with the control sample, diode with pre-annealing shows a lower reverse bias leakage current and tight distribution of leakage current. Wee also found that the ideality factor of the control sample is higher than that of pre-annealed samples, which can be explained by the partial contribution of defect-induced generation current for the control samples.

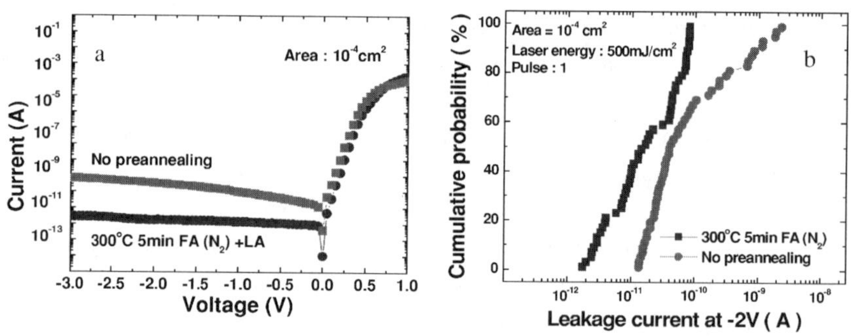

Figure 4. (a) The current-voltage characteristics of the junction diodes. (b) the cumulative probability of junction leakage current measured at –2V. With pre-annealing at a 300°C for 5min, a significant reduction of reverse leakage current was observed.

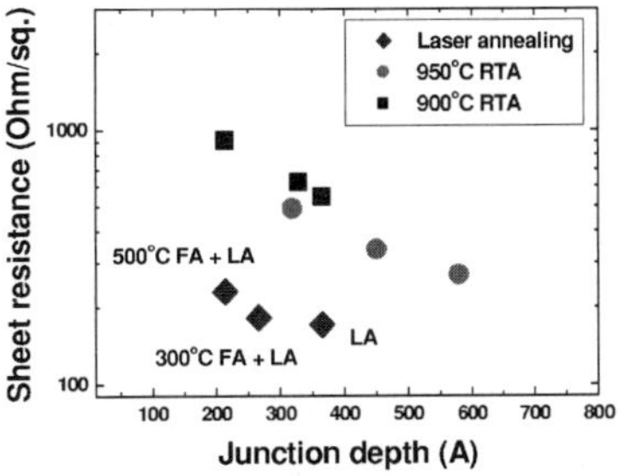

Figure 5. The variation of junction depth vs. sheet resistance. Through the laser annealing, the reduction of junction depth and sheet resistance were obtained.

Figure 5 shows the variation of junction depth (X_j) vs. sheet resistance (R_s) as a pre-annealing temperature. For comparison, the junction depth and sheet resistance obtained by rapid thermal annealing are also indicated. Compared with the rapid thermal annealing (900°C/950°C, 10/30/60sec), the excimer laser annealing with low temperature pre-annealing can attain the reduction of both junction depth and sheet resistance.

CONCLUSIONS

We investigated the effect of low temperature pre-annealing on an ultra-shallow p+/n junction. By employing a low temperature annealing prior to excimer laser annealing, a significant improvement in junction characteristics such as shallow junction depth, low leakage current and a lower defect density was observed. By implementing a low temperature annealing prior to laser annealing, a junction depth as shallow as 28nm and a sheet resistance of 250Ω/sq. can be obtained. The improvement in junction characteristics can be explained by the effective elimination of defects which were generated by the plasma doping process. The reduction in defects was confirmed by TEM analysis and characterization of the reverse bias leakage current. A low temperature annealing before laser annealing represents a promising process for implementing an ultra-shallow junction for sub-100nm MOSFET applications.

ACKNOWLEDGMENT

This research was supported by Brain Korea 21 and System IC 2010 project

REFERENCES

1. A.Agarwal, H.J.Gossmann, D.J.Eaglesham, S.B.Herner, A.T.Fiory, and T.E.Haynes, Appl. Phys. Lett., **74**, 2435 (1999)
2. Y.F.Chong, K.L.Pey, A.T.S.Wee, L.Chan, Y.F.Lu, W.D.Song, and L.H.Chua, Appl. Phys. Lett., **76**, 3197 (2000)
3. B.Yu, Y.Wang, H.Wang, Q.Xiang, C.Riccobene, S.Talwar, and M.Lin, Tech. Dig. Int. Electron Devices Meet., 509 (1999)
4. H.Tsukamoto, H.Yamamoto, T.Noguchi, H.Masuya, and T.Suzuki, Jpn. J. Appl. Phys., part1 **35**, 3810 (1996)

Device Engineering Options

Mat. Res. Soc. Symp. Proc. Vol. 717 © 2002 Materials Research Society C2.1

Optimisation of Junctions formed by Solid Phase Epitaxial Regrowth for sub-70nm CMOS

Richard Lindsay, Bartlomiej J. Pawlak[1], Peter Stolk[1], and Karen Maex
IMEC, Kapeldreef 75, Leuven, B3001 Belgium.
Philips Research Leuven, IMEC, Leuven, B3001 Belgium.

ABSTRACT

For the 70nm CMOS node, it is anticipated that conventional implantation and spike annealing approaches, even with pre-amorphisation and co-implantation, are unlikely to provide pMOS junctions consistent with the ITRS requirements. Here the junction performance is limited by equilibrium solid solubility.

As laser annealing and in-situ doping techniques currently have unsolved integration problems, there is a renewed interest in using solid phase epitaxial regrowth (SPER) to form ultra-shallow metastable junctions. Such junctions have the potential to have an active dopant profile similar to the as-implanted profile. This offers above equilibrium solid solubility and abrupt profiles compatible with 70nm and even 45nm nodes. However there are concerns about residual defects, deactivation, diffusion and uniformity.

In this paper we show how the Ge, F and B implant and SPER anneal can be optimised for abrupt, uniform and highly activated B junctions. There is latitude for higher doses and energies than conventional implants, however results show that this may lead to clustering causing enhanced deactivation and reduced mobility. We give attention to the probing issues involved in characterising partially annealed junctions.

With this approach, p-type junctions having a sheet resistance of 265ohms/sq and depth of 22nm are realised which are compatible with 70nm and potentially 45nm CMOS nodes.

INTRODUCTION

Extending conventional implantation with spike anneal

To optimise the performance of CMOS transistors there are certain restrictions placed on the junctions which extend underneath the gate. These restrictions are dictated by the International Roadmap for Semiconductors (ITRS) and are shown in figure 1 as a function of transistor gate length.

In figure 1 the SEMATECH curve is included to show the limits of using B only and conventional RTA [1]. This curve is a collection of all the best results from literature as of 2001 and these junctions on the graph are almost always formed by high temperature spike anneals (>1050°C for <1s with ramp up rates >75°C/s and ramp down >50°C/s) in <0.1% O_2 concentrations and <1.5keV B implants through <2nm [2] screen oxide. Irrespective of dose, energy, tilt angle, multiple B implants, dose rate, spike temperature, ambient, it seems very difficult to improve much beyond this line with B alone. However results in-house and elsewhere [3] have gained up to 7nm to the left of this curve by using very sharp spike anneals and high B doses.

Co-implanting several species before or alongside the B implant can allow for changing the channeling, diffusivity and activation of the dopant. For pre-amorphisation, Ge, Si or GeF_2 [4,5,6] are the most commonly used. Experiments elsewhere show that C, N, F, Ga, Ge, and In can decrease the B diffusion while Al, Ga, and In can improve its activation [7]. The trouble is that most often these species do not pre-exist in the substrate and require to be co-implanted causing further TED.

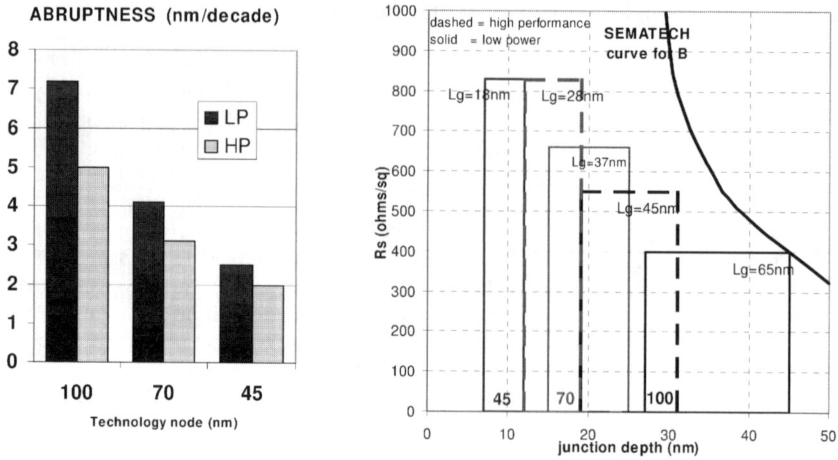

Figure 1. ITRS 2001 specifications on junction depth, abruptness and resistance.

Of all these candidates for co-implantation the best results have been obtained with Ge, F, GeF_2, and C. BF_2 has been shown to offer decreased diffusion and possibly increased activation. In BF_2 however, the position of the F is not optimal for full reduction of TED and experiments have shown that the F should be implanted separately with an energy approximately 10 times that of the B [8]. Deep and shallow C has been shown to strongly reduce TED but concerns remain over its effect on junction leakage [9].

In-house results with B or BF_2 co-implanted with a combination of Ge, GeF_2, F have provided junctions <8nm to the left of the Sematech curve between 30-40nm. The best results were obtained by using the combination of low temperature spike anneal with high doses. The question is whether the Ge and F have more dramatic effects for shallower junctions to reach the Lg=37nm spec for the LP 70nm node, and whether their use can improve the junction abruptness below 4nm/decade. This will be reported on at a later date [10].

The use of non-standard HALO, or pocket, junctions or offset spacers before extension implant may extend the lifetime of conventional ion implantation and spike anneal to the 70nm CMOS node. To implement these though requires significant additional processing and introduces sources of non-uniformity and repeatability. The lifetime of ion implantation itself will depend on the ability to maintain high beam currents without resulting in significant energy contamination. At the moment the levels of contamination are generally much less than the dopant diffusion during anneal, but when junctions <30nm are required limits can appear, especially for SPE regrown junctions, as shown later.

Avoiding deactivation and diffusion of the conventional junctions and HALOs is now essential. This calls for process integration adjustments like low temperature spacers, gate pre-doping or extension-last transistor architectures, and low temperature silicides. Conversely, the activation anneal of the junction can have adverse effects on advanced integration materials like high k gate dielectrics and metal gates. The use of lower thermal budgets for annealing is becoming necessary.

Laser thermal annealing (LTA) and in-situ doping with SiGe offer extremely abrupt and highly activated shallow junctions but they are plagued with integration problems. The main issues with LTA is the poly-Si melting above isolation and pattern effects on absorption. Various capping layers have shown to help but currently there are no known solutions [11,12]. For in-situ doping the main problem is obtaining selectivity on the active B concentrations >1e20atoms/cm^2 and eliminating variations in dopant incorporation caused by active area variations [13]. It is very possible that conventional approaches will not meet the 70nm requirements and integratable solutions for LTA or in-situ doping are not expected to arrive in time for even the 45nm CMOS node.

SPE Regrowth

Recently there has been a renewed interest in taking advantage of the high levels of activation achieved at low temperatures by regrowth of a doped amorphous region. Beforehand higher temperatures were favoured as they annealed all the defects from the junction and there was no need for low temperature device processing nor such stringent requirements on the junction profile. The low temperatures required for SPER of around 550°C-650°C do no allow for significant dopant diffusion in the crystalline Si so the resultant junction looks similar to the as-implanted profile. This lack of diffusion allows for higher implant energies and doses than that allowed for spike annealed junctions. However, these and other parameters can affect the level of activation, mobility, abruptness, uniformity and the amount of residual defects. The positioning of the amorphous region relative to the junction and the introduction of F can affect the dopant redistribution and abruptness. Performing anneals before and after the SPER may affect the residual defects and junction profile.

Much data already exists on several of these influences, especially on post-annealing and defect formation [14-18]. This paper will overview some of the existing knowledge and combine it with the results acquired here on the optimisation of the B dose and energy and the Ge and F implantation for forming junctions compatible with the 70nm and 45nm CMOS nodes.

The need for low temperature post-SPER processing creates difficulties in optimising the junction in integration experiments so the junctions have to be optimised for all Rs, depth, abruptness and defects to evaluate the merits of SPER before integration. The main concern in integration is the residual defects in the channel increasing the off-state leakage. Defects under the extension are normally consumed by the deeper contact junction. Optimised SPER junctions have been integrated into devices here and elsewhere [19] with disposable spacers and will be reported on at a later date.

EXPERIMENT

In n-type 200mm wafers, several experiments were set up to investigate the influence of implant and annealing parameters for SPER junction performance and identify the process window for CMOS integration. Ge was implanted at energies ranging from 8keV to 20keV mostly at 1e15atoms/cm^2. B was implanted between 0.5keV and 1.5keV with doses ranging from

1e15 to 3e15. In some wafers BF_2 or F was implanted with the former at 4.5keV, 3e15/cm^2 and the latter between 3-10keV at 1e15. All implants were angled at 7° tilt and 27° twist on the Applied Materials xR80 LEAP.

Some wafers received a pre-anneal of either 420°C for 30minutes or 500°C for 10minutes. The SPER anneal was tested between 500°C for several minutes up to spikes at 900°C or 1000°C. Post-annealing was done mostly at 900°C for 5s but also for the full range of 700°C to 1000°C. Annealing was either done on whole wafers in the AST-STEAG SHS2800 or on the ASM LEVITOR [20], or in pieces in the HEATPULSE RTA.

Rs measurements were either done by four-point probe (4PP) on the SSM 240 with a probe depth of 25nm or by probe spacing which probes 5nm deep. Probe spacing is typically used in spreading resistance profiling (SRP) on the SSM 150 and is a multiple 2-point measurement. Chemical dopant profiling was done using SIMS on the ATOMIKA 4500 and active dopant profiling by SRP, again on the SSM150. Amorphous and defect imaging was done in by high resolution cross-sectional TEM.

RESULTS AND DISCUSSION

Probing SPER junctions – activation, deactivation and leakage

Figure 2 shows the Rs of various SPER junctions formed at different temperatures for 1 minute when measuring with 25nm or 5nm probes. With spike annealed junctions down to 20nm even measuring with 100nm deep probes typically gives accurate results [21]. The important thing is of course the depth of the carriers at the substrate level, not at 1e18atoms/cm^3, and the depletion width. However the results show that in measuring Rs with 25nm probes on SPER junctions the substrate Rs contributes (and is here typically 200ohms/sq). This is not surprising as the electrical junction is where the original amorphous crystalline interface lay and there are many residual defects causing leakage. We will see that this type of probing comparison may provide some information about the formation of the junction and residual damage.

In both cases shown in figure 2 the Rs decreases coming from 550°C, 1min up to 650°C, 1min. This is due to incomplete regrowth (see HRTEM results later). For the samples with F, regrowth is so slow at these temperatures that no significant junction is formed until 700°C. Correspondingly the Rs is extremely high with the 5nm probe and at the substrate level with the 25nm probe. The formation of the F junction creates the depletion region which isolates the substrate giving an increased Rs. Between 700°C and 750°C the junctions do not diffuse or deactivate then at 800°C the junctions suffer from deactivation. This agrees with that reported elsewhere where a possible reason for the deactivation is the back-flow of Si interstitials coming from the end of range defects at the original amorphous interface [14]. Above 900°C the junctions begin to diffuse significantly and the Rs decreases. Here the pre-amorphised and non-amorphised junctions coincide in Rs (see 3BF4.5) as the equilibrium solid solubility is attained. The reason why the Rs with 25nm dips at 800°C can tell us something about the effect of the defects on junction leakage. Either the deactivation causes the depletion region to enter further into the junction, or perhaps more likely, is that extended defects are formed from the heavily damaged region and the resistance from the junction to the substrate is reduced dramatically. Plan view TEM images shown elsewhere [14,15] show that at temperatures around 800°C the point defects evolve into {311} rod-like defects.

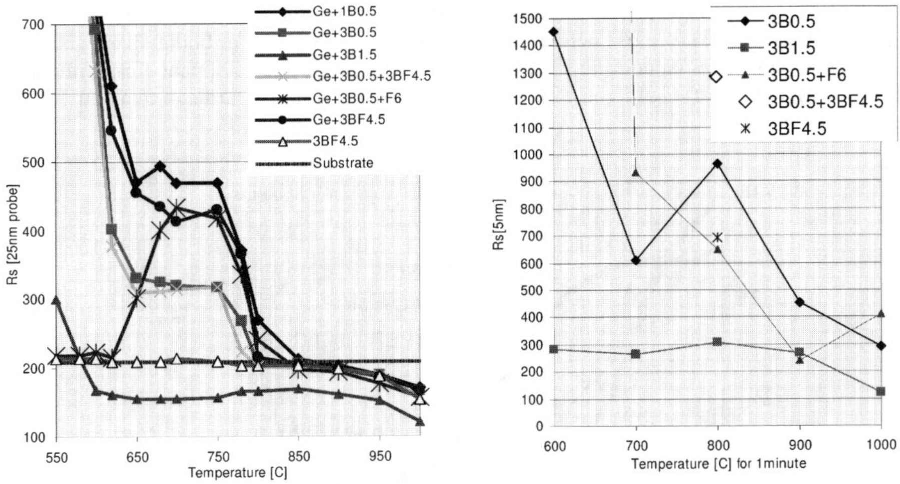

Figure 2. Rs of various SPER junctions measured by 25nm and 5nm probing. Both give an insight to leakage mechanisms in the junction. The nomenclature is, e.g. 3B0.5+F6, indicates 0.5keV B 3e15 + 6keV F 1e15, likewise for the rest.

What is also interesting is that the deactivation is dependent on the energy and dose. For 1.5keV B the Rs only rises by 16% at 800°C whereas for 0.5keV B the increase is almost 60%. The SIMS graphs in figure 3 comparing these junctions regrown at 600°C for 10mins can provide some explanation. The 1.5keV B junction is seen to have a lower peak concentration of B than the 0.5keV case. It is possible that the higher amount of non-activated, or clustered, B can cause enhanced deactivation. If a comparison is made between the 3e15 B and 3e15 B + 3e15 BF_2 it is clear that the additional B is not contributing to the junction activation even though it should make the junction deeper (4.5keV BF2 has a similar B profile as 1keV B). This indicates that having higher inactive B concentrations can degrade the mobility of the carriers. From the 5nm probe data the deactivation is also seen to be worse for the higher B concentration. Figure 4 shows more systematic Rs data on 1keV B with doses of 1e15, 2e15 and 3e15/cm^2 from which it is clear that at low temperatures there is no benefit of going higher than 2e15/cm^2 and at 3e15 there is evidence of reduced hole mobility and greater deactivation. At 900°C for 1min the higher doses diffuse further and the conventional junction Rs trends are seen.

Diffusion during SPER regrowth

Figure 3 shows how the dopant can redistribute even during low temperature processing at 600°C. There are several interesting points to note. Diffusion occurs at around 2e20atoms/cm^3 during SPER. This likely corresponds to the active concentration of the dopants in the SPER region as only substitutional B is subject to diffusion. This agrees with previous observations [14]. The diffusion is far more than expected from thermal diffusion at 600°C so TED seems responsible. The migrating Si interstitials diffuse the dopants towards the end-of-range damage, where they are trapped. For this temperature the dopants do not migrate towards the surface during SPER as the retained dose before and after SPER is very close for both 0.5keV and 1.5keV. Incidentally, self-sputtering seems responsible for some loss of dopant during implant and is 19% for the 0.5keV 3e15 and 11% for the 1.5keV 3e15. This is further explained elsewhere [22].

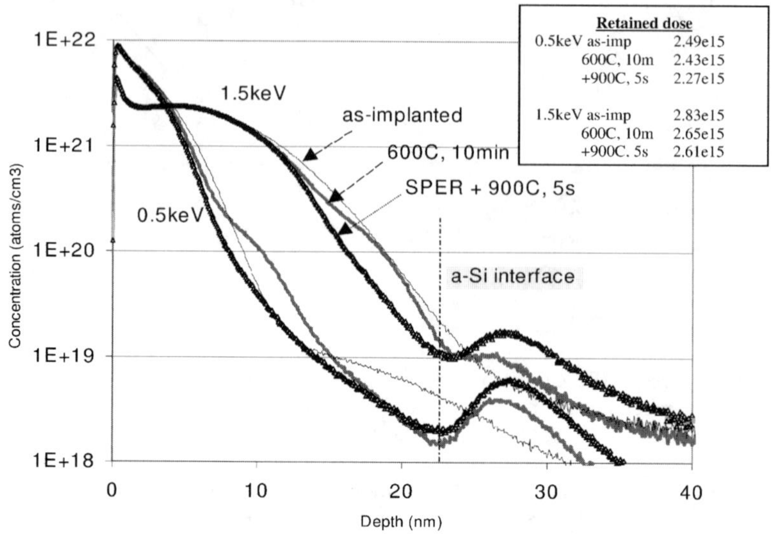

Figure 3. SIMS of 0.5keV and 1.5keV 3e15 B as-implanted and after SPER at 600°C for 10mins and with a post anneal of 900°C for 5s.

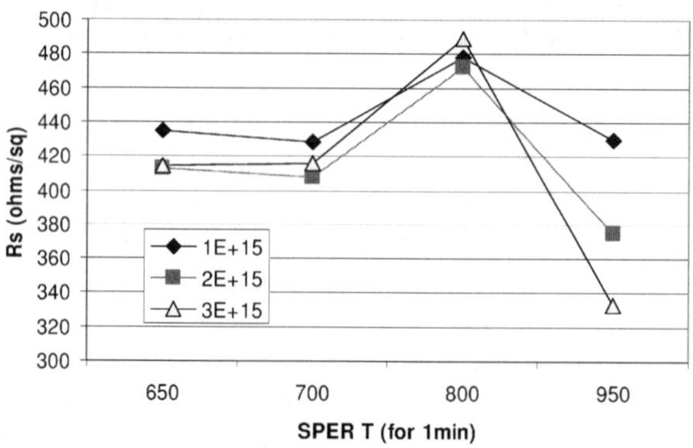

Figure 4. Rs (5nm probe) of B at different doses after various SPER temperatures for 1min.

It is worth pointing out at this moment that energy contamination appeared on the 0.5keV samples. For throughput, these implants were decelerated from 3keV. In house results on 0.5keV 1e15 decelerated from 1,2,3, and 4keV show that the junction depth at 1e18 increases linearly as 40x(% energy contamination). Eliminating energy contamination required a reduction in beam current by a factor of 10. Unless the channel concentration rises above 1e19atoms/cm^3, techniques like plasma doping (PLAD) may become necessary [23].

Central to implementing SPER junctions into transistors is that the leakage mechanism arising from the residual defects is negligible compared to the other influences on off-state leakage. Higher SCEs and gate-to-drain leakage intrinsic to shrinking device dimensions do increase the acceptable leakage from these residual defects. On top of this, shallow junctions may allow easier defect removal underneath the gate in the channel region. It is also important to identify how the nature of the defects affect leakage, cf. point defects, {311} defects and dislocation loops.

Post and pre-annealing – diffusion, deactivation and defects

The use of post-annealing to reduce the defect density has been widely used in laser annealing and SPER [14]. The idea is to anneal the defects without causing deactivation or dopant diffusion. Figure 3 shows the effect of a 900°C, 5s post anneal on the junction profile. Upon post-annealing at 900°C for 5s there is substantial dopant diffusion to the end of range defects and the profiles indicate no uphill diffusion. The release of the interstitials from the end-of-range defects at 900°C seem to be responsible for this dopant redistribution and this agrees with other reports. The small dose change during the 900°C, 5s anneal is most likely due to outgassing in pure N_2. There are several important points with regard to integration here. The result of this dopant diffusion and activation during post annealing is that the junction may be considerably deeper than it was originally (at the a-Si interface) and for the 1.5keV case the abruptness has degraded considerably (from <1nm/decade to >15nm/decade). This is only the case however if there is no activation of the dopants beyond the interface after the 600°C, 10min SPER anneal. Looking at the SRP profiles in figure 5 for the two cases indicate that 600°C does not sufficiently activate the B above a concentration of 1e18/cm^3 beyond the interface and so it is possible that the junction depth is at 22nm and the abruptness is <1nm/decade. SRP also suggests a reduction in mobility at the highly doped surface region. It is noted however that the SRP profiling has some problems in profiling such highly doped SPER junctions where the amount of clustered dopants may affect the mobility used in the calculation and carrier spilling may produce errors in electrical junction depth measurements. With this in mind, further experiments are underway to confirm whether the 600°C SPER does not activate the c-Si tail and allow a <1nm/decade abruptness. Judging by this work and others there seems to be very little margin for post annealing to achieve sufficient defect removal without activating the profile tail or deactivating the bulk junction. The ideal situation would be to remove the defects before SPER.

There are various other approaches that may affect the number of end of range defects that have been researched elsewhere. Among these are wafer temperature during implant [24], beam current dynamic annealing [25], pre-annealing [12], amorphising species, dose and energy, or the use of co-implants to create vacancies for the excess interstitials. For normal device processing there may be operational or device issues with implementing all of these apart from pre-annealing. Pre-annealing at temperatures between 400-500°C for several seconds to minutes has been shown to improve the amorphous interface roughness as it allows for some Si redistribution without inducing SPER. There are also indications that this anneal may bring nearby excess interstitials toward the interface and reduce the number of stacking faults. However it is not clear whether this can affect the end of range defects in low temperature SPER.

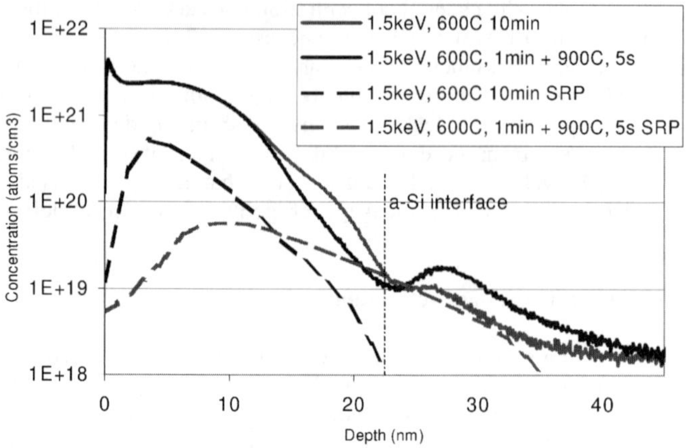

Figure 5. SRP of 3e15 1.5keV B before and after post annealing at 900°C for 5s.

Figure 6 shows the effect on the junction profile and regrowth rate by a pre-anneal of either 420°C for 30mins or 500°C for 10s. As was expected the junction profiles do not change at all. The same amount of B trapping at the end-of-range defects is observed. This indicates that there is no significant reduction in residual defects. However it is worth noting that the Rs consistently takes longer to reach its final value after SPER. This could be due to a reduced regrowth rate as a result of more difficult nucleation. It could also be due to less leakage to the substrate in the Rs measurement, implying possible changes in defect structure. Further investigations are required at this point to understand and optimise the pre-anneal.

Effect of Fluorine

Fluorine is known to reduce B diffusion in most instances but has also been observed to increase its diffusion [26]. Comparisons have also been made elsewhere between B and BF_2 for SPER [14]. In figure 7 the effect of a separate 6keV 1e15/cm^2 F implant before SPER is shown to enhance the diffusion of B in the amorphous state. By looking at the Rs measurements in figure 7 and the XTEM in figure 8 it is evident that the SPER here is not complete in 1 minute at 600°C. This implies that the enhanced B diffusion observed with F occurs during the amorphous phase. This is explained elsewhere as the F trapping the dangling bonds in the amorphous Si thereby allowing B to diffuse more freely [26]. The positioning of the F could be important to allow some 'shoulder' diffusion to obtain a more abrupt profile without implanting deeper and having more B at the end-of range damage. From 5nm PS measurements, F is seen to reduce the Rs due to this enhanced diffusion. F is known to reduce the regrowth rate, as seen comparing 'a' and 'c' in figure 8. Results also show that the F can increase the depth of a pre-existing amorphous region (see figure 7), and may cause and retain more residual defects [14].

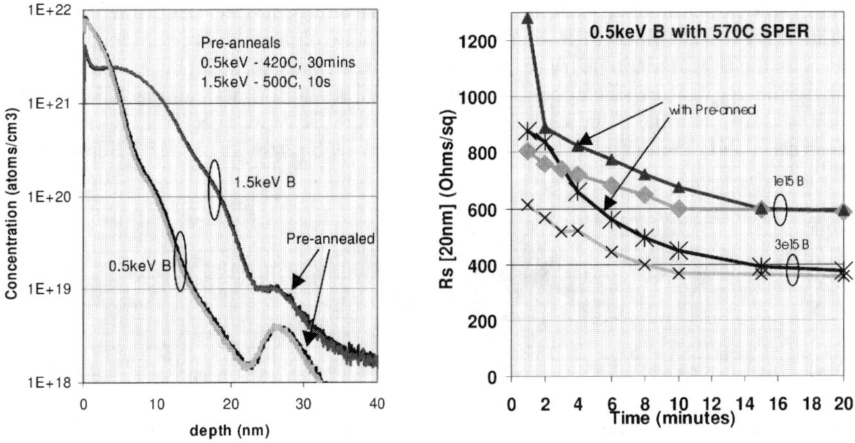

Figure 6: Effect of pre-annealing on diffusion for 0.5keV and 1.5keV B, and Rs (25nm) during SPER.

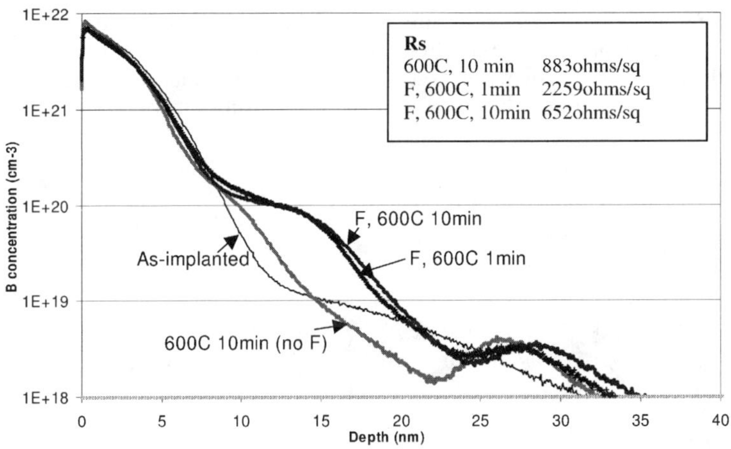

Figure 7. Effect of 6keV 1e15 F on the junction profiles and Rs. SPER occurred at 600°C for 10mins. and the post anneal was 900°C for 5s.

Within-wafer uniformity of SPER junctions

For CMOS integration a major issue as with all shallow junctions is uniformity of the junction in terms of both activation and depth. Since the formation of SPER junction is quite different from that of conventional ones, sources of non-uniformity can be different. Temperature control at these SPER temperatures cannot be obtained by optical methods like pyrometry as the Si wafer can be transparent to the wavelengths used. Figure 9a shows the result of using pyrometry for SPER temperature control where Rs uniformity can reach >15% where the SPER is almost complete due variations in regrowth rates affecting a larger portion of dopants. These, and other, results show the increased growth rate with higher B doping levels.

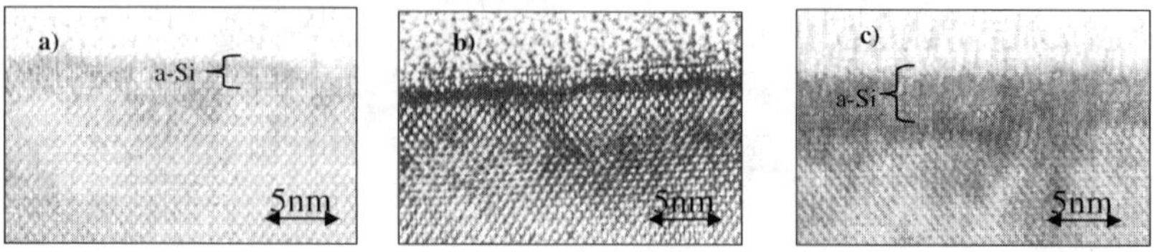

Figure 8. HR-TEM of SPER 0.5keV 3e15 B after a) 600C, 1min, b) 600°C, 10mins and c) with 6keV F 1e15 after 600°C, 1min.

Excellent temperature control, obtained by a conduction based RTA shown in figure 9b can provide junction uniformities better than most spike annealed junctions. The best condition tested here was for 650°C for 1min which fully regrows 22nm amorphous with a uniformity of 0.1%, the same as the implant uniformity (from therma-wave) of the 1.5keV B used. It should be noted that for the lamp-based systems, at these temperatures without an absorbing layer (e.g. silicide) the wafer temperature would normally be measured using a thermocouple

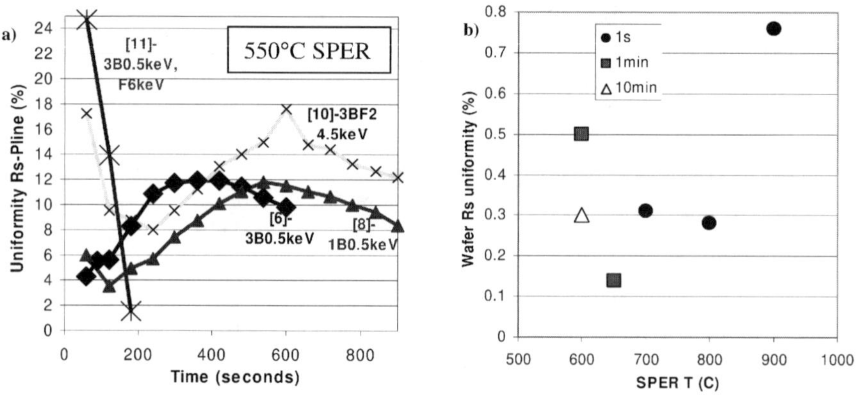

Figure 9. Uniformity of SPER junctions using a) pyrometry and b) thermocouple T control.

SPER and ITRS

Taking the best results from the investigations reported here and comparing them to the ITRS specs for the upcoming technology nodes it is clear that as far as junction performance is concerned, SPER has the potential to stretch to the 45nm node and possibly beyond. Figure 10 shows the Rs, depth and abruptness of the best junctions. Here the abruptness depends on the activation of dopants beyond the original amorphous interface and energy contamination, having the potential to be <1nm decade at 1e18-1e19atoms/cm^3.

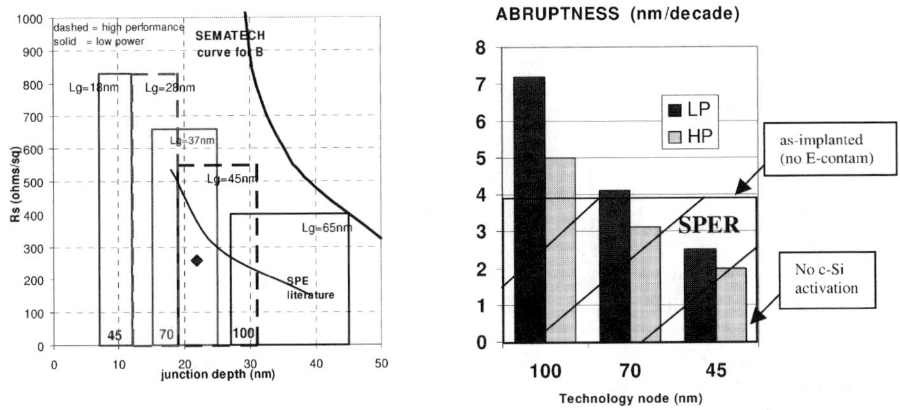

Figure 10. The best SPER junctions against the ITRS specifications for 70nm and 45nm CMOS. Possible improvements can be made with lower B doses with shallower Ge, and F implantation.

CONCLUSIONS

Even with co-implantation of Ge and F, pMOS extension junctions with high temperature spike annealing are not expected to meet the Rs, depth and abruptness requirements for sub-70nm CMOS nodes. The use of offset spacers or super steep HALOs may offer some hope but a lot of fine-tuning and process optimisation will be necessary.

From all the results together it is possible to arrive at an optimum implant and anneal condition for implementing SPER into integration. The results indicate that higher implant energies than conventionally used are preferred along with higher doses, the optimum being around 1-1.5keV B for 15-25nm junctions. However doses above 2e15 can lead to significant clustering, mobility reduction and deactivation upon post thermal treatment. The optimal temperature range of SPER lies between 600°C and 650°C for around 5-10mins as this gives good uniformity and little deactivation. Post annealing can dramatically increase the junction depth and abruptness just by activating the tail and may result in some deactivation if between 750°C and 850°C. Tailoring the F co-implant has the potential to offer broader profiles and reduce Rs. Whether this junction survives the leakage requirements for CMOS is not clear at the moment. There may come some improvements in junction leakage by finding a suitable pre- and post-anneal but results show that the process window for this is narrow. Fortunately off-state leakage specifications are being relaxed for most high performance devices and this may offer some latitude to use SPER even if the defects are not completely annihilated.

ACKNOWLEDGEMENTS

The authors would like to thank Wilfried Vandervorst, Luc Geenen, Ilse Hoflijk, Jens Frühauf, Radu Surdeanu, and Jorge Kittl for characterisation and discussion.

REFERENCES

[1] R. Murto, Proc. 3rd National implant users meeting, Oct 1999
[2] R. Lindsay, A. Lauwers, J. Frühauf, M. De Potter, and K. Maex. Proc. USJ2001, p255
[3] W. Lerch, Mattson, Germany, private communication
[4] A. Al-Bayati, S. Tandon, A. Mayur, M. Foad, D. Wagner, R. Murto, D. Sing, C. Ferguson, and L. Larson, Ion Implantation Technology, 2000. p 54 –61
[5] M. C. Ozturk, J.J Wortman, C. M. Osburn, A. Ajmera, G. A. Rozgonyi, E. Frey,W. –K Chu, C. Lee, IEEE Transactions on Electron Devices, Volume: 35 Issue: 5 , May 1988 pp.659-668
[6] V. Meyssen, P. Stolk, J. Van Zijl, J. Van Berkum, W. Van de Wijgert, R. Lindsay, C. Dachs, G. Mannino, and N. Cowern. Proceedings of Semicon West, San Francisco, April 2001
[7] H-J. Li; P. Kohli, S. Ganguly, T. A. Kirichenko, P. Zeitzoff, K. Torres, S. Banerjee, Electron Devices Meeting, IEDM Technical Digest. International,2000. pp515 –518
[8] L. S. Robertson, P.N. Warnes, M. E. Law, K. S. Jones, D. F. Downey, J. Liu, Ion Implantation Technology, 2000, pp.171 –174
[9] I. Ban, M. C. Ozturk, E. Demirlioglu, IEEE Transactions on Electron Devices, Volume: 44 Issue: 9 , Sept. pp1544 -1551
[10] B.J. Pawlak, R. Lindsay, R. Surdeanu, P. Stolk, K. Maex, submitted to IIT2002-03-25
[11] S. Talwar, S. Felch, D. Downey, Y. Wang, Ion Implantation Technology, 2000. pp175 –177
[12] A. Mayur, AMAT, Santa Clara, private communication
[13] S. Gannavaram, N. Pesovic, M. C. Ozturk, Electron Devices Meeting, 2000. IEDM Technical Digest. International, pp 437 -440
[14] J-Y. Jin, J. Liu, U. Jeong, S. Metha, K. Jones, J. Vac. Sci. Tech. B 20(1) 2002, p422
[15] S. N. Hong, G. A. Ruggles, J. J. Wortman, M. C. Ozturk, IEEE Transactions on Electron Devices, Volume: 38 Issue: 3 , March 1991 pp 476 -486
[16] J. O. Borland, Semiconductor International, April 2000, pp70-80
[17] W. O. Adekoya, M. Hage-Ali, J.C. Muller, P. Stiffert, J. Appl. Phys. 64 (2) 1988, p666
[18] C. M. Osburn, D. F. Downey, S. B. Felch, B. S. Lee, Ion implantation Tech. 1996, p832
[19] K. Tsuji, K. Takeuchi, T. Mogami, IEDM Tech Dig 1999, p9
[20] V. I. Kuznetsov, A.B. Storm, G. J. Snijders, C. de Ridder, T. A. M. Ruijl, J.C.G vd Sanden, E.H.A Granneman, 197th Meeting of Elect. Chem. Soc May 14-18th 2000, p1
[21] T. Clarysse, D. Vanhaeren, W. Vanderworst, J. Vac. Sci. Tech B, 20(1) 2002, p459
[22] A. Agarwal, A. T. Fiory, H.-J. Gossmann, C. Rafferty, P. Frisella, J. Hebb, J. Jackson, Ion Implantation Technology Proceedings, 1998, Volume: 1 , pp22 -25
[23] R. B. Liebert, S. R. Walther, S. B. Felch, Ziwei Fang; Bon-Woong Koo, D. Hacker, The first international workshop on Junction Technology, 2000, p23 –27
[24] Z. Shenjun J. A. V. D Berg, D. G. Armour, S. Whelan, R. G. Goldberg, P. Bailey, T. C. Q. Noakes, Ion Implantation Technology, 2000, pp119 –122
[25] H. Banisaukas, K. Jones, S. Talwar, D. F. Downey, S. Falk, Mat. Sci. Semi. Proc. 4 (2001), p339-343
[26] L. S. Robertson, PhD Thesis, University of Florida 2001

Mat. Res. Soc. Symp. Proc. Vol. 717 © 2002 Materials Research Society C2.2

S/D engineering for sub-100 nm MOSFET using ultra shallow junction formation technique, elevated S/D structure and SALICIDE technique

Kazuya Ohuchi, Kanna Adachi, Akira Hokazono and Yoshiaki Toyoshima
System LSI Research & Development Center, Toshiba Corporation Semiconductor Company,
8 Shinsugita-cho, Isogo-ku, Yokohama, Kanagawa 235-8522, Japan
Phone: +81-45-770-3644, Fax: +81-45-770-3571, E-mail: kazuya.ohuchi@toshiba.co.jp

Abstract

Suppression of short channel effect (SCE) by utilizing the technology of formation of ultrashallow junctions is one of the important issues. The annealing process of implantation-damage that induces transient enhanced diffusion during a subsequent thermal process, such as low-pressure chemical vapor deposition (LPCVD) for gate sidewall spacer, should be optimized. To pursuit high performance of MOSFETs, parasitic resistance must be reduced with scaling. On the other hand, it is difficult to decrease the parasitic resistance in the region of contact junction, which is a function of physical constant such as Schottky barrier height of silicide materials and solid solubility of dopant. The elevated source/drain structure reduces parasitic resistance of contact junction due to reduction of resistance of diffusion beneath salicide materials. Cobalt salicide is widely used till 100nm node. However, cobalt salicide has disadvantage in the thermal budget for shallow junction and quantity of silicon consumption during silicidation. Nickel salicide is one of the candidates for successor of cobalt salicide to 70 nm node or above, because of its characteristics of low temperature formation, low silicon consumption and low contact resistivity on p+ junctions. In this paper, S/D engineering will be discussed from the viewpoint of the process integration of sub-100 nm physical gate length complementary metal-oxide-semiconductor field-effect transistor (CMOSFET) device.

1. Introduction

In order to realize deep sub-100 nm MOSFET, such source/drain engineering techniques as ultrashallow junction (USJ) formation [1-3], elevated source/drain structure[4-6] and low silicon-consumption SALICIDE process are needed. In the case of USJ formation, each of low energy implantation, low thermal budget annealing (spike RTA or laser annealing [7-8]) and thin screen oxide at the silicon surface for preventing dose-loss phenomena, is already adopted. If using low energy implantation, photolithographic resist layer removing process, which is free of etching of surface, is strongly needed since dopants are located at the vicinity of the surface. Capability of dust removing and surface etching during removing resist layer are, unfortunately, in the relation of trade-off. On the other hand, it was newly found that the thermal budget of thermal CVD (around 700°C) of gate sidewall spacer for the mask of contact junction implantation and SALICIDE process seriously affects the dopant diffusion. In the second section, we report detailed study of damage recovery before sidewall spacer CVD [9].

In the view of LSI design, sidewall spacer should be also scaled with grand design rule. This scaling trend should accompany with shallow contact junction to prevent degradation of the characteristics of threshold voltage (V_{th}) roll-off. However, shallow silicided junction leads to unacceptable junction leakage, and increases a contact-junction resistance (R_{co}), where R_{co} consists of silicide/silicon contact resistance (R_c). This means the contact junction needs to be a certain distance away from the silicide/silicon interface. To overcome this trade-off, elevated S/D MOSFETs have been proposed [4-6]. In the 3rd section of this paper, remarks in case of using elevated source/drain structure are mentioned experimentally in the viewpoint of process integration [10].

Although cobalt SALICIDE process is widely and successfully used in the 100 nm SoC process, it includes some issues that cobalt in-diffusion, necessity of relatively high thermal budget owing to transformation from mono-silicide to di-silicide, and large amount of silicon consumption originated in use of disilicide. In the 4th section of this paper, it is demonstrated experimentally that nickel SALICIDE process improves these disadvantages of cobalt SALICIDE [11].

2. Formation of Ultra-shalow Junctions

In order to investigate TED during CVD for sidewall spacer, damage recovery anneal (DRA) before sidewall spacer deposition was studied. Detailed process sequences for the experiment with complementary-MOSFET (CMOS) technology are shown in Fig. 2.1. Before the low-pressure chemical vapor deposition (LPCVD) at 700°C for the sidewall, tilted halo implantation and shallow source/drain extension formation using the low-energy ion implantation at 1-5 keV with a dose of 8×10^{14}-1×10^{15} cm^{-2} BF_2^+ or As and DRA of 750-1000°C for 5-30 s were performed. Three types of sequences for damage recovery were examined in this work. Sequence A as shown in Fig. 2.1, contained simultaneous DRA for n- and p-extension implantation. In sequence B, in the meantime, samples received different temperature DRA for n- and p-extension, separately. Sequence C is highly similar to sequence A except that it includes Ge preamorphization implantation (PAI) before the p-extension implantation. After sidewall formation, n^+/p^- doping of the gate and the diffusion and activation RTA (A-RTA) of 1000°C

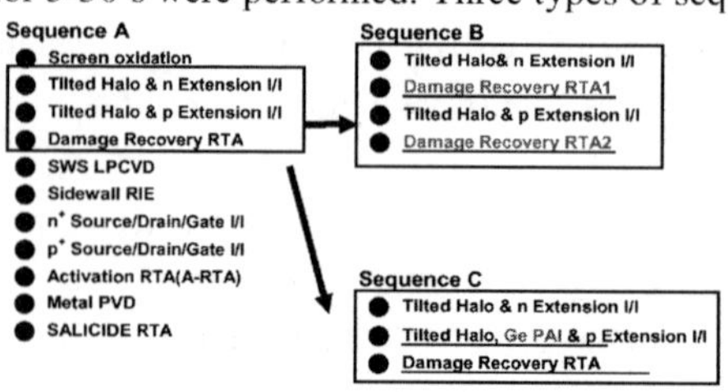

Figure 2.1.
Process sequence is shown. Sequence B or C is substituted for the boxed process module of sequence A.

for 10 s were performed. Profiles of arsenic and boron were measured by secondary ion mass spectroscopy (SIMS) by utilizing an ultralow acceleration energy of primary ions of under 1 keV for USJ analysis.

2.1 Controlling TED for n-type channel MOSFET

The arsenic profile after A-RTA of 900°C DRA is shallower and steeper than that of a sample that received 800°C DRA as shown in Fig. 2.2. In order to clarify this reverse dependence of the arsenic profile on DRA temperature, the evolution of arsenic profiles of the sample that received 800°C DRA were measured by SIMS after each successive process as shown in Fig. 2.3. The as-implanted arsenic profile was as steep as 20 nm at the concentration of $1x10^{18}$ cm^{-3}. The profile after 800°C RTA immediately after ion implantation was almost the same as the as-implanted profile. Rapid diffusion occurred during LPCVD for sidewall formation at the temperature of 700°C. This anomalous diffusion clearly originates in TED assisted by the point defect introduced by ion implantation, according to the diffusion coefficient of arsenic at 700 °C.

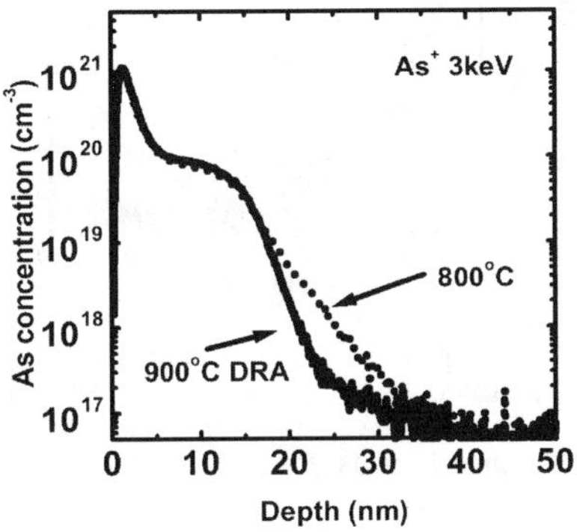

Figure 2.2.
Dependence of arsenic profile after activation RTA on DRA temperature.

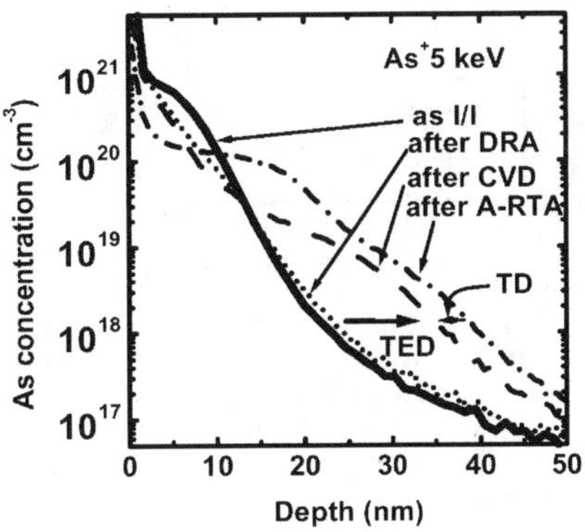

Figure 2.3.
Arsenic profile evolution from as-implanted to final activation RTA for the case of DRA temperature of 800°C. SIMS profile was measured after each process.

The V_{th}-roll-off characteristics of nMOSFET are greatly improved as the DRA temperature increases to approximately 900°C. This phenomenon originates in TED during the chemical vapor deposition (CVD) process and is consistent with arsenic profiles measured by SIMS. On the contrary, a significant degradation of the short-channel effect is observed in the sample that received 1000°C DRA. This is most likely caused by a redistribution of boron as the dopant for the halo implantation. According to a result of the process simulation of nMOSFET with 100 nm physical gate length, the boron concentration of the high-temperature annealed samples (1000°C) is smaller than that of low-temperature annealed samples (960°C). This is due to the drift diffusion of boron in the vicinity of the source/drain extension caused by the electric field brought about by the activation of arsenic by high-temperature DRA.

The dependence of overlap capacitance (C_{ov}) on DRA temperature, which is proportional to lateral diffusion along the MOSFET channel, has a minimum at 900-950°C, as shown in Fig. 2.4. This agrees with SIMS data and V_{th}-roll-off characteristics. Therefore, DRA at 900°C is optimal for nMOSFET. This DRA temperature dependence decreases with decreasing arsenic implantation energy in the region below 900°C. This is, we think, owing to decreasing point defect during arsenic implantation.

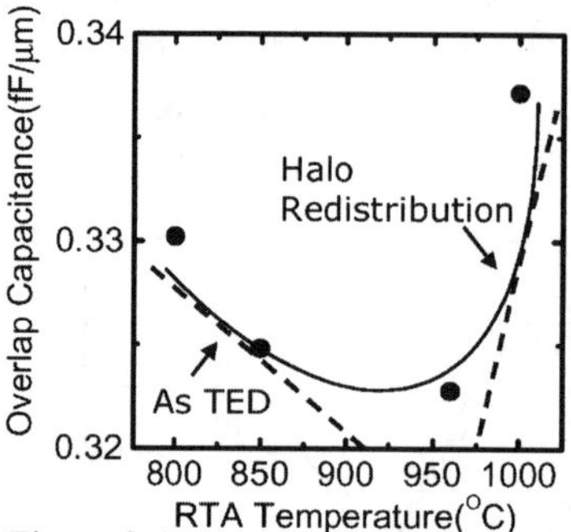

Figure 2.4.
DRA temperature dependence on gate-source overlap capacitance on nMOSFET. All samples received A-RTA. In the low-temperature region, arsenic TED is dominant. In the high-temperature region, boron redistribution is dominant.

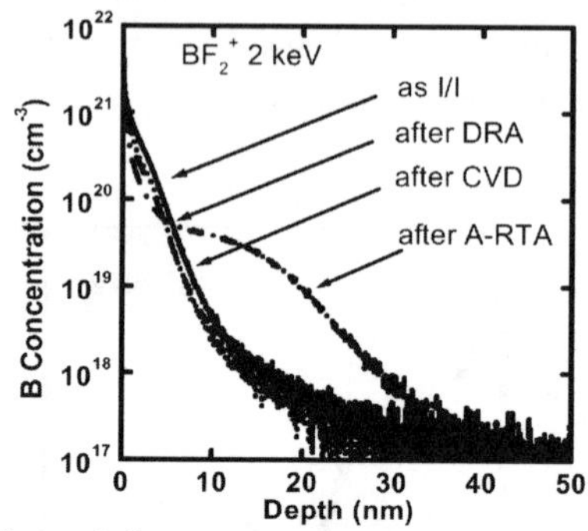

Figure 2.5.
Boron profile evolution from as-implanted to final activation RTA for the case of DRA temperature of 800°C. SIMS profile was measured after each process.

2.2 Controlling TD for p-type channel MOSFET

In contrast to the arsenic case of rapid diffusion during the CVD process, thermal diffusion during the DRA and the A-RTA is the dominant factor determining the junction depth in the pMOSFET. Figure 2.5 shows the profile evolution of boron concentration, where DRA temperature is 800°C. Boron diffusion is scarcely observed during the DRA and CVD processes at 700°C. Significant diffusion is observed during the A-RTA at 1000°C. The boron profile of the sample annealed at 900°C as DRA is deeper than the profile annealed at 800°C. Both profiles are measured after the same A-RTA at 1000°C. This means that the thermal diffusion of boron during DRA is also dominant if the DRA temperature is above 900°C. Therefore, in order to achieve a shallow boron concentration profile, low temperature DRA is preferable.

It was recognized that only the samples, which received DRA at 960 C, exhibited severe V_{th}-roll-off. This means that TD of boron during DRA at below 850°C is negligible. Dependence of the overlap capacitance (C_{ov}) on the DRA temperature is consistent with the SIMS profile and the V_{th} roll-off characteristics of MOSFET, as shown in Fig. 2.6. In contrast to nMOSFET, C_{ov} of pMOSFET increases rapidly with DRA over 850°C.

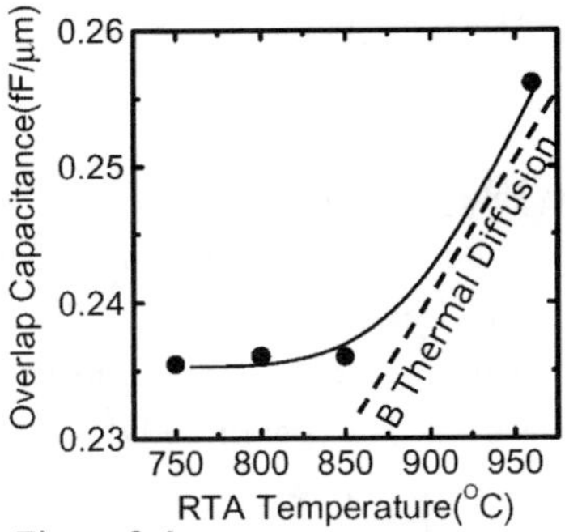

Figure 2.6.
DRA temperature dependence on gate-source overlap capacitance on pMOSFET. All samples received A-RTA. In the high-temperature region, boron TD is dominant.

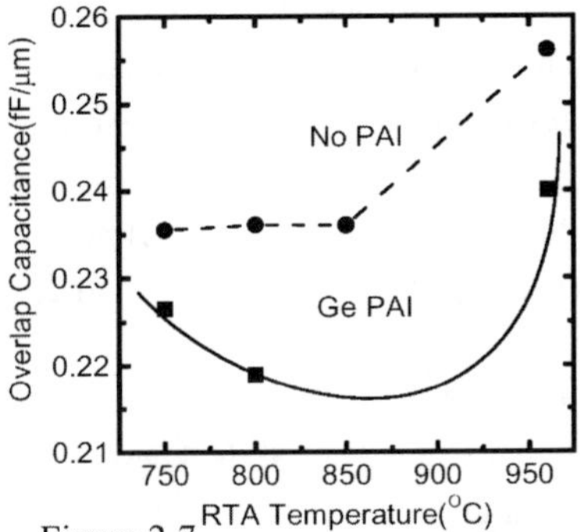

Figure 2.7.
DRA temperature dependence on gate-source overlap capacitance on Ge PAI pMOSFET. Dependence of Ge PAI samples has a optimal DRA temperature and low C_{ov} as compared with samples without PAI.

For USJ formation, samples should receive high-temperature DRA (DRA-1, 900-950°C) after n-extension implantation and low-temperature DRA (DRA-2, 750-800°C) for p-extension implantation, in sequence. In the case of using Ge preamorphization (PAI), the V_{th}-roll-off characteristic is greatly improved in pMOSFET. To investigate the dependence of the junction depth on the DRA temperature, the dependence of C_{ov} of Ge PAI samples on the DRA temperature is plotted, as shown in Fig. 2.7. Utilizing Ge PAI, C_{ov} has a minimum at the temperature of 900°C. Transient enhanced diffusion is, however, the dominant factor below 900°C, as shown in Fig. 2.7 even in the case of p-extension. In the entire range of DRA temperature, C_{ov} of the Ge PAI pMOSFET remains smaller than that of no-PAI samples, as shown in Fig. 2.7. It is speculated that Ge PAI retarded the diffusion of boron. Therefore, it is experimentally confirmed that p- and n-extension DRA can be summed into one DRA, as shown in Fig.2.1, sequence C by means of Ge PAI.

3. Elevated source/drain structure

In this section, we focus on reducing parasitic resistance and junction leakage current by utilizing SEG technique. Moreover, guidelines for application of elevated S/D technologies are described for sub-100 nm CMOS [10].

Before LPCVD layer deposition for sidewall, shallow S/D extension (SDE) formation using low energy ion implantation was performed according to the 2nd section. After sidewall formation, the SEG process was carried out [12]. Thickness of SEG-Si was varied from 0 nm (no SEG as a control) to 70 nm. After SEG process, S/D contact junctions were formed while changing implantation energy of B^+ or As^+ followed by rapid thermal annealing to achieve various doping profiles in the elevated S/D regions. SALICIDE process was applied to source/drain/gate regions of MOSFETs.

3.1 Suppression of the short channel effect

Figure 3.1 (a) and (b) show the dependence of L_{min} on gate sidewall thickness with various SEG-Si thicknesses, where L_{min} is defined as L_{gate} where threshold voltage shift compared to that of long channel MOSFETs' is 0.1 V and 20 mV for n- and pMOS, respectively. As the sidewall thickness decreases, severe SCE is observed in the case of no SEG condition due to lateral diffusion of contact junction. Application of SEG for contact junctions significantly improves SCE with increasing SEG-Si thickness. This means that elevated S/D structure enables sidewall scaling maintaining contact junction depth.

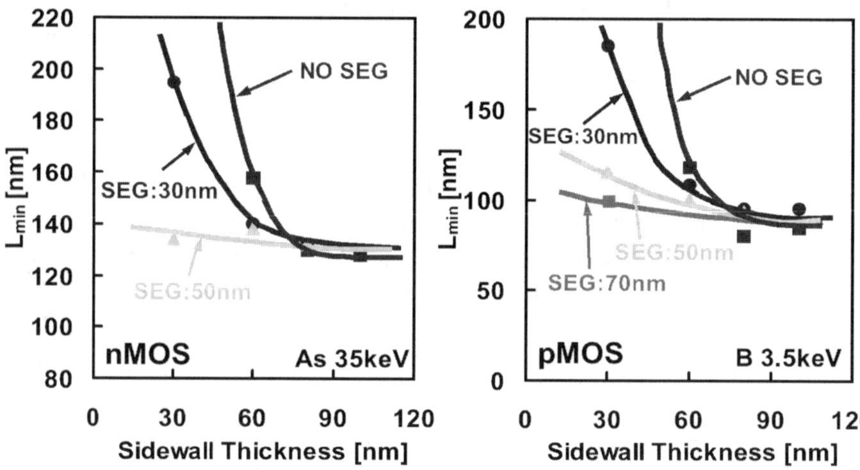

Figure 3.1 Dependence of L_{min} on sidewall thickness with varied SEG-Si thickness, (a) nMOSFET, (b) pMOSFET.

3.2 Consideration of parasitic resistance

We investigated impacts of elevated S/D structures on reducing parasitic resistance of MOSFETs. Firstly, sidewall scaling was studied comparing between elevated S/D MOSFETs with scaled sidewall and conventional MOSFETs with thicker sidewall. As shown in Fig. 3.2, significantly improved current drivability is demonstrated for both n- and pMOSFETs with elevated S/D structures. This means that elevated S/D structure reduces the parasitic resistance originating in SDE region. Furthermore, this improvement is more prominent for pMOSFET than for nMOSFET, due to the higher sheet resistance of SDE of pMOSFET.

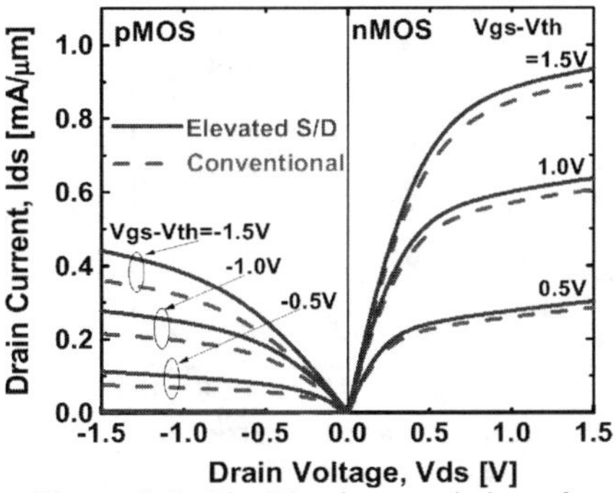

Figure 3.2. *I*ds-*V*ds characteristics of n and pMOSFETs, respectively. Solid line shows elevated source/drain MOSFET. Broken line shows conventional MOSFETs.(SEG=30nm)

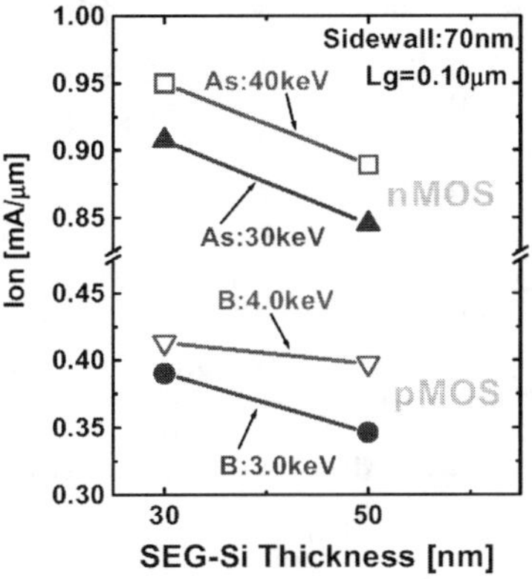

Figure 3.3. Dependence of SEG-Si thickness on parasitic resistance at the same implantation energy.

Contact junction profile should be optimized in elevated S/D structure. Figure 3.3 shows the dependence of current drivability on SEG-Si thickness while changing implantation energy. The elevation of silicon without optimization of contact junction causes the increase of the parasitic resistance, because contact junction is kept away from the gate edge by SEG-Si. Optimized elevated S/D structure also reduces R_{co}, the contact junction portion of the parasitic resistance.

3.3 Suppression of the junction leakage current

Suppression of junction leakage is one of the main concern from the viewpoint of contact junction scaling and motivation of elevated S/D structure introduction. Improved junction leakage characteristics is indicated Fig.3.4 (a) and (b) for n and pMOS, respectively. Cumulative probability plots of n^+ and p^+ junction leakage current are shown. Both contact junction was optimized for each no SEG or SEG sample, respectively in the viewpoint of SCE immunity. Since implantation energy of no SEG sample should be reduced compared to SEG sample, this caused degraded junction leakage distribution.

It should be noted that an additional consideration is required in contact junction and channel doping design with introduction of elevated S/D structure. Figure 3.5(a) shows the p^+ junction leakage current with varying channel dose. In spite of that there is no difference in the case of low channel dose, degraded distributions are observed with the SEG thickness increase in high channel dose case. Figure 3.5(b) shows SEG-Si thickness dependence on the junction capacitance (C_j).

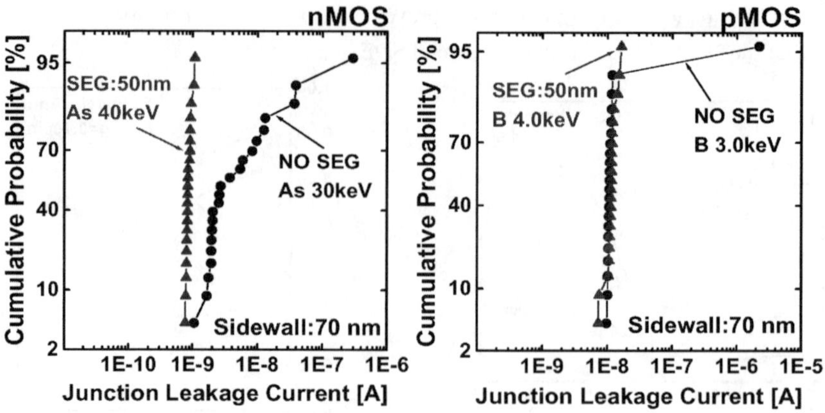

Figure 3.4. Cumulative probability plot of distribution of n+ and p+ junction leakage current. The suppression of leakage current using SEG process is observed.

Corresponding increase in C_j is found in the thick SEG sample with high channel doping. These results indicate that the careful profile design is necessary including channel. Channel profile, therefore, should be designed according to the SEG-Si thickness in order not to increase C_j and control V_{th}.

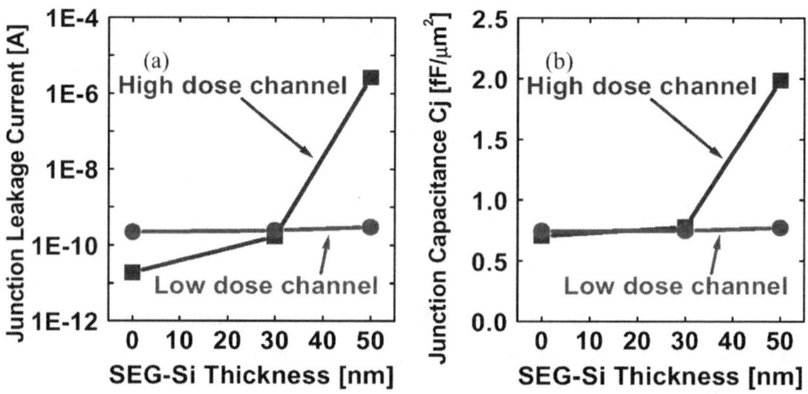

Figure 3.5. SEG thickness dependence of (a) junction leakage and (b) junction capacitance at a varying channel dose.

3.4 LSI application

Practically, imperfection of selectivity of silicon CVD is the most important matter of concern in LSI production. We tried to adopt elevated source/drain structure using selective silicon epitaxial growth to 1 Mb 0.15 μm SRAM and embedded 1 Mb 0.175 μm DRAM, respectively. All-bit-path function yield of SRAM and DRAM, and pause characteristics of DRAM of the samples received selective CVD for elevated source/drain is comparable to controlled samples as shown in Fig.3.6.

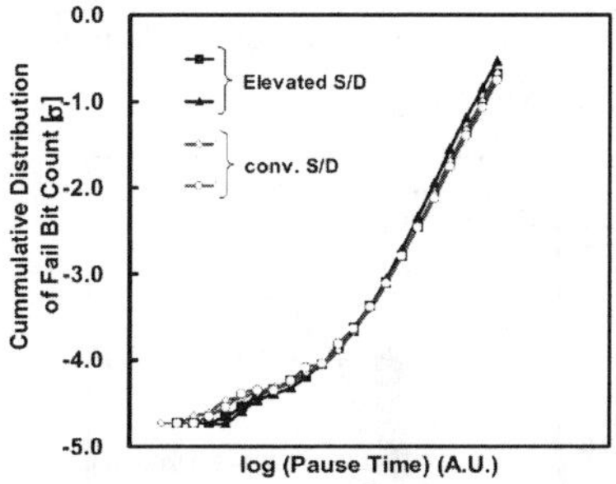

Figure 3.6. Pause characteristics of 0.175 μm embedded 1 M DRAM. No degradation is observed in elevated S/D DRAM.

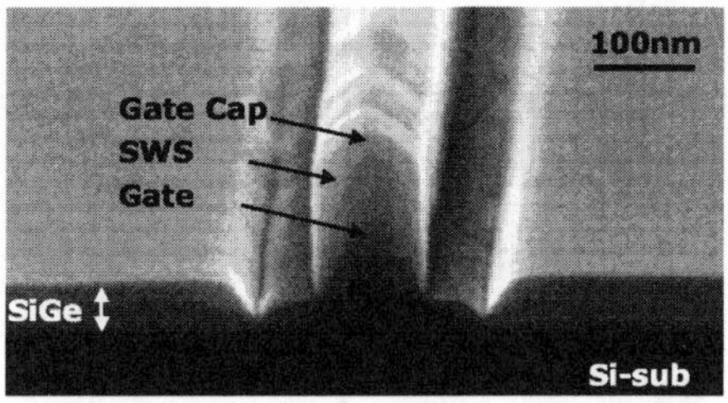

Figure 3.7. SEM image of elevated S/D structure using selective SiGe at the process temperature of 650°C.

3.5 Challenge for low thermal budget

Selective Si CVD, however, has a disadvantage in thermal budget, because its process temperature without breaking selectiveness is higher than 800°C. Unfortunately, this temperature is too high to formation ultrashallow junctions for deep sub-100 nm CMOS. SiGe selective CVD is the most promising candidate for sub-30 nm MOSFETs elevated source/drain structure as shown in Fig.3.7. This process including pre-treatment is achieved at around 650°C for several minutes. Therefore, it is negligible small for ultra shallow junction formation for sub-30 nm CMOS.

4. SALICIDE Process for sub-100 nm

4.1 Low level of metal diffusion and low amount of silicon consumption of nickel silicide

Although cobalt silicide is widely used in CMOS for 100 nm SoC, it is recently reported that cobalt fast in-diffusion during transformation from mono-silicide to di-silicide occurrs and degrades junction characteristics [13]. This in-diffusion will be killer and inevitable defects because contact junction depth will be also scaled with gate length scaling as previously mentioned. According to SIMS analysis comparing the profile of cobalt after first annealing step (475°C for 30 sec) and second annealing step (815°C for 30 sec) in ref [13], this fast diffusion is, we think, originated in transformation from mono-silicide to di-silicide. On the other hand, nickel salicide process needs only one time annealing process for reaction due to its low resistivity of mono-silicide. Cobalt and nickel SIMS profile from the backside of wafer clearly show the different diffusion length as shown in Fig.4.1. The samples of cobalt silicide are received two-step annealing process (475°C for 30 sec & 815°C for 30 sec) and nickel received single step annealing process (450 °C for 30 sec). Therefore, in the viewpoint of metal in-diffusion toward junction, nickel SALICIDE is suitable for deep sub-100 nm CMOS with scaled contact junctions. Moreover, consumption of silicon during nickel silicidation is about 20% less than that of cobalt owing to its stoichiometry and resistivity. Therefore, with NiSi case, we also observed smaller junction leakage current than $CoSi_2$ case in contact junction, though R_{sh} in NiSi case is the same level or below R_{sh} in $CoSi_2$ case as shown in Fig. 4.2.

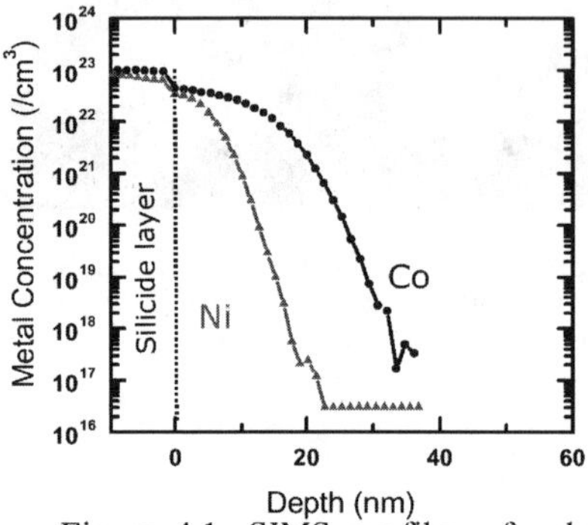

Figure 4.1. SIMS profiles of cobalt and nickel from the backside of silicided sibstrates.

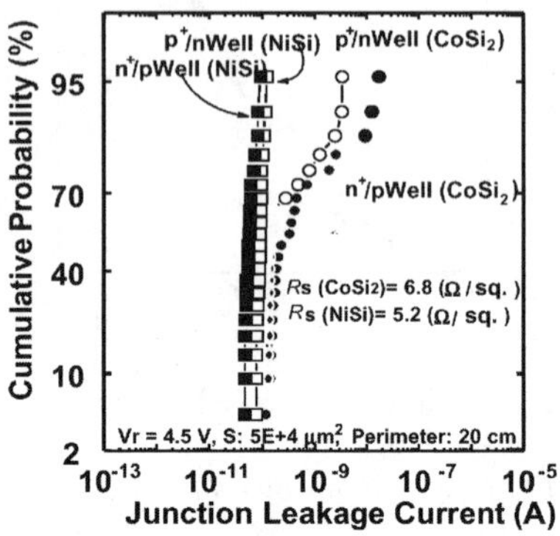

Figure 4.2. Cummulative plot of junction leakage characteristics. Circles or squares show cobalt, nickel samples,respectively. Open symbols show p+ junctions, close show n+ junctions.

4.2 Impact of low temperature silicidation anneal on performance of MOSFET

In addition, NiSi was introduced into these devices instead of $CoSi_2$ to reduce the temperature of silicidation RTA. Maximum process temperature after activation spike anneal was reduced to less than 550°C. According gate voltage dependence of gate capacitance, depletion of gate polycrystalline silicon electrode is improved in case of nickel silicide compared to cobalt silicide case corresponding to about a few % of inversion capacitance with the oxynitrided gate dielectric thickness of 1.6 nm. It is speculated that 2nd step annealing of cobalt silicidataion (815°C) deactivates boron atoms in gate polycrystalline silicon. Moreover, contact resistance between silicide/p+ silicon interface of nickel silicided junction corresponds to one fourth lower than that of cobalt silicided junction. Therefore, especially pMOSFET, MOSFET performance with 70 nm gate length at 1.5 V operation is improved about 5 % if using nickel silicide as shown in Fig.4.3. Moreover, our group has recently reported that nickel salicided CMOS with the gate length of 35 nm at 0.85 V operation was integrated as shown in Fig.4.3 [11]. Current drives of I_{on} = 676 μA / μm in nFET and I_{on} = 272 μA / μm in pFET were achieved with I_{off} = 100 nA / μm at V_{dd} = 0.85 V, respectively. In nFET case, this is the highest current drive at 0.85 V operation in 35 nm gate length CMOS reported to date.

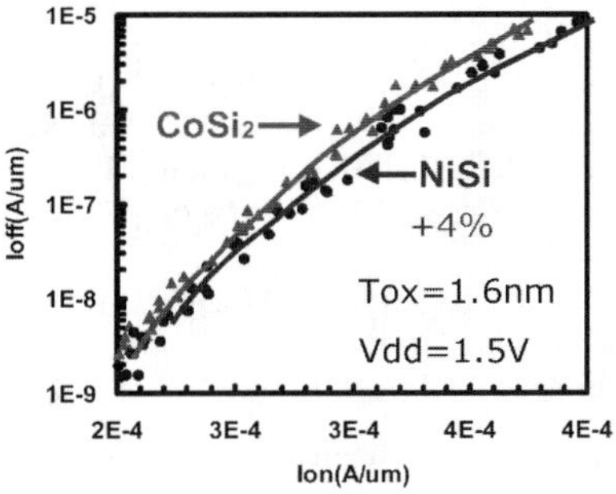

Figure 4.3. *I*on-*I*off characteristics of pMOSFET at *V*d=1.5 V with CoSi2 (triangles) or NiSi (circles).

5. Conclusion

We reviewed source/drain engineering techniques as ultrashallow junction (USJ) formation, elevated source/drain structure and low silicon-consumption SALICIDE process. The annealing process of implantation-damage that induces transient enhanced diffusion during a subsequent thermal process, such as low-pressure chemical vapor deposition (LPCVD) for gate sidewall spacer, should be optimized. The SEG process enables improvement of short channel effect, parasitic resistance and junction leakage. A careful channel profile design for the SEG process is demonstrated. The importance of elevated S/D technology will increase in sub-100nm CMOS technology. Cobalt salicide is widely used till 100nm node. However, cobalt salicide has disadvantage in the thermal budget for shallow junction and quantity of silicon consumption during silicidation. Nickel salicide is one of the candidates for successor of cobalt salicide to 70 nm node or above, because of its characteristics of low temperature formation, low silicon consumption.

Acknowledgements

Authors would like to thank Y. Okayama, H. Takato, K. Yasumoto and T. Yoshida for support of memory fabrication and evaluation. K. Miyano and I. Mizushima would be also acknowledged for selective CVD process. N. Aoki, T. Kanemura, A. Murakoshi, T. Iinuma, S. Inaba, T. Ohguro and K. Suguro would be thanked for useful discussion. We express our thanks to T. Furuyama for his support. We are also deeply grateful to H. Ishiuchi for his continuous encouragement.

REFERENCES

[1]K. Goto, Y. Sambonsugi and T. Sugii: *Symp. VLSI Technology Digest of Technical Papers* (Business Center for Academic Societies Japan, Tokyo, 1999) p.49.

[2]K. Tsuji, K. Takeuchi and T. Mogami: *Symp. VLSI Technology Digest of Technical Papers* (Business Center for Academic Societies Japan, Tokyo, 1999) p. 9.

[3] E. Morifuji, A. Ohishi, K. Miyashita, H. Kawashima, T. Nakayama, H. Yoshimura and Y. Toyoshima: *Int. Electron Device Meeting Technical Digest* (IEEE Service Center, New Jersey, 1999) p. 649.

[4]S. S. Wong, D. R. Bradbury, D. C. Chen, and K. Y. Chiu: *Int. Electron Device Meeting Technical Digest*, (IEEE Service Center, New Jersey, 1984) p. 634.

[5]H. Wakabayashi, T. Yamamoto, T. Tatsumi, K. Tokunaga, T. Tamura, T. Mogami and T. Kunio: *Int. Electron Device Meeting Technical Digest*, (IEEE Service Center, New Jersey, 1997), p. 99.

[6] A. Waite , S. Kubicek, D. Howard, M. Caymax, K. DeMeyer and A. Evams: *Proc. The 28th European Solid-State Dev. Res. Conf.*, (Editions Frontieres, Paris, 1998) p. 148.

[7]S. Shishiguchi, A. Mineji, T. Hayashi and S. Saito: *Symp. VLSI Technology Digest of Technical Papers* (Business Center for Academic Societies Japan, Tokyo, 1997) p. 89.

[8] B. Yu, Y. Wang, H. Wang, Q. Xiang, C. Riccobene, S. Talwar and M.-R. Lin: *Int. Electron Device Meeting Technical Digest* (IEEE Service Center, New Jersey, 1999) p. 509.

[9]K. Ohuchi, K. Adachi, A. Murakoshi, A. Hokazono, T. Kanemura, N. Aoki, M. Nishigohri, K. Suguro and Y. Toyoshima: *Jpn. J. Appl. Phys.* **40** (2001) 2701 Part 1.

[10]A. Hokazono, K. Ohuchi, K. Miyano, I. Mizushima, Y. Tsunashima and Y. Toyoshima., *Int. Electron Device Meeting Technical Digest* (IEEE Service Center, New Jersey, 2000) p. 243.

[11]S. Inaba, et al., *Int. Electron Device Meeting Technical Digest* (IEEE Service Center, New Jersey,2001) p. 641.

[12]K. Miyano, I. Mizushima, K. Ohuchi, A. Hokazono and Y. Tsunashima: *Jpn. J. Appl. Phys.* **38**. (1999), p2419.

[13]M. Tsuchiaki and K. Ohuchi: *Ext. Abst. 2001 Int. Conf. Solid State Dev. Mat.*,(Business Center for Academic Societies Japan, Tokyo,2001) p. 216.

Mat. Res. Soc. Symp. Proc. Vol. 717 © 2002 Materials Research Society C2.4

Characterization and monitoring of silicon-on-insulator fabrication processes by high-resolution x-ray diffraction

G. M. Cohen, P.M. Mooney, H. Park, C. Cabral, Jr., and E.C. Jones
IBM T. J. Watson Research Center, P.O. Box 218, Yorktown Heights, NY 10598, U.S.A

Abstract

High-resolution x-ray diffraction (HRXRD) was used to monitor silicon-on-insulator (SOI) processing steps. The use of HRXRD is attractive since it is non-destructive and can be applied directly to product wafers. We show the usefulness of this technique for the characterization of amorphizing implants for shallow junctions, solid phase re-crystallization of implanted junctions, cobalt-silicide formation, and oxidation; all are critical processes for CMOS fabrication on SOI.

Introduction

The idea behind the use of HRXRD to monitor SOI processing is straightforward. Oxidation, silicide formation and amorphizing implants reduce the single crystal SOI film thickness, which is easily and accurately measured by HRXRD [1]. For example, oxidation would convert the upper part of the SOI film into SiO_2. Similarly, silicidation reacts a metal (such as Co) with the top portion of the SOI film to form a silicide (such as CoSi and $CoSi_2$). The phase of the silicide that is formed can be determined from the amount of Si consumed. In the case of amorphizing implants, which are routinely used to form the source and drain region of a device [2], a portion of the SOI film is converted from single crystal silicon to amorphous silicon. Post-implant annealing, typically carried out by rapid thermal annealing (RTA), re-crystallizes the amorphous portion of the film and restores the SOI film thickness. Accurate measurement of the SOI film thickness is essential for tight monitoring of all these processes.

The change in the SOI layer thickness is not the only parameter obtained by HRXRD. The HRXRD data of amorphizing and non-amorphizing Ge, Si or boron implanted wafers, is further used to extract the strain profile and the implant damage profile in the SOI film. Moreover, in the case of boron-implanted wafers a change in the sign of the strain is recorded when the boron assumes a substitutional site following the RTA anneal.

Experiment

The x-ray diffraction measurements were carried out with a Philips materials research diffractometer (MRD), which is equipped with an asymmetrically cut four-crystal Ge(220) monochromator and mounted on a rotating Cu anode x-ray source. ω is the angle between the incident beam and the wafer surface and 2θ is the angle between the incident beam and the detector (reflected beam); therefore the angle of the incident or reflected beam with respect to the analyzed atomic planes, the Bragg angle, is $\theta_B = 2\theta/2$. ϕ denotes a rotation in the plane of the wafer. The 004 reflections were measured in the triple-axis (TA) mode [3], i.e. by taking $\omega/2\theta$ scans with an analyzer crystal installed in front of the detector. The 113 reflections were measured using a large detector aperture and standard rocking curves (ω scans). The 113 reflections are taken with a low angle incidence beam. This makes them surface sensitive and useful for measurements of thin films.

Figure 1 demonstrates how the cobalt silicide formation process is monitored on an SOI wafer. The process begins with a sputter deposition of a 7 nm thick cobalt layer over the SOI film. The cobalt layer is capped with a 20 nm thick TiN film to prevent the cobalt from oxidizing during annealing. A first anneal at a temperature of 520°C is performed to react all the cobalt with the underlying silicon to form the mono-silicide phase (CoSi). The TiN cap and unreacted cobalt on the regions of the wafer having no exposed Si (e.g. dielectric isolation regions) are then selectively etched off, and a second anneal at a temperature of 750°C is performed to form the disilicide phase ($CoSi_2$). Since the silicide reaction is self-limited, the nominal amount of silicon consumed from the SOI layer is known. The formation of the CoSi phase requires 1.82 nm of silicon per 1 nm of metal and the formation of the $CoSi_2$ phase takes 3.64 nm of silicon for each nanometer of metal [4]. By monitoring the change in the SOI thickness, which is equivalent to the amount of silicon reacted to form the silicide, it is possible to confirm the formation of the desired silicide phases.

We note that it is critical to form the lowest resistivity silicide phase for the manufacturing of high-speed devices. To reduce the parasitic resistance of a device the $CoSi_2$ phase is used since it exhibits a lower sheet resistance than CoSi. The lowest resistivity phase is not always the disilicide phase, however. For example, nickel mono-silicide (NiSi) has a lower resistivity than the disilicide $NiSi_2$.

Figure 1(a) shows a 004 triple-axis scan of the as-deposited cobalt on a bonded SOI wafer, where the diffraction peak of the SOI layer is offset from that of the substrate [1]. The trace is recorded with the incident angle, ω, centered at the SOI layer diffraction peak. The SOI layer thickness is accurately measured from the spacing in the Pendellosung fringes (also known as thickness fringes). Figure 1(b) shows a scan following the first anneal which forms the monosilicide phase, and figure 1(c) shows a scan following the second anneal which forms the disilicide phase. As is clearly seen, the fringe spacing increases when the silicide phase becomes more silicon rich. By measuring the change in the SOI thickness the silicide phase is identified as indicated on figures 1(b) and 1(c).

Ion implantation is another key process for semiconductor device fabrication. To obtain a shallow junction by implanting a light atom dopant such as boron, a pre-amorphizing implant of Ge is used. The amorphized surface prevents boron channeling, which would lead to a tail in the boron profile. The thickness of the amorphized portion of the SOI film is easily obtained by measuring the thickness of the remaining crystalline portion of the film.

To investigate amorphizing implants by HRXRD we implanted Ge into a SIMOX wafer using different implant conditions. For SIMOX wafers, the diffraction peak from the SOI layer overlaps that from the Si substrate. Figure 2 shows three rocking curves that correspond to the initial SOI wafer and two different implant conditions. The first trace (solid line) of figure 2 is a measurement of a region with no implant. The second trace (dashed line) corresponds to a Ge dose of $3x10^{14}$ cm^{-2} and implant energy of 15 keV, and the third trace (dash-dot line) corresponds to an implant with the same dose but a higher energy of 30 keV. TEM images (not shown) confirm that the top portion of the SOI film is amorphized by the implant. A 15 keV implant resulted in a 20.5 nm thick amorphous layer, while the 30 keV-implant amorphized about 37.2 nm of single crystal silicon. The HRXRD measurements of figure 2 provide the same thickness data. The implanted samples show a larger fringe spacing than of the non-implanted sample, because the top portion of the single crystal SOI film was amorphized. The traces from the implanted samples also show an asymmetric spectrum, having pronounced thickness oscillations for negative ω and few fringes for positive ω. This asymmetry is due to the strain and damage introduced by the implant into the crystalline portion of the SOI, as explained below.

Figure 3 shows a rocking curve of a SOI film implanted with a Ge dose of $3x10^{13}$ cm^{-2} and energy of 50 keV, and figure 4 shows a TEM cross-section of the same sample. Due to the lower Ge dose, the implant does not amorphize the SOI film. Hence the fringe spacing is the same as that of the non-implanted SIMOX sample. To better explain the asymmetry in the rocking curve, we simulated the diffraction pattern using the wave summation approach outlined in reference 5. The Ge implant leads to strain and damage in the SOI film, which were incorporated into the simulation in the following way: The SOI film was sectioned into N laminae each having two parameters: strain and static atomic disorder. The strain distribution is represented by a set of coherently diffracting laminae. The laminae all have the same in-plane lattice parameter, but the out-of-plane lattice parameter of each one varies according to an assigned uniform strain value chosen to approximate the strain profile introduced by the implant. In addition to the coherent atomic displacements (strain), static random displacements (damage) are treated through their effect on the mean structure factor in each lamina [6]. The static atomic displacement factor is also known as the Debye-Waller factor, which accounts for the displacements of atoms from their lattice position. The higher the static atomic disorder value assigned to a lamina the less diffractive the lamina will be.

We further assumed that a Gaussian distribution could approximate the strain and damage profiles. Both the center position and width of each distribution were changed to obtain the best fit with the measured data. The maximum positions of the strain distribution and of the atomic disorder distribution obtained from simulation are in excellent agreement with the position of the maximum damage found by TEM. The simulated rocking curve, shown in Figure 3, predicts well all the features in the measured rocking curve. We note that the measured intensity is greater than the calculated intensity away from the center of the scan because of high background scattering when data is taken without a receiving slit to reduce the detector aperture. We also think that an improved prediction can be obtained by not fixing the distributions to an analytical form (such as a Gaussian distribution), but rather allowing the distribution to be calculated using a minimization procedure as discussed in [7].

Figures 5 and 6 demonstrate how HRXRD can be used to monitor RTA processing of implanted wafers. Figure 5 shows the rocking curves of a SIMOX sample that was implanted with a Si dose of $5x10^{14}$ cm^{-2} and energy of 23.8 keV and then annealed at 1000 °C for 2 seconds by RTA. The rocking curve clearly shows that the Si implant amorphized the top portion of the SOI film, and induces damage which cause the film to strain compressively, similarly to the observation made with Ge implants. During RTA the amorphous silicon film re-crystallizes by solid phase epitaxy and the thickness fringe spacing is restored to its pre-implant value (shown in figure 2 by the solid line). The post-RTA curve shows a slight asymmetry, which is due to residual damage that could not be removed by RTA.

Figure 6 shows the rocking curve of a sample that was implanted with a boron dose of $3x10^{14}$ cm^{-2} and energy of 3.8 keV. The thickness fringe spacing does not change since the light-mass boron atom does not amorphize the film. The rocking curve indicates that the post-implanted film is compressively strained, due to implant damage and the boron mostly assuming interstitial sites. As is seen from the layer peak shift, following an anneal at 1000 °C for 2 seconds by RTA, the boron atoms assume substitutional sites and the strain in the SOI film becomes tensile, since boron is a smaller atom than silicon.

Conclusions

We have demonstrated that HRXRD can be used to monitor critical SOI device fabrication steps. This technique is non-destructive and can, therefore, be applied to product wafers. While this method primarily provides the thickness change in the SOI film due to the processing, additional data such as the ion implantation damage and the strain distribution in the film was also obtained by simulating the x-ray diffraction spectrum.

References:

1. G. M. Cohen, P. M. Mooney, E. C. Jones, K. K. Chan, P. M. Solomon, and H-S. P. Wong, Appl. Phys. Lett., **75**(6), 787, (1999).
2. S. Wolf and R. N. Tauber, *Silicon Processing for the VLSI Era*, Vol. 1, Lattice Press, (1986).
3. P.F. Fewster, Semicond. Sci. Technol. **8**, 1915 (1993).
4. S. P. Murarka, Silicide for VLSI Applications, Academic Press New York, 1983.
5. E. Zolotoyabko, J. Appl. Cryst. **31,** 241 (1998).
6. V. S. Speriosu, J. Appl. Phys. **52**(10), 6094, (1981).
7. A. Parisini, S. Milita, and M. Servidori, Acta Cryst. **A52**, 302, (1996).

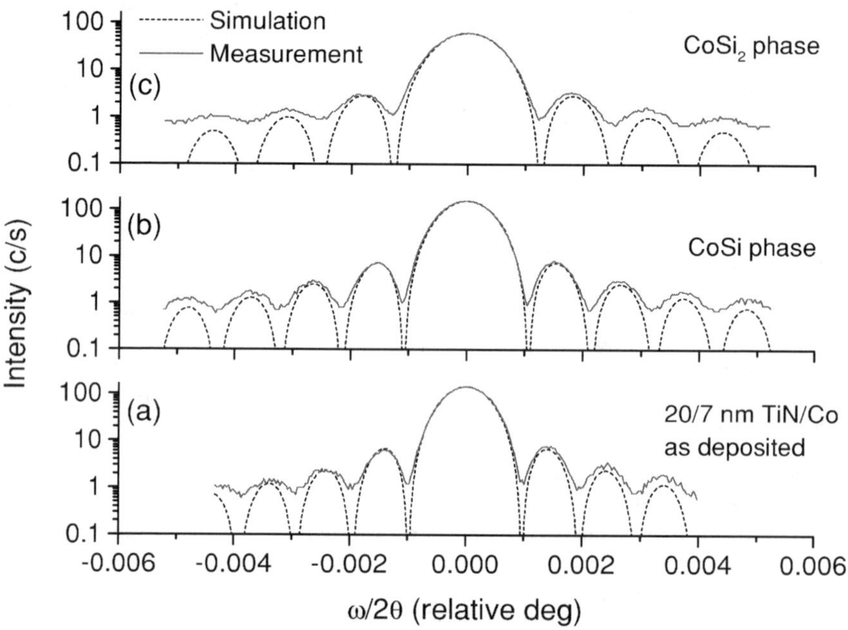

Figure 1: An 004 triple-axes scan of a bonded SOI wafer, showing only the diffraction pattern obtained from SOI film. (a) With as deposited cobalt. (b) After first RTA that forms cobalt monosilicide. (c) After second anneal that form the cobalt disilicide.

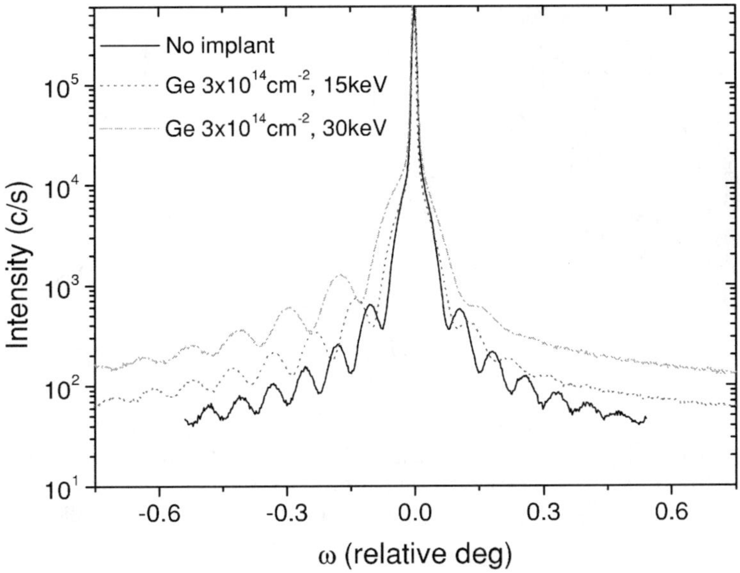

Figure 2: 113 rocking curves of a SIMOX wafer. The solid line is a measurement of a region with no implant. The dashed line corresponds to a Ge dose of $3x10^{14}$ cm^{-2} and implant energy of 15 keV, and the dash-dot line corresponds to an implant with the same dose but a higher energy of 30 keV.

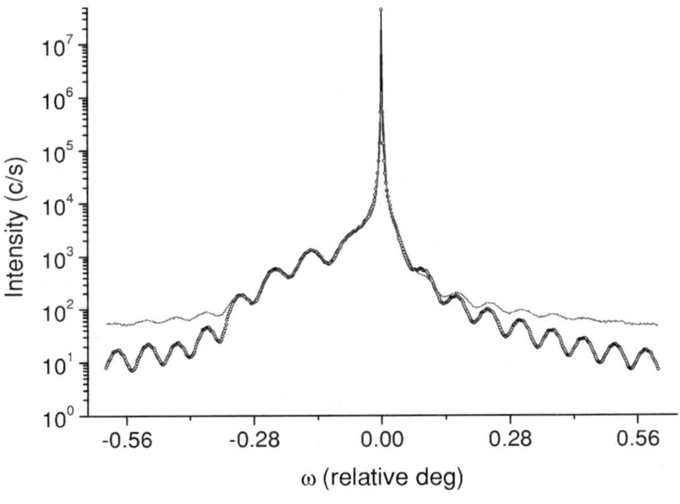

Figure 3: A 113 rocking curve of a SOI film (shown in Fig. 4) implanted with a Ge dose of $3x10^{13}$ cm^{-2} and energy of 50 keV (solid line). The open circles show a simulation obtained by fitting the strain and damage distribution in the film.

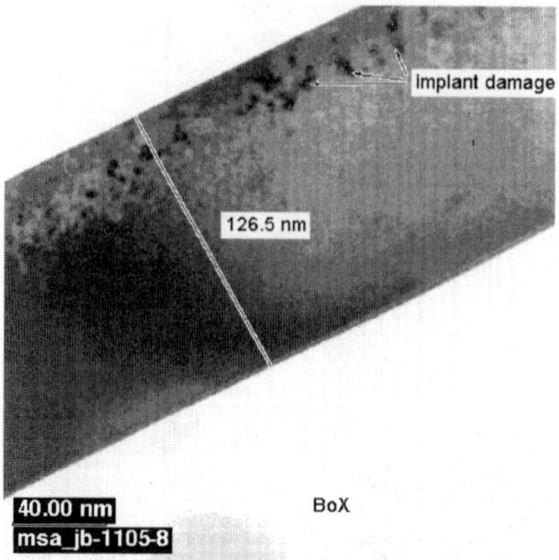

Figure 4: A TEM image of a SOI film implanted with a Ge dose of $3x10^{13}$ cm^{-2} and energy of 50 keV. Note the implant damage.

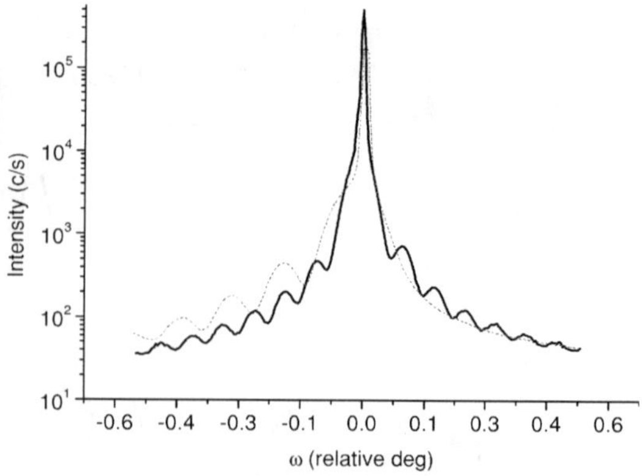

Figure 5: 113 rocking curves of a SIMOX sample that was implanted with a Si dose of $5x10^{14}$ cm^{-2} and energy of 23.8 keV (dashed line) and then annealed at 1000 °C for 2 seconds by RTA (solid line).

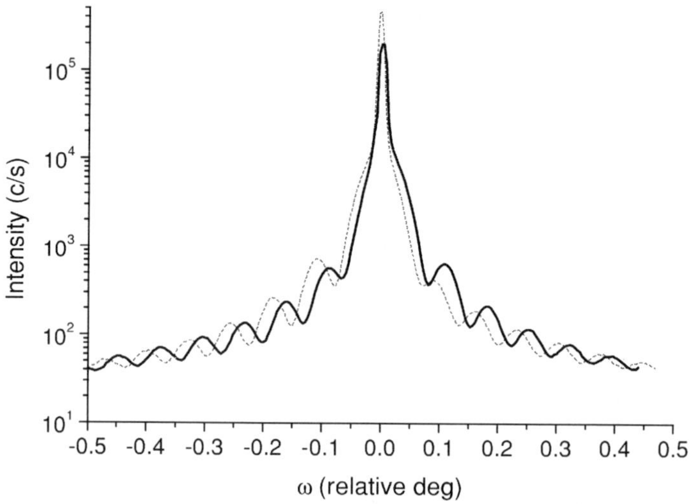

Figure 6: 113 rocking curve of a sample that was implanted with a boron dose of $3x10^{14}$ cm^{-2} and energy of 3.8 keV (dashed line) and then annealed at 1000 °C for 2 seconds by RTA (solid line).

Silicon Self-Interstitial Cluster Formation and Dissolution in SOI

A. Saavedra, J. Frazer, D. Wrigley, K. Jones
Department of Materials Science and Engineering, University of Florida, Gainesville, FL

I. Avci, S. Earles, M. Law
Department of Electrical and Computer Engineering, University of Florida, Gainesville, FL

E. Jones
IBM Research Division, T.J. Watson Research Center, Yorktown Heights, NY

Abstract

Silicon-on-insulator (SOI) is a promising alternative to bulk silicon as ultra shallow junction depths have begun to shrink below 50 nm. This study examined the effect of the SOI surface silicon/buried oxide interface on {311} defect evolution after Si^+ ion implantation. SOI wafers were produced such that the surface silicon thickness varied from 300Å to 1600Å. Non-amorphizing Si^+ implants at 5 and 20 keV with a dose of $2x10^{14}$ cm^{-2} were performed into SOITEC SOI wafers. Furnace anneals were done at 750°C from 5 minutes to 4 hours and quantitative transmission electron microscopy (QTEM) was used to study the implant damage evolution. At 5 keV, the dissolution behavior of the SOI was very similar to that of the bulk. However, the extended defects in the 300 Å SOI did not nucleate the same as those observed in the bulk or thicker SOI. Similar results were seen at 20 keV for the 700 Å SOI, but a slight decrease in the concentration of trapped interstitials was observed due to interface recombination as a result of the increased projected range of the implant. It is concluded that the surface Si/BOX interface does not significantly affect recombination of interstitials trapped in extended defects unless the interstitial profile is close to or truncates the interface. However, the interface does appear to affect the stability of zig-zag {311} defects and dislocation loops in thin SOI at lower implant energies.

Introduction

While the advantages of silicon-on-insulator (SOI) over bulk silicon have been well documented over the years, only recently has it begun to make its way into mainstream CMOS technology.[1] In order to scale SOI devices in the future, significant research must be done to understand the diffusion processes that make SOI different from bulk Si. Transient enhanced diffusion (TED) is exhibited by numerous dopants and is one of the main issues affecting ultra shallow junction formation.[2] One of the sources of TED are extended defects such as {311} defects and dislocation loops, which regulate the release of excess interstitials.[3] It is via these interstitials that many dopants, most notably B and P, diffuse throughout the silicon lattice.[4] Thus, in order to understand TED in SOI, it is imperative to discern the role of the surface Si/buried oxide (BOX) interface on the evolution of extended defects. The current study has attempted to understand this role by viewing the evolution of extended defects created by Si^+ ion implantation using quantitative transmission electron microscopy (QTEM).

A few studies have been aimed at trying to understand how the surface Si/BOX interface affects dopant diffusion in SOI. Boron was shown to pile-up at the surface Si/BOX interface in thinner (530 Å) SOI films, while phosphorus and arsenic exhibited a depletion at the interface.[5] However, the P depletion was not seen in shallow implants that were closer to the screen oxide. Crowder et. al modeled the enhanced diffusion of B and P in SOI under oxidation and inert ambients.[6] They extracted the effective surface recombination velocity and found that at temperatures below 925°C the surface Si/BOX interface retarded the diffusion, but the opposite effect occurred above 925°C. It is clear from these studies that the presence of the surface Si/BOX interface plays a crucial role in dopant diffusion in SOI.

As Si interstitials approach the surface Si/BOX interface it is important to understand how they interact with the interface. Interstitial recombination at Si/SiO_2 interfaces has been undertaken by a few groups in previous years, but there is still significant debate over the mechanisms that take place. At non-oxidizing Si/SiO_2 interfaces, the recombination flux of single silicon interstitials is expressed as

$$F_s = \sigma_{eff}(C_I - C_I^*) \qquad (1)$$

where σ_{eff} is the effective surface recombination velocity, C_I is the concentration of silicon interstitials and C_I^* is the equilibrium concentration of interstitials.[7] Tsamis et. al modeled the recombination velocity of Si interstitials as a function of time at non-oxidizing Si/SiO_2 interfaces.[8] They were able to determine the surface recombination velocity as a function of time, which is typically assumed to be independent of time. Law et. al were able to model data from oxidation enhanced diffusion (OED) studies by proposing recombination at the interface took place via a di-interstitial mechanism.[9] This is certain to be an area that will continue to be discussed in the future as the Si/SiO_2 interface becomes more and more important.

Experimental

In the experiment, 200 mm p-type (001) SOITEC and Czochralski (CZ) wafers having resistivities of 14 – 22 Ω cm were used. SOITEC wafers were used, instead of SIMOX, because of the integrity that is maintained in the active layer, as well as to avoid effects from other extended defects created by the high dose O^+ ion implantation process.[10] Prior to implantation the SOI wafers were thinned using oxidation and etching in HF. The thickness of the surface Si layer was monitored using a Rudolf dual wavelength ellipsometer. The final thickness obtained for the SOI wafers were 299 Å – 305 Å for the 300 Å target and 670 Å – 688 Å for the 700 Å target. The thickest SOI wafer, which did not receive oxidation/etching, was 1600 Å and all the BOX were 4000 Å. Wafers were then implanted at room temperature with non-amorphizing $^{28}Si^+$ ions at energies of 5 and 20 keV with a dose of $2x10^{14}$ cm^{-2}. Inert ambient furnace anneals in N_2 were performed at 750°C for times ranging from 5 minutes up to 4 hours. Plan-view transmission electron microscopy (PTEM) specimens were then cored and lapped using a 15 μm Al_2O_3 slurry. Etching of the PTEM specimens was done using HF/HNO_3 1:3. In addition, the SOI PTEMs received a buffered oxide etch (BOE) 6:1 to assist in removing the BOX. A JEOL 200CX TEM operating at 200 kV was used to image the PTEM specimens by weak beam dark field (WBDF) microscopy under $\mathbf{g}_{220}$ two-beam diffraction conditions. Finally, quantitative transmission electron microscopy (QTEM) was used to calculate the number of trapped interstitials in extended defects.

Results and Discussion

The Si interstitial profiles from UT-Marlowe are shown in figure 1 for each thickness of SOI. Implant energies were designed such that the projected range of the implant was strategically placed at a given depth within the surface Si layer. The projected range from SRIM was calculated to be 100 Å and 308 Å for the 5 keV and 20 keV $2x10^{14}$ cm^{-2} implants, respectively. It is important to realize that a certain portion of the dose was lost when the implant profile overlapped into the BOX as the SOI was thinned. Thus, it would be expected that this portion of the dose could not contribute to the formation of extended defects in the surface Si layer. Figure 2 shows the percent of dose retained and the actual dose confined to the surface Si layer for 5 to 40 keV Si^+ implant energies. The 1600 Å SOI essentially received the same dose as the bulk Si for both energies, while the 700 Å SOI suffered a 6% dose loss at 20 keV. The 300 Å SOI lost 6% and 45% at 5 and 20 keV, respectively.

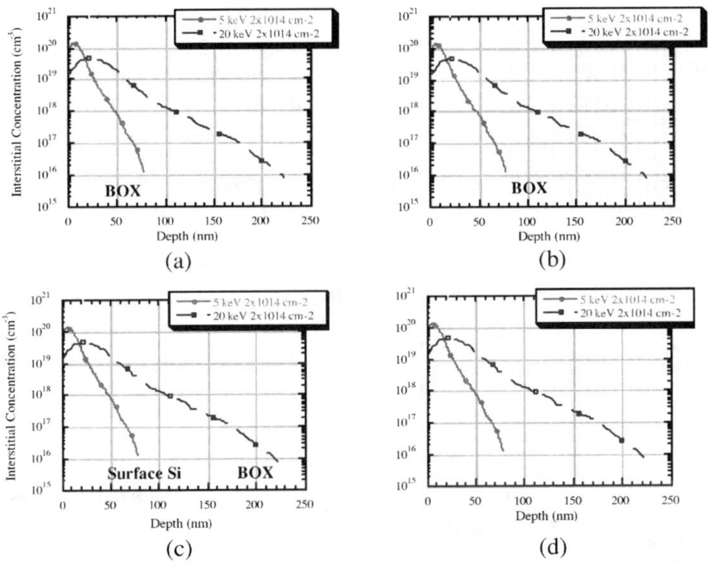

Figure 1. Interstitial profiles from UT-Marlowe for (a) 300 Å, (b) 700 Å, (c) 1600 Å, and (d) bulk Si.

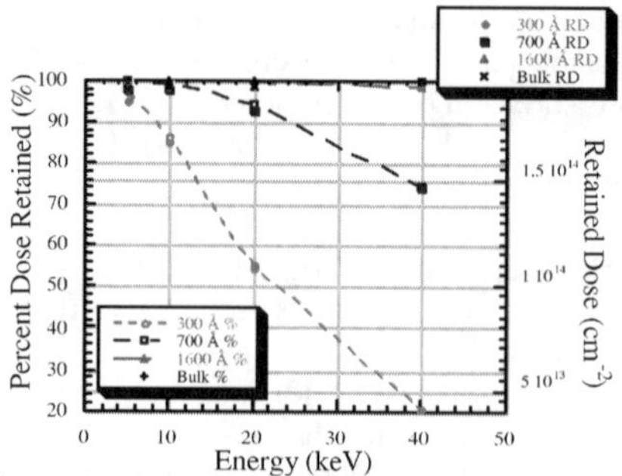

Figure 2. Percentage of dose retained and effective retained dose in surface Si layer of SOI for Si^+ implants from 5 keV to 40 keV, $2x10^{14}$ cm^{-2}.

Figure 3 shows PTEM micrographs for the 5 keV implants illustrating the extended defect evolution for the 300 Å and bulk Si. After annealing for 5 minutes, an assortment of small interstitial clusters, which preclude the formation of the rodlike {311}s[11], formed in all the specimens. Between 10 and 15 minutes the small clusters evolved into zig-zag {311} defects, which are {311} defects that unfault and then continue growing along another {311} plane[12]. Zig-zag {311}s have previously been observed at low evergies[12], and were stable until after approximately 2 hours. Dislocation loops then formed in the 700 Å, 1600 Å, and bulk, and were still present after annealing for 4 hours. However, the zig-zag {311} defects did not appear to evolve into loops in the 300 Å SOI. The small interstitial clusters were still present in the 300 Å SOI after 4 hours and appeared to be smaller and more numerous than at early anneal times. Figure 4 shows the concentration of trapped interstitials, Si_I, as a function of time obtained from the QTEM measurements for 5 keV. Data from the self-interstitial experiments done by Agarwal et. al for 5 keV, $1x10^{14}$ cm^{-2} and $3x10^{14}$ cm^{-2} are included for comparison.[12] The dissolution rate for the bulk specimens compares favorably with that done in their study. All the specimens had essentially identical dissolution behavior

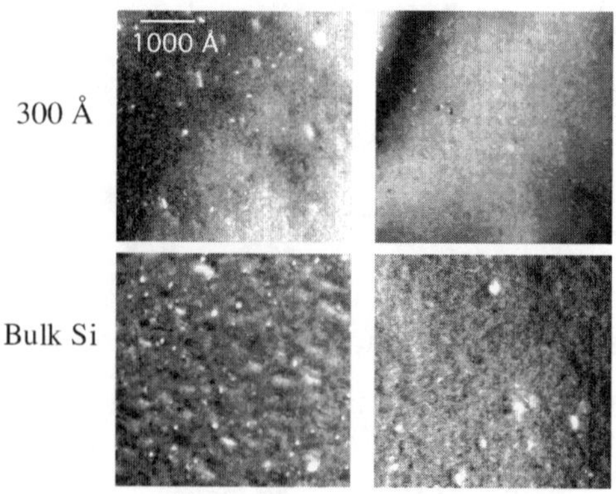

Figure 3. Plan-view TEM micrographs showing defect evolution in 300 Å SOI and bulk Si for 5 keV, $2x10^{14}$ cm^{-2} after annealing at 750° C.

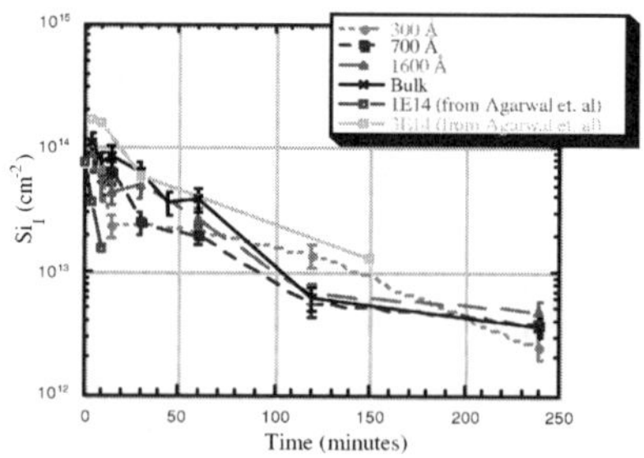

Figure 4. Concentration of trapped interstitials in extended defects for 5 keV, $2x10^{14}$ cm^{-2} after annealing at 750° C.

despite the differences observed in the PTEM images from figure 3. The fact that the 300 Å SOI trapped more interstitials after 120 minutes indicates a high density of small interstitial clusters existed in the 300 Å and did not nucleate the same as the extended defects in the other specimens.

Figure 5 shows the defect evolution at 20 keV for the 700 Å and bulk. It should be noted that extended defects did not form in the 300 Å SOI due to the dose loss to the oxide, so QTEM could not be performed. Once again, an assortment of small interstitial clusters appeared at early times and precluded the formation of {311} defects. After approximately 15-30 minutes the {311} defects began to unfault and form dislocation loops, which coarsened with time. The loops, which become more stable at higher implant energies[13], were still present after annealing as long as 8 hours. It can clearly be seen in the micrographs that the 700 Å SOI had fewer trapped interstitials due to the smaller loops that formed compared to those in the bulk. Figure 6 shows a plot of Si_I versus time at 20 keV for the 700 Å, 1600Å and bulk. Like 5 keV, the dissolution rate appeared to be the same in SOI as bulk Si. There was not a reduction in trapped interstitials with time due to the stability of dislocation loops at this energy and annealing temperature. The 1600 Å and bulk Si behaved the same, while there was a near constant reduction of ~35 - 40% in Si_I in the 700 Å for each of the times. This is slightly more than can be attributed to the dose loss of 6% in the 700 Å SOI.

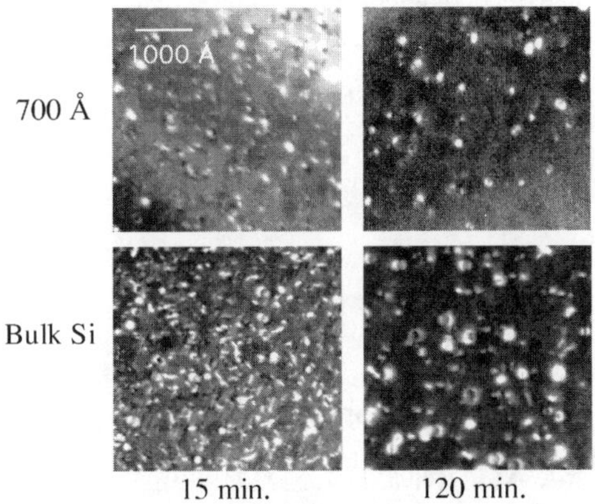

Figure 5. Plan-view TEM micrographs showing defect evolution in 700 Å SOI and bulk Si for 20 keV, $2x10^{14}$ cm^{-2} after annealing at 750° C.

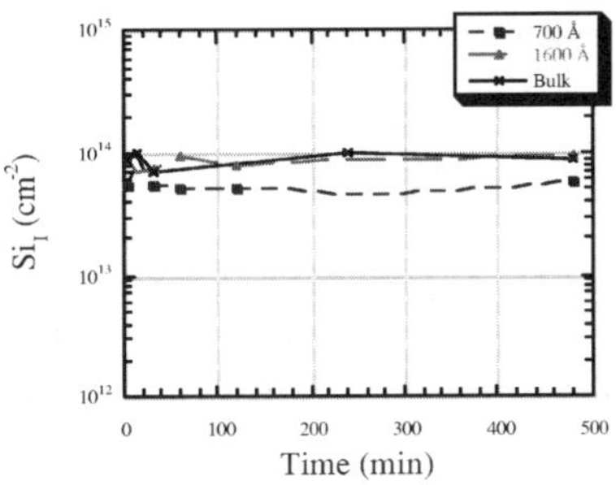

Figure 6. Concentration of trapped interstitials in extended defects for 20 keV, $2x10^{14}$ cm^{-2} after annealing at 750° C.

The proximity of the projected range to the surface accounts for the dissolution of the extended defects at 5 keV. The presence of small interstitial clusters throughout the annealing cycle in the 300 Å SOI indicates that something inhibits the stablility of {311} defects and dislocation loops in thin SOI at low implant energies. The slight difference in Si_I between the 700 Å and bulk at 20 keV is not solely a result of dose loss, but is attributed to recombination at the surface Si/BOX interface. However, the fact that the 1600 Å SOI behaved the same as the bulk in both cases indicates that trapped interstitials do not diffuse great distances in order to recombine at the interface. It is now proposed that the surface Si/BOX interface does not serve as a significant sink for interstitials trapped in extended defects unless the interstitial profile is close to or overlaps significantly into the BOX. Thus, it would be expected that the effect of the interface would be observed in thicker SOI at higher implant energies and with increased straggle.

Conclusions

It was observed that the surface Si/BOX interface did not increase the dissolution rate of extended defects in SOI after implantation at 5 and 20 keV, $2x10^{14}$ cm^{-2}. The proximity of the projected range to the surface accounts for the dissolution of extended defects at the 5 keV implant energy. The presence of the small interstitial clusters throughout the annealing cycle in the 300 Å SOI indicates that something inhibits the stability of defects in thin SOI at low implant energies. At 20 keV a slight decrease in the concentration of trapped interstitials was observed in the 700 Å SOI. This difference is attributed to interface recombination as the projected range and straggle of the Si^+ implant increased. The fact that the 1600 Å SOI behaved the same as the bulk in both cases indicates that trapped interstitials do not diffuse great distances in order to recombine at the interface. It is concluded that the surface Si/BOX interface does not serve as a significant sink for interstitials trapped in extended defects unless the interstitial profile is close to or overlaps significantly into the buried oxide.

Acknowledgements

This project was funded by the Semiconductor Research Corporation under task 819.001.

References

[1] G. Shahidi, et. al, IEEE International SOI Conference, 1 (1999).
[2] N. E. B. Cowern, K. T. F. Janssen, and H. F. F. Jos, J. Appl. Phys. **68**, 6191 (1990).
[3] D. J. Eaglesham, P. A. Stolk, H. -J. Gossmann, and J. M. Poate, Appl. Phys. Lett. **65**, 2305 (1994).
[4] P. M. Fahey, P. B. Griffin, and J. D. Plummer, Rev. Mod. Phys. **129**, 2292 (1989).
[5] H. Park, E. C. Jones, P. Ronsheim, C. Cabral, Jr., C. D'Emic, G. M. Cohen, R. Young, and W. Rausch, IEDM, 337 (1999).
[6] S. W. Crowder, C. J. Hsieh, P. B. Griffin, and J. D. Plummer, J. Appl. Phys. **76**, 2756 (1994).
[7] S. M. Hu, Appl. Phys. Lett. **27**, 165 (1974).
[8] C. Tsamis and D. Tsoukalas, J. Appl. Phys. **84**, 6650 (1998).
[9] M. E. Law, Y. M. Haddara, and K. S. Jones, J. Appl. Phys. **84**, 3555 (1998).
[10] A. Plößl and G. Kräuter, Solid-State Electronics **44**, 775 (2000).
[11] S. Coffa, S. Libertino, and C. Spinella, Appl. Phys. Lett., 76, 321 (2000).
[12] A. Agarwal, T. E. Haynes, D. J. Eaglesham, H. -J. Gossmann, D. C. Jacobson, J. M. Poate, Yu. E. Erokhin, Appl. Phys. Lett., **70**, 3332 (1997).
[13] J. Li and K. S. Jones, Appl. Phys. Lett. **73**, 3748 (1998).

Doping Activation

Mat. Res. Soc. Symp. Proc. Vol. 717 © 2002 Materials Research Society C3.1

Current Understanding and Modeling of Boron–Interstitial Clusters

Peter Pichler
Fraunhofer–Institut für Integrierte Bauelemente, Bauelementetechnologie
Schottkystrasse 10, 91058 Erlangen, Germany

ABSTRACT

Scaling of devices requires not only shallow junctions but also high levels of dopant activation. For boron as the main p–type dopant, the latter requirement is especially problematic since small clusters of boron atoms and self–interstitials, known also as boron–interstitial clusters (BICs), were found to deactivate and immobilize large fractions of the implanted atoms during post–implantation annealing. In this article, the properties of BICs are reviewed and their influence on semiconductor processes are highlighted.

INTRODUCTION

Since many years, ion implantation is the standard technique of dopant introduction in ULSI fabrication. However, the impinging ions undergo several collisions with host atoms before they are sufficiently slowed down and come to rest. In these collisions, the host atoms may be displaced from their lattice sites and large concentrations of defects are generated. Annealing of the defects and activation of the implanted ions requires a post–implantation annealing step at elevated temperature.

A major effect during post–implantation annealing, reported already in 1973 by Wittmaack et al. [1] and Hofker et al. [2, 3], is a considerably enhanced diffusion of the implanted ions. After time–resolved studies were possible by heating systems based on incoherent–light sources, investigations like that of Michel et al. [4] showed that the diffusion enhancement decreases rapidly with time. At the same time it was found that not all of the implanted atoms, especially in the case of boron, may be activated under certain conditions. Both effects are highly undesired considering the ongoing trend to shallower junctions which require also high dopant concentrations to fulfill the requirements of sufficiently low sheet resistivities.

The incomplete activation has triggered a notable number of investigations since then and small clusters of boron atoms and silicon self–interstitials, called boron–interstitial clusters or, shorter, BICs, were identified as the main reason. This article aims at summarizing especially the experimental evidence about composition and dynamics of BICs obtained from a variety of experiments, including studies based on implantation of boron followed by annealing, the use of samples with special marker–layer structures, and electron irradiation.

IMMOBILIZATION OF BORON

After ion implantation and short–time annealing by incoherent light sources came into use, Michel et al. [4] reported that rapid diffusion of boron occurs only below a concentration which correlates more with the intrinsic carrier concentration than with the solid solubility of boron. Similar investigations before and thereafter established values for the threshold concentration below which boron is mobile during post–implantation annealing [3–9]. These values are summarized in Figure 1. Also shown in the figure is the solid solubility limit of

$$C_B^{sol} = 9.25 \cdot 10^{22} \cdot \exp\left(-\frac{0.73\text{eV}}{k \cdot T}\right) \text{cm}^{-3} \qquad (1)$$

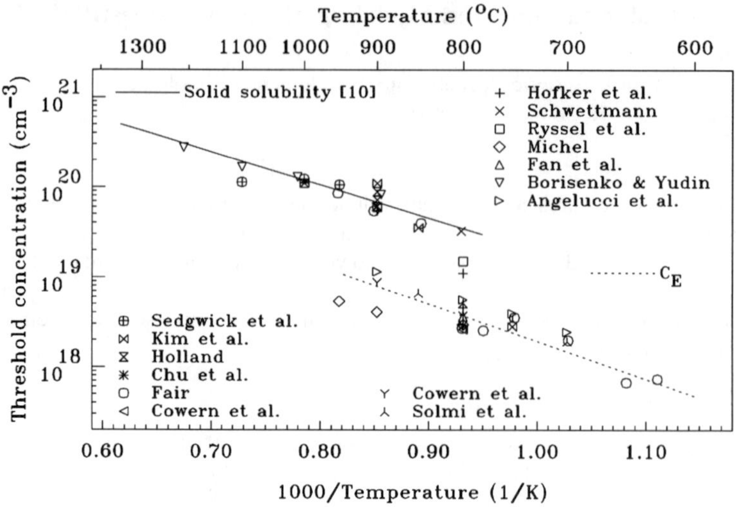

Figure 1: Threshold concentration below which boron is mobile during post–implantation annealing.

determined by Armigliato et al. [10]. It is apparent that the values above 950 °C follow the solubility limit while the values below 800 °C are described better by the expression

$$C_E = 2.82^{+10.4}_{-2.21} \cdot 10^{22} \cdot \exp\left(-\frac{0.827 \pm 0.137\text{eV}}{k \cdot T}\right) \text{cm}^{-3} \tag{2}$$

obtained by a regression analysis of the data points associated. In the range of intermediate temperatures higher threshold values are usually observed for higher thermal budgets. In the investigations of Hofker et al. [3], Cowern et al. [8], and Solmi et al. [9] it was even shown that the threshold concentration increases with process time towards the solid solubility. The time constants characteristic for this process increase from about 20 s to 30 min at 900 °C [8] to about two hours at 850 °C [9] and about 8 hours at 800 °C [9]. Complementary measurements of the carrier concentrations indicated that the immobile peak is largely electrically inactive [6, 8, 9, 11].

Major insights about transient diffusion and the immobile and electrically inactive fraction of boron atoms came also from a different class of experiments. They were based on the creation of implantation damage by a surface–near silicon implantation and the characterization of its effects via the diffusion of boron spikes. In such experiments, Eaglesham et al. [12] and Stolk et al. [13, 14] observed that significant fractions of the boron in the spikes were immobile. This effect diminished the further remote the spikes were from the surface. For the boron spikes nearest to the surface, the concentration below which the boron was mobile agreed closely to the threshold concentration (2) found from annealing implanted boron. But in contrast to the implanted profiles, significant immobile fractions were observed here also in regions remote from the original implantation damage. The straightforward conclusion from these observations was that self–interstitials originating from the implant damage caused the clustering in the boron peaks.

The observation of Stolk et al. that the immobile fraction reduced at 790 °C on a timescale of about 10 hours agreed well with similar results from ion–implanted profiles. The immobilization of boron in spikes remote from the implantation damage was observed also by Simpson et al. [15] who reported that the immobile part of the boron profile is electrically inactive. The experiments indicated that the immobilization reduces with distance from the implant damage even when only nominally undoped silicon is in between. The electrical inactivity of the immobile fraction was also confirmed by the experiments of Castell et al. [16] who characterized the electrically active boron concentration by secondary electron imaging of the workfunction.

THE DAWN OF BORON–INTERSTITIAL CLUSTERS

To account for the transient diffusion and the reduced mobility threshold observed for implanted boron profiles quantitatively, Cowern et al. [7,8,17] assumed the formation of immobile boron clusters with the participation of intrinsic point defects. Concrete numbers given in [7] favor B_2I clusters. At this point it is necessary to introduce the generally established nomenclature for boron–interstitial clusters. Their composition, denoted in the form B_mI_n, refers to m boron atoms and n silicon atoms which occupy m regular lattice sites. This notation does not include full structural information and, e. g., two boron atoms in a split configuration and an interstitial boron atom next to a substitutional boron atom are both denoted by B_2I. Formation of a B_2I cluster was also included in the model of Hane and Matsumoto [18] to reproduce incomplete activation and the threshold concentration for diffusion. The parameters describing the formation of B_2I in equilibrium, including the binding energy of 1.61 eV, were taken from the work of Mathiot and Pfister [19] who estimated them from the diffusion of boron at high concentrations. Similar to the work of Hane and Matsumoto, a B_2I cluster was included in the model of Baccus and Vandenbossche [20, 21] which aimed at simulating transient diffusion effects at high concentrations. Although details of its properties were not given, it was noted that a B_2I cluster would be sufficient to reproduce short annealing times while they are not sufficient to simulate long annealings at 900 °C [21]. These problems, on the other hand, were related to boron precipitation rather than the formation of boron–interstitial pairs.

Long after the effects were reported for the first time, the models contained hardly more than conventional clusters, precipitates, or B_2I complexes. Inspired by the concept of "intermediate defect configurations" introduced by Tan et al. [22], Cowern et al. [8] suggested that a certain number of substitutional boron atoms and self–interstitials might form such an intermediate defect configuration. A similar explanation was given by Stolk et al. [13] to explain the immobile fraction of boron atoms in marker layers remote from the original implantation damage. However, the real breakthrough of the concept of BICs came with the work of Pelaz et al. [23] and Caturla et al. [24]. Pelaz et al. [23, 25] presented a physically motivated model in which individual reactions between BICs and boron interstitials and silicon self–interstitials were explicitly taken into considerations. The reaction path suggested is shown schematically in Figure 2. It starts with a reaction of substitutional boron with self–interstitials to immobile BI and further to BI_2. During annealing, the BICs are assumed to grow via the capture of mobile boron interstitials. At a later stage, when the oversaturation of self–interstitial is reduced, self–interstitials may be liberated from the BICs which implies a composition change towards B_mI. Dissolution finally is a very slow process since it requires the repeated dissociation of boron interstitials and capture of self–interstitials. Using a model based on a similar reaction scheme, Caturla et al. [24] took the energy differences resulting

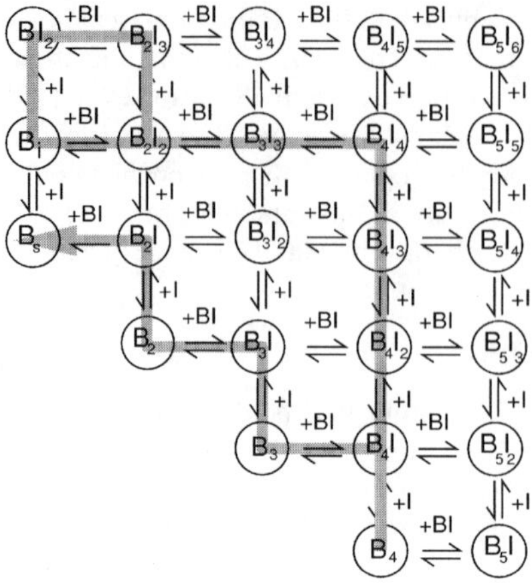

Figure 2: Reaction path suggested by Pelaz et al. [25] for the formation of boron–interstitial clusters.

from the reactions from *ab–initio* calculations. The two approaches — adaption of parameters based on a physical concept and taking them from *ab–initio* simulations — mark the boundaries of basically all model–development activities since then.

EXPERIMENTAL EVIDENCE

Complementary to the diffusion studies, there were various investigations in which individual BICs were identified. From infrared (IR) spectroscopic investigations after electron irradiation, Laithwaite et al. [26] concluded that interstitial boron can react at room temperature with silicon self–interstitials. The first reaction product would be BI_2, but it eluded the observation by IR spectroscopy. As an explanation, the authors assumed that various complexes of the composition BI_n may form so that the absorption is spread throughout a wide spectral range.

After electron irradiation, Newman and Smith [27] observed absorption bands at 552.5, 560.0, and 570.0 cm^{-1} (77 K) which were labeled P1 to P3. The concentrations were found to reflect the abundances of boron isotopes in substitutional B_2 pairs. Pairs of boron atoms with equal masses are D_{3d}–symmetric with respect to the $\langle 111 \rangle$–oriented bond between the boron atoms. For pairs composing of different boron isotopes, the symmetry reduces to C_{3v}. Similar conclusions were presented by Tsvetov et al. [28] and Bean et al. [29].

From its quadratic dependence on the boron concentration, Drevinsky et al. [30] assigned a hole trap at $E_v + 0.3$ eV to a B_iB_s complex. The same level was observed by Khan et al. [31] to correlate in irradiation studies with the dissolution of B_iO_i complexes as well as with the concentration of substitutional boron atoms. The tentative identification as B_iB_s complex is supported by the absence of this level in gallium–doped samples and by the independence of oxygen. Based on an ion–beam analysis after electron irradiation and annealing, Swanson [32] suggested two boron atoms to assume a $\langle 100 \rangle$ split–interstitial configuration in which they are separated by a distance of 1.6 Å. Near–edge x–ray absorption fine–structure spectroscopy investigations carried out by Vailionis et al. [33] were also consistent with a $\langle 100 \rangle$–oriented boron pair around a substitutional site. An entirely different configuration was suggested for B_2I by Thonke et al. [34] to explain their photoluminescence investigations of the 1.080 eV I_2 line. They concluded from isotope experiments and stress studies that the atomic arrangement consists of two substitutional boron atoms with a silicon self–interstitial in between which is in the (110) plane through the two boron atoms, but distorted out of the bond–centered position.

In their IR investigations on electron–irradiated, boron–doped samples, Tipping and Newman [35] observed the formation of absorption bands at 903, 917, 928, 599, and 603 cm^{-1} (77K). These lines were labeled S1 to S5 and suggested to reflect the different isotopes and vibration modes of B_iB_i pairs of boron interstitials. Break–up of the defects was observed around 270 K. In addition, the existence of a metastable precursor was suggested since the concentration of B_iB_i pairs increased quadratically with time during the initial annealing stage while the concentration of boron–interstitial complexes decreased linearly.

After implantation of boron with doses up to 10^{17} cm^{-2}, Mizushima et al. [36] observed by Hall–effect measurements that the number of holes increased approximately by one for about six implanted boron atoms. Complementary X–ray photoemission spectroscopy and Fourier–transform–infrared spectroscopy measurements revealed spectra known from icosahedral B_{12} clusters. From size considerations it was tentatively suggested that icosahedral boron clusters replace tetrahedral structures of five silicon atoms and bond to the surrounding silicon atoms without dangling bond. Taking the known properties of icosahedral B_{12} into account, it was further suggested that this structure acts as a double acceptor, explaining the observed increase of holes with the total number of boron atoms. *Ab–initio* simulations of Yamauchi et al. [37] confirmed the icosahedron to be more stable than a cubo–octahedron B_{12} cluster. While they did not find acceptor levels in the band–gap, the generation of holes found experimentally was ascribed to an unoccupied level near the edge of the valence band. Later, Yamauchi et al. [38] showed that the computed oscillation frequencies for an icosahedral B_{12} cluster are compatible to the measured ones.

After boron–interstitial complexes were identified as the reason for the incomplete activation during post–implantation annealing, several dedicated experiments provided insight into dynamics and composition of boron–interstitial clusters.

An important information about the size of BICs came from high–resolution transmission–electron–microscopy (TEM) analyses where, explicitly noted, e. g., by Stolk et al. [13] and Zhang et al. [39], no indication of BICs was ever found. This is generally interpreted in terms of clusters containing only a few boron atoms.

Accessible to TEM studies, on the other hand, are agglomerates of self–interstitials generated by sub–amorphizing implants of boron. In such an experimental set–up, Haynes et al. [41] investigated the formation of $\{113\}$ defects generated by silicon implantation to a dose of 10^{14} cm^{-2} and annealing at 740 °C for 15 min as a function of the boron background concentration. They found that the fraction of self–interstitials in the $\{113\}$ defects decreases by about one order of magnitude when the boron concentration is increased to 10^{19} cm^{-3}. From the concentration dependence the authors also concluded that the effect is in agreement with the formation of B_2I clusters. These experiments were extended by Lilak et al. [42]. However, under similar conditions as in the work of Haynes et al., significant effects of the boron concentration on the formation of $\{113\}$ defects were observed already at one tenth of the boron concentration. The experiments indicated quite clearly that boron is indeed able to capture self–interstitials. But the information was insufficient for a quantitative analysis since neither self–interstitial agglomerates with small sizes nor the BICs could be quantified.

To acquire additional information, marker layers and electrical measurements were used in certain experiments to quantify the oversaturation of self–interstitials and the BICs, respectively.

Spatially separated marker layers with different peak concentrations were used already by Griffin et al. [40] to study the effects of surface–near implant damage. They found that the diffusivity of the marker layer increases when its peak concentration decreases. This effect was described quantitatively by assuming the formation of BI_2 complexes. Combining their results with an estimate of the number of electrically inactive boron atoms from a comparison of secondary ion–mass spectroscopy (SIMS) and spreading–resistance profiling (SRP) measurements, Lilak et al. [42]. estimated upper limits for the ratio of self–interstitials to boron atoms in the BICs which formed after annealing at 740 °C for 10 and 20 minutes. Assuming these complexes to be partially electrically active, B_3I or B_4I were suggested to be responsible for the deactivation observed. It has to be noted that this result would also be compatible to an inactive B_2I cluster. Comparing silicon implants into samples with and without a boron–doped layer within the damage region and a remote boron marker layer to monitor the oversaturation of self–interstitials, Mannino et al. [43] concluded that BICs are energetically more stable than small interstitial clusters and that they form during the very early stage of annealing, probably already during ion implantation. Correlating the number of immobile boron after silicon implantation with a dose of $2 \cdot 10^{13}$ cm^{-2} and an anneal of 2 h at 600 °C with the number of self–interstitials which were captured during this time at the BICs they estimated that one self–interstitial is captured per boron atom at the already nucleated BICs. Using a similar set–up, Solmi et al. [44] confirmed that BICs are more stable than small interstitial clusters and $\{113\}$ defects. Using two different boron background concentrations, they estimated that the ratio of boron atoms and self–interstitials in the BICs formed after silicon implantation with a dose of 10^{14} cm^{-2} and annealing at 740 °C for 15 min is about 1.5. Characterizing the concentration of clustered boron atom by SRP measurements, Mannino et al. [45] observed that the concentration of electrically active boron atoms after silicon implantation is lowest for the shortest annealing time (2 s at 740 °C) and increases for prolonged annealing times. A comparison of several silicon implant doses showed that the concentration of clustered boron atoms increases with the silicon implant dose but saturates when the hole concentration drops to about 70% of the boron background concentration. Finally, the authors also estimated from their experiments that one implanted silicon atom may deactivate several boron atoms.

Estimates of compositions of boron–interstitial clusters were also obtained from reproducing diffusion profiles. Cowern et al. [46] and Cacciato et al. [47] suggested from their simulations that a maximum of about one silicon self–interstitial is trapped per clustered boron atom. Since they started with substitutional boron, their analysis indicated B_mI_m complexes or complexes with less self–interstitials. Using more advanced models, Uematsu [48] also claimed that one self–interstitial would be captured per boron atom in the cluster. However, since he considered reactions of boron interstitials with silicon self–interstitials, his analysis rather indicates boron–interstitial clusters of composition $B_mI_{2 \cdot m}$. Based on an extensive matrix of experiments, Mokhberi et al. [49] suggested by applying a simple cluster model that BI_2 complexes would form in the initial stages of an annealing step while later larger, boron–rich BICs are likely to form.

The importance of charge states of BICs is still a point of considerable uncertainties. *Ab–initio* simulations, as discussed below, predict that some BICs are ionized in their minimum–energy configuration. Experimental evidence in favor or against is rather scarce. Comparing SRP measurements with the number of substitutional atoms measured by ion–beam analysis, Kobayashi et al. [50] found an almost linear relationship. Since only few BICs are assumed to be negatively charged, this might indicate that the dominating BICs might be electrically inactive.

There are various indications in the literature that boron–interstitial clustering occurs very early in the implant and anneal process. In one such experiment, Kyllesbech Larsen et al. [51] and Cowern et al. [52] implanted homogeneously boron–doped samples and samples containing epitaxially grown boron spikes with a non–amorphizing dose of silicon atoms. From SRP measurements they found that the implant deactivated the boron even in depths far beyond the range of the implanted silicon atoms. In a different experimental set–up, Jones et al. [53] amorphized samples with grown boron spikes. Comparing the diffusion of spikes inside and beyond the regrown layer, similar diffusion enhancements were found. Clustering, however, was found only in the spikes outside the regrown layer. This can be taken as an indication that BICs or their precursors formed already before the recrystallization front swept over the position of the marker layer inside the amorphized layer.

While it seems generally accepted that BICs or, at least, their precursors, form already during implantation, the dominant species is still in debate. Because of reasons discussed below, in most models the formation of an immobile boron–interstitial complex BI is assumed as the first step. Further reactions with self–interstitials lead then to BI_2 complexes. However, as also outlined below, the assumption of BI being immobile may not be justified so that reactions with substitutional boron atoms may rather lead to B_2I.

The experiments of Jones et al. [53] mentioned already above contributed also to elucidate the formation of BICs in crystalline and amorphous silicon. Using different implant energies, a structure with marker layers was amorphized to different depths prior to annealing at 800 °C. While no clusters were found in a marker layer 170 nm shallower than the interface between amorphous and crystalline silicon, BICs were found to form around a marker which was only about 65 nm shallower. It can thus be concluded that BICs can be expected to form when the amorphous layer is not sufficiently deep. A quantitative prediction, on the other hand, is likely to exceed the capabilities of contemporary models.

THEORETICAL INVESTIGATIONS

The experimental work which was hardly conclusive was soon complemented by *ab–initio* simulations which contributed considerably to the clarification of certain phenomena, but also to confusions. Already the structure of the most simple boron–interstitial complex was a matter of long–lasting debate. In one of the first studies, Zhu et al. [54] suggested that an immobile pair of a self–interstitial next to a substitutional boron atom transforms with an activation energy of 1.0 eV to a boron interstitial. For its migration an activation energy of 0.3 eV was given while 0.6 eV would be needed for the return to the initial B_sI configuration. Taking charge states into account, Zhu [55] confirmed his results for positively charged boron–interstitial complexes with barriers of 1.1, 0.2, and 0.4 eV for the initial kick–out reaction, diffusion, and final return to the stable configuration. As a consequence, most of the models discussed below included an immobile BI complex in addition to a mobile boron interstitial. However, Zhu's results were not reproduced in the later investigations of Windl et al. [56], Sadigh et al. [57], and Jeong and Oshiyama [58]. All of them found that the energetically most favorable configuration of a self–interstitial next to a substitutional boron atom is assumed again after each diffusive jump of the boron atom.

Qualitative discrepancies with respect to the structure of the BICs and between the formation energies of specific B_mI_n complexes exist between the theoretical work based on tight–binding [59] and *ab–initio* methods [24, 60–65]. Characteristic for the accuracy of such computations is that in the work of Liu, Windl, and Masquelier [62, 64] differences of more than 1 eV were reported for some BICs between the binding energies computed by the local density approximation and the generalized gradient approximation while Lenosky et al. [61] reported that the difference between the two approaches is insignificant. Similar discrepancies of nearly 1 eV exist between the work of Liu, Windl, and Masquelier [62, 64] and the work of Lenosky et al. [61] despite the fact that the same code was used in both investigations. This is most likely associated to different structures found for the BICs, but also to the general problems of *ab–initio* schemes with excited states.

Despite some differences, the *ab–initio* calculations of Liu, Windl, and Masquelier [62, 64] and Lenosky et al. [61] agree that the formation of B_m complexes from substitutional boron atoms increase the system energy. There is also agreement that the BICs may exist in several charge states and that the most stable ones correspond to B^-, BI^+, $B_2^=$, B_2I^0, $B_2I_2^0$, B_3I^+, and $B_4I^=$. All of the BICs containing self–interstitials cause a decrease of the system energy upon formation. Usually, this decrease becomes larger for larger numbers of boron atoms and self–interstitials in the BIC. An exception is $B_4I^=$ which was found to be less stable than B_3I^-.

To make use of the binding energies extracted from theoretical calculations, they were implemented either in Monte–Carlo diffusion schemes [23–25, 63, 66] or in continuum models [62, 64, 67]. Depending on the study, agreement with experiments was rated from fair to excellent for BIC parameters taken directly from theory. Considering the large discrepancies between the investigations, this is more than one could have hoped for. In addition, moderate adaptions of some of the parameters were reported in some of the publications. The models, probably because of the inherent assumptions of an immobile BI complex, confirmed that the BIC dominating at low thermal budgets is BI_2. Extended anneals were found favor the formation of more boron–rich clusters like B_3I or B_4I_2.

Unlike for Monte–Carlo diffusion schemes, a high number of equations to be solved may be a considerable problem for daily process simulation. Therefore, a reduction of the complexity to a minimum set of BICs was frequently suggested. As an example, Figure 3 shows a drastically reduced model from the work of Stiebel [68] which, as shown in Figure 4, allows in combination with a model for self–interstitial agglomerates to solve a variety of experimental profiles.

CONCLUSIONS

Modeling of transient dopant diffusion and activation has matured to a point where a qualitative and, in many cases, quantitative knowledge allows the prediction of technological processes. Incomplete activation was found to be related to the formation of complexes comprising boron atoms and silicon self–interstitials. Despite major progress in the quantitative description of their kinetics, owing especially to the application of *ab–initio* simulations, significant discrepancies still exist. A convergence of the models would be highly desirable.

REFERENCES

[1] K. Wittmaack, J. Maul, and F. Schulz in Ion Implantation in Semiconductors and Other Materials ed. by B. L. Crowder, New York: Plenum Press, 119 (1973).

[2] W. K. Hofker, H. W. Werner, D. P. Oosthoek, and H. A. M. de Grefte, ibid., 133 (1973).

[3] W. K. Hofker, H. W. Werner, D. P. Oosthoek, and H. A. M. de Grefte, Appl. Phys. **2**, 265 (1973).

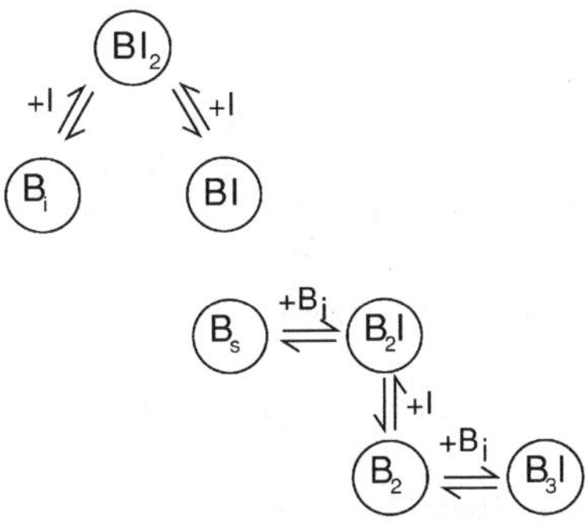

Figure 3: Simplified model for the formation of boron–interstitial clusters.

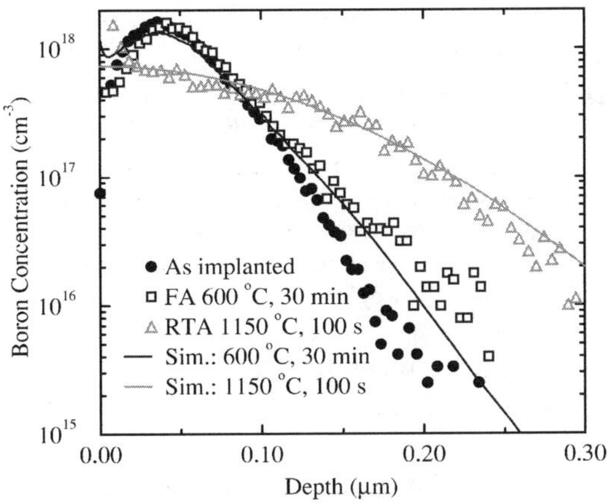

Figure 4: Simulation of boron profiles using a combined model for self–interstitial agglomerates and BICs.

[4] A. E. Michel in Mat. Res. Soc. Symp. Proc. **52**: Rapid Thermal Processing, edited by T. O. Sedgwick, T. E. Seidel, and B.-Y. Tsaur, 3 (1986); A. E. Michel, W. Rausch, P. A. Ronsheim, and R. H. Kastl, Appl. Phys. Lett. **50**, 416 (1987); Appl. Phys. Lett. **51**, 487 (1987).

[5] F. N. Schwettmann, J. Appl. Phys. **45**, 1918 (1974); H. Ryssel, K. Müller, K. Haberger, R. Henkelmann, and F. Jahnel, Appl. Phys. **22**, 35 (1980); D. Fan, J. Huang, R. J. Jaccodine, P. Kahora, and F. Stevie, Appl. Phys. Lett. **50**, 1745 (1987). V. E. Borisenko and S. G. Yudin, phys. stat. sol. (a) **101**, 123 (1987); R. Angelucci, R. Cembali, P. Negrini, M. Servidori, and S. Solmi, J. Electrochem. Soc. **134**, 3130 (1987); T. O. Sedgwick, A. E. Michel, V. R. Deline, S. A. Cohen, and J. B. Lasky, J. Appl. Phys. **63**, 1452 (1988); Y. Kim, H. Z. Massoud, and R. B. Fair, J. Electronic Materials **18**, 143 (1989); O. W. Holland, Appl. Phys. Lett. **54**, 798 (1989); R. B. Fair, J. Electrochem. Soc. **137**, 667 (1990).

[6] W. K. Chu, M. Z. Numan, J. Z. Zhang, and G. S. Sandhu, Nuclear Instruments and Methods in Physics Research B **37/38**, 365 (1989).

[7] N. E. B. Cowern, J. F. F. Jos, K. T. F. Janssen, and A. J. H. Wachters in Mat. Res. Soc. Symp. Proc. **163**: Impurities, Defects and Diffusion in Semiconductors: Bulk and Layered Structures, edited by D. J. Wolford, J. Bernholc, and E. E. Haller, 605 (1990).

[8] N. E. B. Cowern, K. T. F. Janssen, and H. F. F. Jos, J. Appl. Phys. **68**, 6191 (1990).

[9] S. Solmi, F. Baruffaldi, and R. Canteri, J. Appl. Phys. **69**, 2135 (1991).

[10] A. Armigliato, D. Nobili, P. Ostoja, M. Servidori, and S. Solmi in Electrochem. Soc. Proc. **77-2**: Semiconductor Silicon 1977, edited by H. R. Huff and E. Sirtl, 638 (1977).

[11] W. K. Hofker, H. W. Werner, D. P. Oosthoek, and N. J. Koeman, Appl. Phys. **4**, 125 (1974).

[12] D. J. Eaglesham, P. A. Stolk, H.-J. Gossmann, and J. M. Poate, Appl. Phys. Lett. **65**, 2305 (1994).

[13] P. A. Stolk, H.-J. Gossmann, D. J. Eagleshm, D. C. Jacobson, J. M. Poate, and H. S. Luftman, Appl. Phys. Lett. **66**, 568 (1995).

[14] P. A. Stolk, H.-J. Gossmann, D. J. Eaglesham, and J. M. Poate, Nuclear Instruments and Methods in Physics Research B **96**, 187 (1995).

[15] T. W. Simpson, R. D. Goldberg, I. V. Mitchell, and J.-M. Baribeau in Mat. Res. Soc. Symp. Proc. **438**: Materials Modification and Synthesis by Ion Beam Processing, edited by D. E. Alexander, N. W. Cheung, B. Park, and W. Skorupa, 15 (1997); T. W. Simpson, I. V. Mitchell, and J.-M. Baribeau in Mat. Res. Soc. Symp. Proc. **469**: Defects and Diffusion in Silicon Processing, edited by T. Diaz de la Rubia, S. Coffa, P. A. Stolk, and C. S. Rafferty, 277 (1997).

[16] M. R. Castell, T. W. Simpson, I. V. Mitchell, D. D. Perovic, and J.-M. Baribeau, Appl. Phys. Lett. **74**, 2304 (1999).

[17] N. E. B. Cowern, H. F. F. Jos, and K. T. F. Janssen, Materials Science and Engineering B **4**, 101 (1989).

[18] M. Hane and H. Matsumoto, IEDM Techn. Digest, 701 (1991); IEEE Trans. Electron Devices **40**, 1215 (1993).

[19] D. Mathiot and J. C. Pfister, J. Appl. Phys. **55**, 3518 (1984).

[20] B. Baccus and E. Vandenbossche in Simulation of Semiconductor Devices and Processes **5**, edited by S. Selberherr, H. Stippel, and E. Strasser, Wien: Springer-Verlag, 133 (1993).

[21] B. Baccus and E. Vandenbossche, Microelectronics Journal **26**, 235 (1995).

[22] T. Y. Tan, H. Föll, and W. Krakow in Inst. Phys. Conf. Ser.: Microscopy of Semiconducting Materials, 1981, edited by A. G. Cullis and D. C. Joy, 1 (1981).

[23] L. Pelaz, M. Jaraiz, G. H. Gilmer, H.-J. Gossmann, C. S. Rafferty, D. J. Eaglesham, and J. M. Poate, Appl. Phys. Lett. **70**, 2285 (1997).

[24] M.-J. Caturla, T. Diaz de la Rubia, J. Zhu, and M. Johnson in Mat. Res. Soc. Symp. Proc. **469**: Defects and Diffusion in Silicon Processing, edited by T. Diaz de la Rubia, S. Coffa, P. A. Stolk, and C. S. Rafferty, 335 (1997).

[25] L. Pelaz, G. H. Gilmer, H.-J. Gossmann, C. S. Rafferty, M. Jaraiz, and J. Barbolla, Appl. Phys. Lett. **74**, 3657 (1999).

[26] K. Laithwaite, R. C. Newman, and D. H. J. Totterdell, J. Phys. C **8**, 236 (1975).

[27] R. C. Newman and R. S. Smith, Physics Letters **29A**, 671 (1967); in Localized Excitations in Solids, edited by R. F. Wallis, New York, NY, USA: Plenum Press, 177 (1968).

[28] V. Tsvetov, W. Allred, and W. G. Spitzer, ibid., 185 (1968).

[29] A. R. Bean, S. R. Morrison, R. C. Newman, and R. S. Smith, J. Phys. C **5**, 379 (1972).

[30] P. J. Drevinsky and H. M. DeAngelis in Thirteenth International Conference on Defects in Semiconductors, edited by L. C. Kimerling and J. M. Parsey, Jr., The Metallurgical Society of AIME, 807 (1985); P. J. Drevinsky, C. E. Caefer, S. P. Tobin, J. C. Mikkelsen, Jr., and L. C. Kimerling in Mat. Res. Soc. Symp. Proc. **104**: Defects in Electronic Materials, edited by M. Stavola, S. J. Pearton, and G. Davies, 167 (1988).

[31] A. Khan, M. Yamaguchi, Y. Ohshita, N. Dharmarasu, K. Araki, T. Abe, Hisayoshi Itoh, T. Ohshima, M. Imaizumi, and S. Matsuda, J. Appl. Phys. **90**, 1170 (2001).

[32] M. L. Swanson, Vacuum **39**, 87 (1989).

[33] A. Vailionis, G. Glass, P. Desjardins, D. G. Cahill, and J. E. Greene, Phys. Rev. Lett. **82**, 4464 (1999).

[34] K. Thonke, J. Weber, J. Wagner, and R. Sauer, Physica **116B**, 252 (1983); K. Thonke, N. Bürger, G. D. Watkins, and R. Sauer in Thirteenth International Conference on Defects in Semiconductors, edited by L. C. Kimerling and J. M. Parsey, Jr., The Metallurgical Society of AIME, 823 (1985).

[35] A. K. Tipping and R. C. Newman, Semicond. Sci. Technol. **2**, 389 (1987).

[36] I. Mizushima, M. Watanabe, A. Murakoshi, M. Hotta, M. Kashiwagi, and M. Yoshiki, Appl. Phys. Lett. **63**, 373 (1993).

[37] J. Yamauchi, N. Aoki, and I. Mizushima, Phys. Rev. B **55**, R10245 (1997).

[38] J. Yamauchi, N. Aoki, and I. Mizushima, phys. stat. sol. (b) **210**, 273 (1998).

[39] L. H. Zhang, K. S. Jones, P. H. Chi, and D. S. Simons, Appl. Phys. Lett. **67**, 2025 (1995).

[40] P. B. Griffin, R. F. Lever, R. Y. S. Huang, H. W. Kennel, P. A. Packan, and J. D. Plummer, IEDM Techn. Digest, 295 (1993).

[41] T. E. Haynes, D. J. Eaglesham, P. A. Stolk, H.-J. Gossmann, D. C. Jacobson, and J. M. Poate, Appl. Phys. Lett. **69**, 1376 (1996).

[42] A. D. Lilak, S. K. Earles, M. E. Law, and K. S. Jones, Appl. Phys. Lett. **74**, 2038 (1999).

[43] G. Mannino, N. E. B. Cowern, F. Roozeboom, and J. G. M. van Berkum, Appl. Phys. Lett. **76**, 855 (2000).

[44] S. Solmi, M. Bersani, M. Sbetti, J. Lundsgaard Hansen, and A. Nylandsted Larsen, J. Appl. Phys. **88**, 4547 (2000).

[45] G. Mannino, S. Solmi, V. Privitera, and M. Bersani, Appl. Phys. Lett. **79**, 3764 (2001).

[46] N. E. B. Cowern, A. Cacciato, J. S. Custer, F. W. Saris, and W. Vandervorst, Appl. Phys. Lett. **68**, 1150 (1996).

[47] A. Cacciato, J. G. E. Klappe, N. E. B. Cowern, W. Vandervost, L. P. Biró, J. S. Custer, and F. W. Saris, J. Appl. Phys. **79**, 2314 (1996).

[48] M. Uematsu, J. Appl. Phys. **84**, 4781 (1998).

[49] A. Mokhberi, P. B. Griffin, and J. D. Plummer in 2000 International Conference on Simulation of Semiconductor Processes and Devices, Piscataway, NJ, USA: IEEE, 163 (2000).

[50] H. Kobayashi, I. Nomachi, S. Kusanagi, and F. Nishiyama in Mat. Res. Soc. Symp. Proc. **669**: Si Front-End Processing - Physics and Technology of Dopant-Defect Interactions III, edited by K. S. Jones, M. D. Giles, P. Stolk, J. Matsuo, and E. C. Jones, J.5.3.1 (2001).

[51] K. Kyllesbech Larsen, V. Privitera, S. Coffa, F. Priolo, S. U. Campisano, and A. Carnera, Phys. Rev. Lett. **76**, 1493 (1996).

[52] N. E. B. Cowern, H. G. A. Huizing, P. A. Stolk, C. C. G. Visser, R. C. M. de Kruif, K. Kyllesbech Larsen, V. Privitera, L. K. Nanver, and W. Crans, Nuclear Instruments and Methods in Physics Research B **120**, 14 (1996).

[53] K. S. Jones, R. G. Elliman, M. M. Petravic, and P. Kringhøj, Appl. Phys. Lett. **68**, 3111 (1996).

[54] J. Zhu, T. Diaz de la Rubia, L. H. Yang, C. Maillhot, and G. H. Gilmer, Phys. Rev. B **54**, 4741 (1996).

[55] J. Zhu in Mat. Res. Soc. Symp. Proc. **469**: Defects and Diffusion in Silicon Processing, edited by T. Diaz de la Rubia, S. Coffa, P. A. Stolk, and C. S. Rafferty, 151 (1997); Comput. Mater. Sci. **12**, 309 (1998).

[56] W. Windl, M. Bunea, R. Stumpf, S. T. Dunham, and M. P. Masquelier in 1999 International Conference on Modeling and Simulation of Microsystems MSM'99, Cambridge, MA, USA: Computational Publications, 369 (1999); W. Windl, M. M. Bunea, R. Stumpf, S. T. Dunham, and M. P. Masquelier in Mat. Res. Soc. Symp. Proc. **568**: Si Front-End Processing - Physics and Technology of Dopant-Defect Interactions, edited by H.-J. L. Gossmann, T. E. Haynes, M. E. Law, A. Nylandsted Larsen, and S. Odanaka, 91 (1999); Phys. Rev. Lett. **83**, 4345 (1999).

[57] B. Sadigh, T. J. Lenosky, S. K. Theiss, M.-J. Caturla, T. Diaz de la Rubia, and M. A. Foad, Phys. Rev. Lett. **83**, 4341 (1999).

[58] J.-W. Jeong and A. Oshiyama in Proceedings of the 25th International Conference on the Physics of Semiconductors, edited by N. Miura and T. Ando, Berlin, Germany: Springer, 1381 (2001).

[59] W. Luo, P. B. Rasband, P. Clancy, and B. W. Roberts, J. Appl. Phys. **84**, 2476 (1998); W. Luo and P. Clancy, J. Appl. Phys. **89**, 1596 (2001).

[60] M. Okamoto, K. Takayanagi, and S. Takeda in Mat. Res. Soc. Symp. Proc. **510**: Defect and Impurity Engineered Semiconductors II, 343 (1998).

[61] T. J. Lenosky, B. Sadigh, S. K. Theiss, M. J. Caturla, and T. Diaz de la Rubia, Appl. Phys. Lett. **77**, 1834 (2000).

[62] X.-Y. Liu, W. Windl, and M. P. Masquelier, Appl. Phys. Lett. **77**, 2018 (2000).

[63] S. K. Theiss, J. M. Caturla, M. D. Johnson, J. Zhu, T. Lenosky, B. Sadigh, and T. Diaz de la Rubia, Thin Solid Films **365**, 219 (2000).

[64] W. Windl, X.-Y. Liu, and M. P. Masquelier, phys. stat. sol. (b) **226**, 37 (2001).

[65] P. Alippi, P. Ruggerone, and L. Colombo in Solid State Phenomena **82-84**: Gettering and Defect Engineering in Semiconductor Technology GADEST 2001, edited by V. Raineri, F. Priolo, M. Kittler, and H. Richter, 163 (2002).

[66] M. J. Caturla, M. D. Johnson, and T. Diaz de la Rubia, Appl. Phys. Lett. **72**, 2736 (1998); M.-J. Caturla, S. K. Theiss, T. J. Lenosky, and T. Diaz de la Rubia, IEDM Techn. Digest, 489 (1998); S. K. Theiss, M.-J. Caturla, T. J. Lenosky, B. Sadigh, and T. Diaz de la Rubia in 2000 International Conference on Simulation of Semiconductor Processes and Devices, Piscataway, NJ, USA: IEEE, 18 (2000).

[67] A. D. Lilak, S. K. Earles, K. S. Jones, and M. E. Law, IEDM Techn. Digest, 493 (1997); S. Chakravarthi and S. T. Dunham in Simulation of Semiconductor Processes and Devices 1998, edited by K. De Meyer and S. Biesemans, Wien: Springer-Verlag, 55 (1998); D. Stiebel, P. Pichler, and H. Ryssel in Mat. Res. Soc. Symp. Proc. **568**: Si Front-End Processing - Physics and Technology of Dopant-Defect Interactions, edited by H.-J. L. Gossmann, T. E. Haynes, M. E. Law, A. Nylandsted Larsen, and S. Odanaka, 141 (1999); S. Chakravarthi and S. T. Dunham in 2000 International Conference on Simulation of Semiconductor Processes and Devices, Piscataway, NJ, USA: IEEE, 167 (2000); J. Appl. Phys. **89**, 3650 (2001).

[68] P. Pichler and D. Stiebel, Nuclear Instruments and Methods in Physics Research B **186**, 256 (2002).

Mat. Res. Soc. Symp. Proc. Vol. 717 © 2002 Materials Research Society C3.4

Electrical Activity of B and As Segregated at the Si-SiO_2 Interface

Jens Frühauf[1,2], Richard Lindsay[3], Andreas Bergmaier[2], Wilfried Vandervorst[3], Georg Tempel[1], Karen Maex[3], Günther Dollinger[2] and Frederick Koch[2]
[1] Infineon technologies AG, affiliated to IMEC vzw, Kapeldreef 75, B-3001 Leuven, Belgium
[2] Technische Universität München, Physik-Department, Garching, Germany
[3] IMEC vzw, Leuven, Belgium
Corresponding author: Jens.Fruehauf@imec.be

ABSTRACT

During spike annealing of ultra-shallow junctions, large fractions of the dopants form a partially active pile-up at the interface between silicon and the screening oxide layer. In this paper, we show results of sheet resistance, SIMS and high resolution Elastic Recoil Detection measurements to investigate the physical and electrical behaviour of B and As dopant atoms at the interface.

Our results show that the fraction of dopants segregated at the interface is as high as 30-50% for B, but is dependent on dose and the type of screening oxide. Concentrations of up to 3e20cm^{-3} and more of active dopants are found on the Si side of the interface. The presence of nitrogen in the oxide at the interface causes a higher and sharper pile-up. Results indicate that a similar peak is expected for As, with active concentrations above 6e20cm^{-3}. In an HF dip, the pile-up is removed together with the oxide or deactivated during native oxide regrowth.

Further experiments show that immediately after removing the screening oxide in an HF dip the sheet resistance for B decreases sharply due to carrier accumulation, then raises to about 6-9% above the initial level depending on the oxide and dopant species. The sharp decrease in resistance is not observed for As.

INTRODUCTION

To form ultra-shallow junctions, dopants are usually implanted with projected ranges <5nm through a screening oxide layer. Other work [1,2] already pointed out that during spike anneal, a large fraction of the dopants diffuses against the concentration gradient towards the Si/SiO_2 interface and form a pile-up. To avoid an increase in sheet resistance to meet the requirements of the ITRS roadmap, the activation of these dopants needs to be optimised. Therefore it is important to understand the behaviour of dopants at the interface and to avoid their removal during subsequent process steps.

EXPERIMENTS

All experiments used blanket 8" wafers. After growing a thermal screening oxide (O) or oxynitride (NO) they were implanted with B (1keV), BF_2 or As (both 5keV) with different doses using an Applied Materials XRLEAP implanter. In case of the NO, only a small layer of oxide next to the interface is nitrided.

The wafers were spike annealed in an RTP process for 1s at 1070ºC in 133ppm O_2, with ramping rates of 100ºC/s (up) and ~70ºC/s (down), using an AST-Steag SHS-2800.

Table I – Data extracted from ERD and SIMS (500eV) profiles. Integrated dose over the top 5 nm is shown to compare ERD and SIMS accuracy. Data for an as-implanted sample are shown.

	implant	dose	oxide	HF dip	oxide thickness	dose (total) ERD	dose (total) SIMS	%dose in pile-up	FWHM pile-up	dose (top 5nm) ERD	dose (top 5nm) SIMS
A	5keV BF_2	1e14	O	--	1.89 nm	8.43e13	7.60e13	31%	1.71 nm	3.36e13	3.24e13
B	5keV BF_2	4e14	O	--	1.77 nm	2.14e14	2.21e14	31%	1.95 nm	7.44e13	8.52e13
C	5keV BF_2	4e14	NO	--	1.65 nm	2.81e14	2.52e14	54%	1.66 nm	1.73e14	1.27e14
D	5keV BF_2	1e15	O	--	1.69 nm	4.46e14	4.98e14	46%	1.84 nm	2.24e14	1.98e14
E	5keV BF_2	4e14	O	30s	0.86 nm	1.77e14	1.90e14	--	--	5.16e13	5.08e13
F	5keV BF_2	4e14	NO	30s	0.97 nm	1.42e14		--	--	3.63e13	
G	1keV B	4e14	O	30s	0.95 nm	2.46e14	2.68e14	(20%)	(1.62nm)	7.81e13	6.77e13
H	1keV B	4e14	NO	30s	1.07 nm	1.77e14	1.99e14	(15%)	(0.94nm)	4.71e13	4.88e13
	5keV BF_2	4e14	O	--	2.3 nm	4.05e14		as-implanted sample			

As analysis method, the sheet resistance (R_s) of all samples has been measured with the four-point-probe method, using a Tencor RS75. Secondary Ion Mass Spectroscopy (SIMS, with an Atomika 4500) with a sputtering energy of 500eV was performed on selected samples. Spreading Resistance Profilometry (SRP, on a SSM-150) was used to determine the degree of activation.

High-resolution Elastic Recoil Detection (ERD) measurements were done on several wafers, using the tandem accelerator and Q3D spectrograph at Munich University of Technology. Using 40MeV Au ions as primary beam, the distribution of dopants as well as of the oxygen and nitrogen close to the surface was investigated with a depth resolution <1nm [4]. The thickness of the surface oxide was determined using ERD data and Ellipsometry (Tencor ASET-F5).

RESULTS & DISCUSSION

Elastic Recoil Detection (experiment 1)

On different wafers, BF_2 was implanted through a 2nm thermal oxide or oxynitride (cf. Table I). The wafers were spike annealed and measured with high-resolution ERD. For comparison, one wafer was measured as implanted. The signal-to-noise ratio of ERD is very good for concentrations >1e20cm^{-3}. Even though the tail of the junction has a far lower concentration, ERD still resolves the total dose within 10% error, compared to SIMS.

The B dose seen in ERD and SIMS is only 45% to 85% of the nominally implanted dose (Table I). This is due to outdiffusion of dopants during spike anneal. For the 1e15cm^{-2} sample, also self-sputtering during implant is expected to have a slight effect.

Fig. 1 and 2a show a clear pile-up of dopants at the Si/SiO_2 interface. For NO, the interface pile-up is higher and narrower, and contains a higher fraction of the dopants than in the O case (Table I). With a FWHM of 1.65nm, the measured pile-up is almost consistent with the model of a monolayer of B at the interface. For pure oxide, the pile-up is >0.3nm thick and does extent into the oxide and/or silicon.

Comparing BF_2 implants with different doses, fig. 2a shows that for doses of 1e14cm^{-2} and 4e14^{-2}, around 30% of the observed dopants have segregated to the interface pile-up. For the 1e15cm^{-2} implant however, this number increases to 46%. SIMS does not resolve the pile-up in detail (cf. [3]), but confirms the dose within 10%, when integrating over the top 5nm. Only for sample C (with oxynitride), a 30% dose difference is seen between ERD and SIMS.

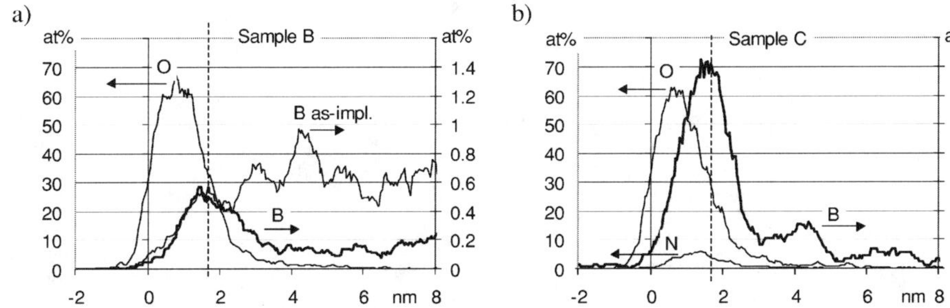

Figure 1 – ERD depth profiles of samples B, C and E (cf. Table I). The concentrations are given on a linear scale in atomic %, on the left scale for the O and N content, on the right scale for the B content. The dashed line indicates the average interface depth. For sample B, the as-implanted profile is shown for comparison.

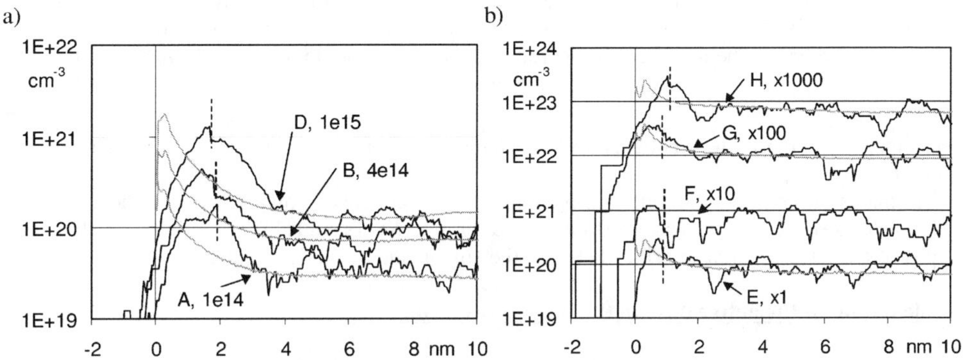

Figure 2 – ERD (black) and SIMS (gray) profiles of different implants. a) Comparison of different BF_2 doses implanted through pure oxide. b) BF_2 and B implants through O or NO. The oxide has been etched in a 30s HF dip and native oxide has regrown prior to profiling. (no SIMS profile available for sample F) – The dashed lines indicate the depth of the Si/SiO_2 interface. The small concentration steps seen at the interface in every graph are due to the rescaling from atomic % to concentration values and are therefore artificial.

Pile-up investigation by surface treatment

Figure 2b shows ERD profiles of BF_2 (5keV, 4e14) and B (1keV, 4e14) implanted samples, on which the implant oxide was removed with a 30s HF dip, and which have been given time to develop a native oxide. For B implants, the remains of a pile-up can be seen, but for BF_2, no significant peak is left.

The HF dip removes most of the interfacial dopant pile-up. As the HF dip cannot remove dopants from within the silicon, these results show that in all cases most of the pile-up dopants reside at or in immediate vicinity to the interface. This is more pronounced in the presence of

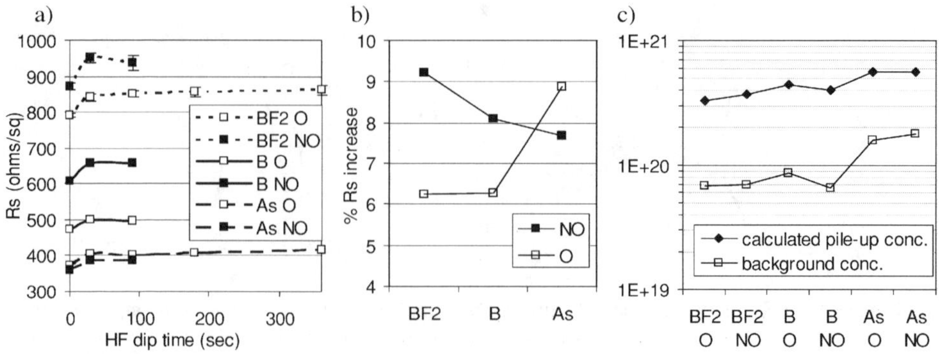

Figure 3 – a) Sheet resistance measurements after each sample had been etched for a different time in a 2% HF solution. The "0s" wafers were not etched at all. b) Increase in sheet resistance after the 30s HF dip and native oxide regrowth. c) Calculated concentration of dopants within the first ~0.5 nm below the interface, compared to the background doping, as it is measured by SIMS in 5-10nm depth.

fluorine, which therefore is believed to cause a higher and narrower pile-up, as it was already seen for nitrogen.

Remains of the pile-up are incorporated into the subsequently growing native oxide and are thereby deactivated. In the case of B implanted through NO (sample "H", fig. 2b), a small pile-up is observed at the new interface between the silicon and the native oxide.

Calculation of the active dose in the pile-up (experiment 2)

On different wafers, a 2nm thermal oxide (O) or oxynitride (NO) was grown, then B (1keV, 4e14), BF_2 (5keV, 4e14) or As (5keV, 8e14) were implanted. After anneal, the wafers were etched for different times (30s, 90s, 180s, 360s or not at all) in 2% diluted HF. Native oxide was given 2 weeks' time to grow, after which the growth process is not completely finished, but has reached a reasonably stable level. As a sheet resistance comparison shows very little difference between wafers etched for 30s and for 360s (fig. 3a), the etch rate of doped silicon in diluted HF is estimated to be <0.3nm/min. Another experiment (not presented here) determined the etch rate of 2% HF to 9nm in 30sec. We can therefore assert that a 30s HF dip removes all existing oxide, but does not attack the silicon.

To calculate the amount of active dopants next to the interface, it is first assumed that all carriers have the same mobility. SRP measurements (not shown here) confirmed 90-100% of the dopants to be active in depths >5nm.

After etching for 30s and native oxide regrowth, the sheet resistance increased by 6% to 9% (fig. 3b). Since the total number of dopants in the bulk part of the junction is know from SIMS, the number of active dopant atoms lost during the process can be calculated. Because all of the lost dopants were located within the layer of silicon consumed during native oxide growth, an average active dopant concentration can be calculated for this layer (typically ~0.5nm thick).

Values of 3-6e20cm^{-3} are obtained. In all cases, the active concentration near the interface is 3 to 6 times higher than the background concentration found in 5-10nm depth (fig. 3c). These concentrations correspond to active doses between 1.1e13cm^{-2} and 2.8e13cm^{-2} within the investigated layer. For BF_2, these are only 16% (O) or 9% (NO) of the total dose seen by ERD in the peak.

Discussion and estimation of accuracy

- Due to the high concentrations, the mobility in the interfacial pile-up is reduced. The mobility at concentrations of >5e20cm^{-3} is 10-15% lower than at 6e19cm^{-3} [5], indicating an active dose which is by the same percentage higher than the calculation.
- An average dose over the top ~0.5nm below the interface was calculated. Since the pile-up is believed to be smaller, its actual active concentration is higher (active concentrations of up to 1e21cm^{-3} (B, BF_2) or 2e21cm^{-3} (As) are expected if all pile-up dopants are confined to one monolayer)
- Interface defects could trap carriers and thereby deplete a small layer of the silicon. Their density is estimated to be 1-5e11cm^{-2}, which would reduce the number of carriers by less than 1% and therefore does not contribute significantly to the observed effects.
- Mobility close to the interface could be reduced because of charged interface impurities, increased roughness or other effects in the less-than-perfect native oxide. The influence of this will be subject to further investigation.

Since no As implanted samples could be measured in ERD until the publication deadline, no direct evidence for an As pile-up at the interface is shown. However, the calculation (fig. 3c) shows a very similar behaviour for As and B/BF_2 implanted samples and therefore strongly suggests the existence of a similar pile-up. Active concentrations of more than 6e20cm^{-3} are expected, when taking the above mentioned accuracy estimations into account.

Investigation of the resistance increase after an HF dip (experiment 3)

Three wafers, implanted with 5keV BF_2, 1keV B and 5keV As through 2nm oxide and annealed in a spike anneal, were etched with a 30s HF dip. Sheet resistance and oxide thickness were measured before and several times after the HF dip. Fig. 4 shows the increase in sheet resistance as a function of oxide thickness. The first data point was measured ~30min after the HF dip, the last one after 79 days (thus much longer than in experiment 2).

All three samples show a linear dependence, which is explained by a continuous incorporation of dopants into the growing oxide. For the BF_2 sample, an initial decrease in sheet resistance after the HF dip is observed. Linear extrapolation towards the HF dip (dashed lines) show the same for the B sample.

This initial R_s drop is believed to be due to charged surface defects causing band bending next to the surface and thereby creating an accumulation layer. The latter is strong enough to also compensate for the carrier loss due to pile-up removal during the HF dip (as described in exp. 1 and 2). The growing native oxide finally heals the defects, removes the accumulation layer and increases the sheet resistance by consuming dopants.

For As, a similar, but less pronounced effect seems possible. However, the observed behaviour could also be explained by an initial R_s increase due to pile-up removal. After that, the native oxide could push dopants into the junction and thereby retard any further R_s increase.

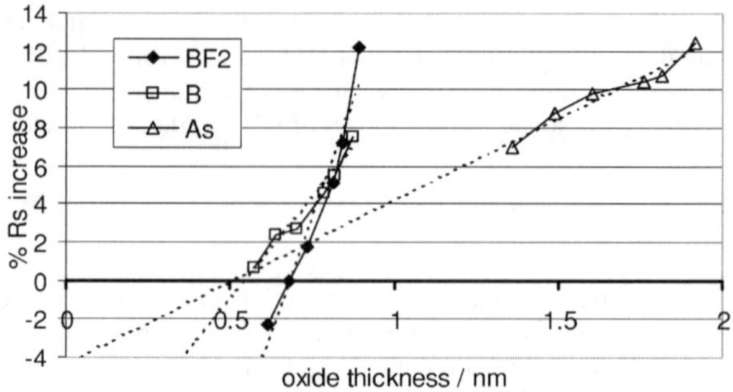

Figure 4 – Sheet resistance increase as a function of native oxide thickness after an HF dip. The first measurement was made ~30min after the HF dip, the last one after 79 days. Linear extrapolation is indicated with dashed lines.

CONCLUSIONS

When implanting ultra-shallow junctions with low-energy B, BF_2 or As and annealing the wafers with a 1070ºC, 1sec spike anneal, up to 50% and more of the originally implanted dose are lost during anneal. Furthermore, 30-50% of the remaining dopants form a pile-up at the interface between the silicon junction and the screening oxide layer. Only 10-15% of the pile-up dopants are active, but still a very high active concentration of >3e20cm^{-3} (boron) or >6e20cm^{-3} (arsenic) was found averaging over the top ~0.5nm below the surface oxide. When taking the oxide off with an HF dip, most of the pile-up is removed. In the case of a BF_2 implant, an initial decrease in sheet resistance is observed after an HF dip.

REFERENCES

[1] A. Shima, T. Jinbo, N. Natsuaki, J. Ushio, J.-H. Oh, K. Ono and M. Oshima, "A model for the segregation and pileup of boron at the SiO_2/Si interface during the formation of ultrashallow p$^+$ junctions", J. Appl. Phys., Vol. 89, No. 6 (2001) 3458-3463

[2] H. C. Wang, C.-C. Wang, C.-S. Chang, T. Wang, P. B. Griffin and C. H. Diaz, "Interface induced uphill diffusion of boron: An effective approach for ultrashallow junction", IEEE El. Dev. Lett., Vol. 22, No. 2 (2001) 65-67

[3] W. Vandervorst, T. Janssens, R. Loo, M. Caymax, I. Peytier, R. Lindsay, J. Frühauf, A. Bergmaier and G. Dollinger, "An (un)solvable problem in SIMS : B-interfacial profiling", Proc. SIMS-XIII (Nara, 2001), Appl. Surf. Sci. (accepted for publication)

[4] P. Neumaier, G. Dollinger, A. Bergmaier, I. Gechnev, L. Görgens, R. Fischer, C. Ronning and H. Hofsäss, "High-resolution elastic recoil detection utilizing Bayesian probability theory", Nucl. Instr. and Meth. in Phys. Res. B 183 (2001) 48-61

[5] "Standard practice for conversion between resistivity and dopant density for boron-doped and phosphorous-doped silicon", ASTM Standard F 723-88, 1988

Mat. Res. Soc. Symp. Proc. Vol. 717 © 2002 Materials Research Society C3.6

The Local Structure of Antimony in High Dose Antimony Implants in Silicon by XAFS and SIMS.

M. Alper Sahiner[1], Steven W. Novak[1], Joe C. Woicik[2], Yayoi Takamura[3], Peter B. Griffin[3], James D. Plummer[3]

[1]Evans East, East Windsor, New Jersey 08520
[2]National Institute of Standards and Technology, Gaithersburg, Maryland 20899
[3]Center for Integrated Systems, Laboratory for Advanced Materials, Stanford University, Stanford, CA 94305

ABSTRACT

One of the important challenges in semiconductor industry is to sustain high concentration of dopant atoms electrically active in very small areas. In investigating the optimum post implantation treatment methods that will help to attain these conditions, the local structural information around the dopant atom is crucial. In this study, we have used secondary ion mass spectroscopy (SIMS) and x-ray absorption fine structure spectroscopy (XAFS) to obtain the concentration depth profiles and the local structural information around the Sb atom in laser thermal annealed (LTA) Sb implants in Si wafers. The Sb implant doses used in this work are $6.4x10^{15}/cm^2$ and $2.0x10^{16}/cm^2$. The XAFS results for the $6.4x10^{15}/cm^2$ Sb dose sample do not exhibit any rhombohedral-Sb precipitation as the Fourier Transformed (FT) data can be fit successfully using only substitutional-Sb in the Si lattice sites. However, a multi-shell analysis of the Fourier Transformed (FT) data for the $2.0x10^{16}/cm^2$ Sb dose sample clearly indicates there is a substantial contribution from the Sb-Sb scattering, which is a signature of precipitated form of Sb.

INTRODUCTION

Doping intrinsic Si with n or p-type dopant atoms above the solid solubility limits to increase the charge carrier concentration, and sustaining high level carrier concentration by eliminating any inactive structures (clusters or precipitates) in terms of annealing mechanisms is one of the main challenges of the semiconductor industry. Information on the local structure around the impurity atom is of crucial importance to face with these requirements. XAFS, being capable of probing the short range order in crystalline and amorphous materials provides information on the coordination number, near-neighbor distances around the impurity atom. This information can be used to identify the nature and amount of dopant-related inactive structures in the material.

In reducing the number of these inactive structures in high dose implants in Si, laser annealing (LA) is one of the important post implantation techniques. Laser annealing has been shown to be very useful in repairing the damage to the crystal due to implantation and recovering the active carrier concentration in many systems[1]. In this study, Sb implanted Si wafers at two different doses were laser annealed and subsequently the local structure around the Sb atom was probed by XAFS. The Sb doses for this study are $6.4x10^{15}$ atoms/cm^3 (will be referred as the low dose) and $2.0x10^{16}$ atoms/cm^3 (will be referred as the high dose). The high dose sample was subjected to further thermal annealing after LA. Sb doping in Si above solid solubility limit may lead to substitutional-Sb (antimony at the Si crystal sites), precipitates in the form of metallic (rhombohedral) antimony, and/or Sb clusters.[2]. Previously, donor-pair type of defects were also suggested by Chadi *et al.*, [3]. Using Secondary Ion Mass Spectroscopy (SIMS), the Sb plateau concentrations were determined for the low dose and high dose samples. We have used simulated-XAFS-standards for metallic-Sb (rhombohedral) and substitutional form Sb in Si, to probe the local structural changes in the Sb complexes with increasing dose.

EXPERIMENT

Czochralski (CZ) Si(100) wafers were pre-amorphized with 55 keV, $1\text{x}10^{15}$ atoms/cm^2 Si implants. The wafers were then implanted with Sb with doses of $6.4\text{x}10^{15}$ and $2.0\text{x}10^{16}$ atoms/cm^2 at 60 keV implant energy. After the implantation the wafers were laser annealed (LA) with laser fluences of 0.80 and 0.73 J/cm^2 for the low and the high dose Sb implants, respectively. The high dose sample was then thermally annealed at 600°C for 40 minutes in nitrogen environment. The secondary ion mass spectroscopy results were performed using 3 keV Cs^+ primary ion beam conditions. Antimony K-edge x-ray absorption fine structure spectroscopy (XAFS) experiments were performed at the National Institute of Standards and Technology's (NIST) beamline (X23A2) at the National Synchrotron Light Source (NSLS) at Brookhaven National Laboratory (BNL). XAFS data were acquired in the fluorescence detection mode with the x-ray angle of incidence was set approximately to twice the Si critical angle during the measurements.

RESULTS AND DISCUSSION

Fig. 1 shows the SIMS concentration depth profiles for both the low (solid line) and high dose (dashed line) samples. The profiles exhibit boxed shape dopant distributions as expected from a successful laser annealing process. The average plateau concentrations for the low and high dose samples, $6.8\text{x}10^{20}$ and $2.5\text{x}10^{21}$ atoms/cm^3, are well above the solid solubility limits of Sb in Si. The junction depths obtained from these SIMS depth profiles are 110 nm for the high dose and 140 nm for the low dose sample, respectively.

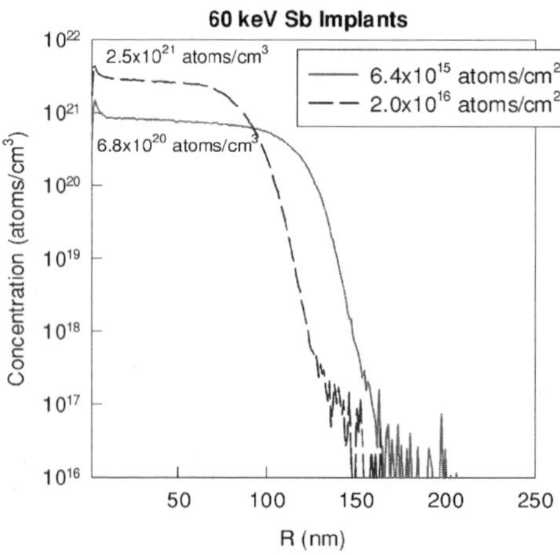

Figure 1. SIMS concentration depth profiles for the laser annealed $6.4\text{x}10^{15}$ atoms/cm^2 and $2.0\text{x}10^{16}$ atoms/cm^2 Sb dose samples. The average plateau concentrations are noted on the plot.

In XAFS data analysis, the energy dependent absorption coefficient was normalized to the absorption edge after the atomic absorption was subtracted [4]. The Fourier transformation of the resulting XAFS function, $\chi(k)$, provides the pseudo radial distribution function around the main absorbing Sb atom. The Fourier transformed (FT) data obtained from the XAFS function using a Gaussian window for the [2.5-10.5 $Å^{-1}$] k-range for these Sb implanted laser annealed samples are shown in Fig. 2. The peaks appearing in the FT data are due to back-scattering from the near-neighbors of the main absorbing Sb atom and the positions and intensities are related to near-neighbor distances and coordination numbers of the surrounding atoms. The peak distances are shifted from the actual near-neighbors distances because of the additional phase shift due to backscattering from the surrounding atoms.

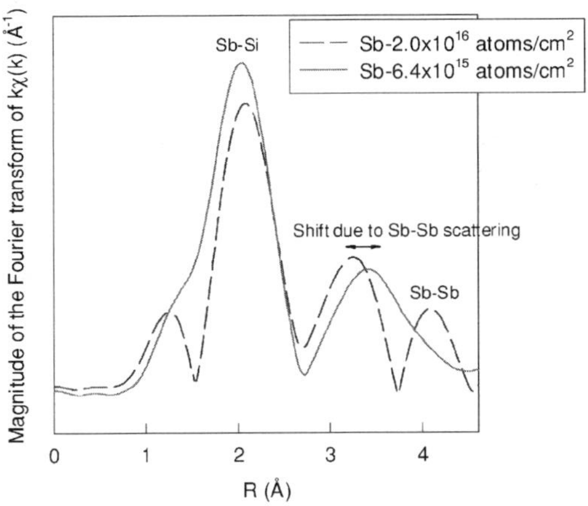

Figure 2. Fourier transformed XAFS data for the $6.4x10^{15}$ atoms/cm^2 and $2.0x10^{16}$ atoms/cm^2 samples.

Compared to the previous data on similar laser annealed samples [5], the FT transformed data shown in Fig. 2 exhibit much better long range order, (*i.e.*, robust peaks well beyond the first and second near-neighbors), which is an indicator of the full re-crystallization of the samples after annealing. A comparison the FT data for the high dose and low dose samples in Fig. 2, shows that an additional peak emerges for the high dose sample just beyond 4 Å. We have calculated all the multi-shell scattering contributions corresponding to two different Sb local structure configurations; 1- Sb in Si lattice sites (substitutional-Sb) and 2- Sb in rhombohedral form (precipitated Sb) using University of Washington's multiple scattering analysis code, FEFF7.0 [6]. The peak in the FT data just above 4 Å corresponds to Sb-Sb scattering from the 5th and 6th near-neighbors and are signatures of the precipitated (rhombohedral) form of antimony in the material. The region between 2.90 Å and 4.00 Å in the FT data has scattering contributions both from the Sb-Sb and the Sb-Si (substitutional form) scattering, where the Sb-Sb peak is located at slightly lower R values compared to Sb-Si peak. The shift of the peak to the lower R values for the high dose sample indicates that Sb-Sb scattering is gaining weight relative to the Sb-Si scattering. Finally the transfer of intensity from the Sb-Si 1st shell peak ~2 Å to the second peak where Sb-Sb scattering is present is further proof that there is an emergent Sb-Sb rhombohedral structure in the system.

In trying to determine the coordination numbers and near-neighbor distances from the least-squares fits to the FT data for the low dose and the high dose samples, we have assumed all the inactive structure in the material is metallic–Sb in the rhombohedral form. For the low dose sample the FT data can be successfully fit using only substitutional form of Sb as it is shown in Fig. 3. The corresponding scattering contributions to the peaks are listed on the plot. Being able to fit the whole range of the FT data using only Sb-Si scattering provides us confirmation that the only form of Sb in the system is substitutional-Sb.

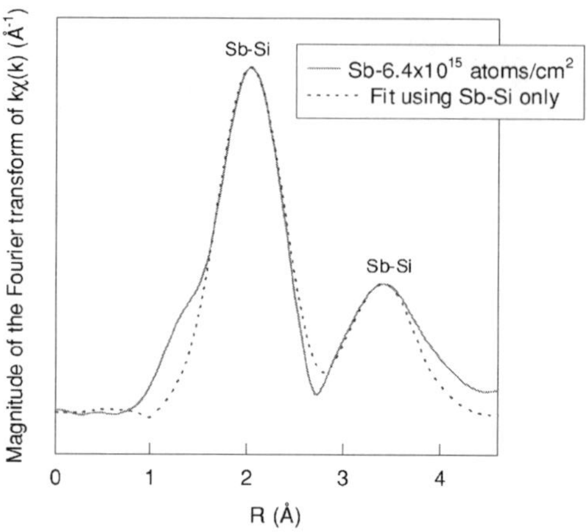

Figure 3. Fourier transformed XAFS data for the $6.4x10^{15}$ atoms/cm^2 sample and the least-square non-linear fit to the data. The contributing scattering mechanisms to the peaks are noted on the plot.

However, for the high dose sample this is not completely true. Fig. 4 is a plot of the FT data for the high dose sample with the least square fit plotted in the dotted-lines. There is a significant contribution from the Sb-Sb rhombohedral structure.

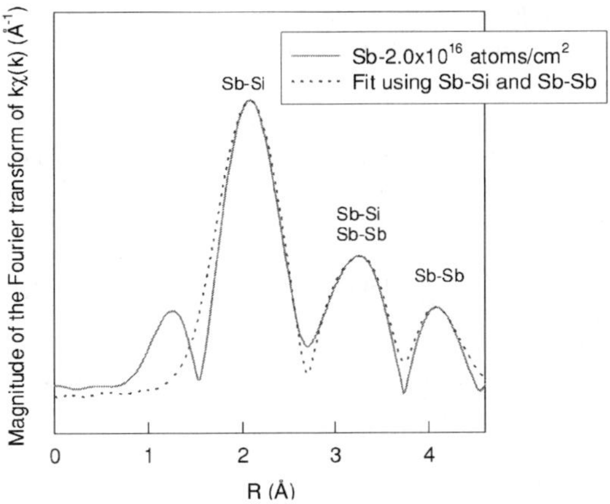

Figure 4. Fourier transformed XAFS data for the $2.0x10^{16}$ atoms/cm^2 sample and the least-square non-linear fit to the data. The contributing scattering mechanisms to the peaks are noted on the plot.

The fit results for these two FT data are summarized in Table I. In order to perform these fits, two different structures: substitutional Sb, and metallic-Sb in the rhombohedral have been assumed to be present [7], and all the possible scattering amplitudes and phases up to 4.50 Å have been created for these two different configurations. The combined scattering amplitudes and phases are then sorted according to their R values and scattering contributions from the six near-neighbors are included in the final calculated XAFS function. This analysis method was proved to be very successful interpretation of the FT data even in much more complex structures like the layered cuprate superconductors [8].

Table I. Least-square fit results to the FT data.

	Near-Neighbors											
	1st (Sb-Si)		2nd (Sb-Sb)		3rd (Sb-Sb)		4th (Sb-Si)		5th (Sb-Sb)		6th (Sb-Sb)	
Sample	**N**	**R (Å)**	**N**	**R (Å)**	**N**	**R (Å)**	**N**	**R (Å)**	**N**	**R (Å)**	**N**	**R (Å)**
Sb-$6.4x10^{15}$ (/cm^2)	4.00	2.54	-	-	-	-	11.85	4.13	-	-	-	-
Sb-$2.0x10^{16}$ (/cm^2)	3.71	2.55	0.80	2.91	3.58	3.35	5.00	4.13	6.80	4.31	4.00	4.51

For the low dose sample, the least squares fit to the FT data results in N = 4.00 ± 0.10 and R = 2.54 ± 0.02 Å for the first shell and N = 11.85 ± 0.10 Å and r =4.12 ± 0.02 for the second shell. As it is listed on Table I. there is no Sb-Sb scattering contribution needed to successfully fit the FT data for this sample. On the other hand, for the high dose sample, mere Sb-Si (substitutional–Sb configuration) scattering is not sufficient for a successful fit to the FT data. The coordination numbers for the Sb-Si scattering is observed reduced in the fit of the high dose sample. The Sb-Sb scattering, at R = 2.91, 3.35, and 4.51 A introduces an inevitable contribution to the XAFS data. If there were only these two assumed structures in the system, the coordination numbers would be sufficient in determining the relative weight of these two structures.

Previously, in the heavily doped arsenic in Si case, using just the linear combinations of the pure substitutional and the precipitated form of As standards, the XAFS data for the intermediate dose could be expressed and the relative weight of the precipitated versus the substitutional As could be determined [9]. Applying the same procedure to Sb, the XAFS data for the high dose sample was attempted to be reproduced using a linear combination of the low dose sample (substitutional-Sb) and the Sb-foil, which has a rhombohedral structure. Although the best trial was obtained using a 20% rhombohedral and 80 % subtitutional-Sb as shown in Fig. 5, the match was not perfect indicating the presence of other Sb structures in addition to the two assumed form of Sb. High resolution x-ray diffraction (HR-XRD) analysis on the same samples indicates the presence of new inactive structures, which are larger than substitutional-Sb but residing in the Si lattice sites [10].

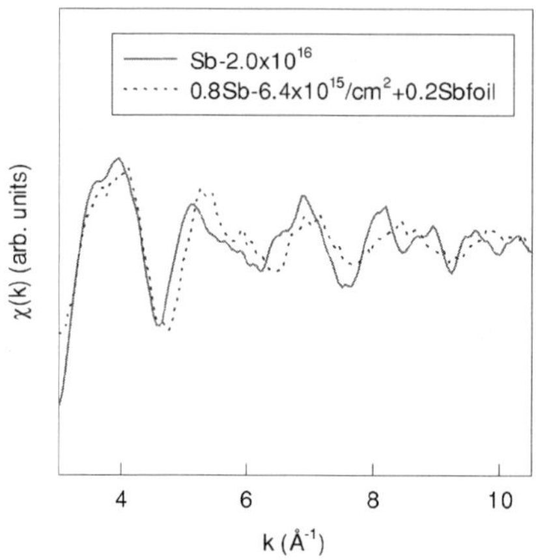

Figure 5. XAFS function $\chi(k)$ of the high dose sample (solid line) and a linear combination of the low dose sample (pure substitutional-Sb) and the Sb-foil (pure Sb-precipitates) (dotted-line).

In order to confirm that there are other forms of inactive structures in addition to precipitated and substitutional form of Sb in the high dose samples, the local crystal configuration of these possible forms should be incorporated in the calculated XAFS functions, which will be the main task of our future studies. By this way, the relative weight of these structures can be calculated.

CONCLUSIONS

Using XAFS, the local structural differences around the Sb atom were probed for laser annealed Sb implant in Si at two different doses. The XAFS data for the low dose ($6.4\times10^{15}/cm^2$) sample can be fully explained with substitutional-Sb configuration. However, unlike the previous studies a dominant signature of precipitated form of Sb have been found for the high dose ($2.0\times10^{16}/cm^2$) sample using multi-shell XAFS analysis. The relative weight of the precipitated form of Sb appears to be around 20 %, however, the lack of a perfect match of the XAFS function for the high dose sample with the one expressed as a combination of the precipitated and substitutional form of Sb leads to the presence of different Sb structures in the system. In our future studies, we will attempt to identify other existing structures, by incorporating the scattering contributions from all the possible Sb configurations to the XAFS function.

REFERENCES

1. Y. Takamura, S. Jain, P. B. Griffin, J. D. Plummer, MRS Proceedings, **669**, J7.3.6 (2001), and references therein.
2. J. L. Allain, A. Bourret, J. R. Regnard, and A. Armigliato, Appl. Phys. Lett. **61**, 264 (1992).
3. D. J. Chadi, P. H. Citrin, D. L. Adler, M. A. Marcus, and H. –J. Gossmann, Phys. Rev. Lett. **79**, 4834, (1997).
4. A. Bianconi, *X-ray Absorption: Principles, Applications, Techniques of EXAFS, SEXAFS, and XANES*. John Wiley & Sons, New York, (1988).
5. A. Armigliato, F. Romanato, A. Drigo, A. Carnera, C. Brizard, J. R. Regnard, and J.L. Allain, Phys. Rev. B **52**, 1859 (1995).
6. S. I. Zabinsky, A. Ankudinov, J. J. Rehr, and R. C. Albers, Phys. Rev. B **52**, 2995 (1995).
7. C. Revenant-Brizard, J. R. Regnard, S. Solmi, A. Armigliato, S. Valmorri, C. Cellini, and F. Romanato, J. Appl. Phys. **79**, 9037 (1996).
8. A. Sahiner, D. Crozier, D. T. Jiang, and R. Ingalls, Phys. Rev. B **59**, 3902 (1999).
9. M. A. Sahiner, S. W. Novak, J. C. Woicik, J. Liu, and V. Krishnamoorty, MRS Proceedings, **669**, J5.8.1 (2001).
10. Y. Takamura, A Vailionis, A. Marshall, P. B. Griffin, J. D. Plummer, will be published in the Proceedings of this symposium, (MRS Spring Meeting 2002 Proceedings).

Mat. Res. Soc. Symp. Proc. Vol. 717 © 2002 Materials Research Society C3.7

Modeling of Diffusion and Activation of Low Energy Arsenic Implants in Silicon

Srinivasan Chakravarthi, Chidambaram P.R., Charles Machala,
Amitabh Jain, and Xin Zhang
Silicon Technology Development, Texas Instruments, Dallas, TX 75243.

Ultra-shallow n^+ source/drain extensions require low energy arsenic implants combined with rapid thermal anneals in order to minimize diffusion. However, annealing of low energy arsenic implants exhibits anomalous activation/diffusion. Accurate models are needed to enable TCAD driven transistor design/optimization. The actual extent of arsenic extension diffusion/activation is critical to modeling *n*MOS transistors. In this work, we develop a physics-based model for arsenic that successfully models both diffusion and activation/deactivation during both front-end and back-end processing. An As_4V cluster coupled with a arsenic precipitation model was used to model anomalous behavior of low energy arsenic.

INTRODUCTION

Transient enhanced diffusion (TED) has been the dominant effect in determining junction depths for the past decade and will continue to be important. However, the use of ultra-low energy implants and short, high temperature Rapid Thermal Processing (RTP) annealing has greatly diminished the importance of TED in ultra-shallow arsenic junctions. Reducing the implant energy is particularly effective as it puts the damage closer to the surface where it can be more readily annihilated, thus reducing the time period over which TED is present. Recent work shows that TED can be nearly eliminated for low implant energies.[1] However, ultra-shallow arsenic data suggests other effects controlling arsenic diffusivity/activation.[2] This paper focuses on understanding and modeling these effects.

EXPERIMENTS

Arsenic was implanted at energies of 500 to 5000 eV and doses from 5×10^{14} to 2×10^{15} cm^{-2} in 200mm *p*-type wafers. All implants were carried out on an Applied Materials LEAP-II. Implants into bare silicon were capped with an oxide before annealing. Anneals were carried out on a Centura XE+ (also Applied Materials). The RTP anneals were carried out in pure nitrogen (5 ppm O_2). Secondary Ion Mass Spectroscopy (SIMS) analysis was carried out by Evans Texas using a 1 keV Cs beam. Electrical sheet resistance measurements were performed on a Prometrix Rs-75 instrument using a type-H four-point in-line probe.

MODELS

Initial Damage

Each implanted ion creates a defect cascade, producing a large number of interstitials and vacancies. Fig. 1(a) shows a typical set of defect and dopant profiles from UT-MARLOWE, a Monte Carlo ion implantation simulator. Although the large initial interstitial and vacancy profiles

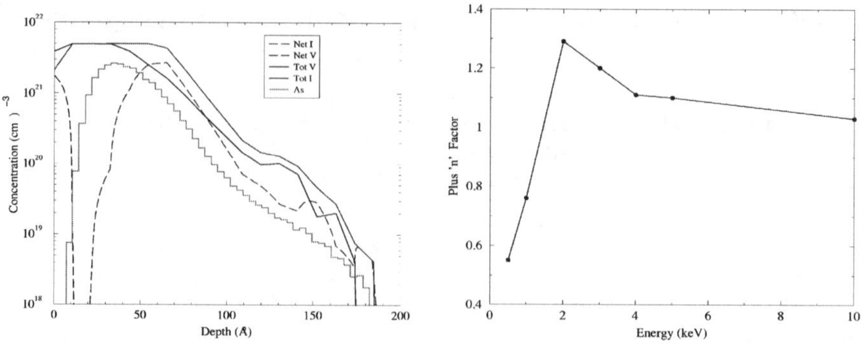

Fig. 1: (a) Simulations from UT-Marlowe for a 2keV 1×10^{15} arsenic implant showing total interstitial and vacancy profiles. The net profiles are also plotted. (b) Calculated 'plus n' factors from UT-Marlowe damage cascade for $1 \times 10^{15}\text{cm}^{-2}$ arsenic implanted at 0.5–10 keV.

are nearly equal, subtracting them reveals that the surface is vacancy rich while interstitials are knocked deeper into the substrate, with the integrated net I-V dose approximately equal to the i implant dose. Further, high dose implants (as used in our experiments) lead to a formation of a surface amorphous layer. Hence, we use the net I-V in the non-amorphous region as the starting interstitial profile. To estimate the enhancement from the interstitials in the non-amorphous region, we calculate a net 'plus n' factor as:

$$+n = \frac{\int x(C_\text{I} - C_\text{V})dx}{\int x C_\text{As} dx} \tag{1}$$

Fig. 1(b) shows the effective 'plus n' factor obtained for $1 \times 10^{15}\text{cm}^{-2}$ arsenic implanted at 0.5–10 keV. With decrease in implant energy, more of the arsenic layer is amorphized leading to a decrease in expected net TED enhancements from the implant.

Coupled Diffusion

It is well known that dopants migrate and interact with point-defects by coupled diffusion. We use a three-stream approach for our simulation of dopant diffusion as implemented in the process simulator T-SUPREM4.[3] As TED is reduced, coupled diffusion effects become more dominant.[4]

Arsenic Clusters

Figs. 2(a) and (b) show experimental SIMS profiles representative of different implant energies. To capture the expected junction depth from normal diffusion, we plot simulations (dashed lines) using only a simple solid solubility model. The solubility was adjusted to match the chemical concentration for a 5keV arsenic annealed at 950°C + 1050°C. Evidently, it is not possible to account for the SIMS data satisfactorily with such an approach. Also, note the high active concentrations needed to match the chemical concentrations. For instance, the measured sheet resistance for the 5keV implant with a junction depth of 285Å is 316 Ohms/Sq, whereas the calculated value is 249 Ohms/Sq. From these experiments and simulations, we can make the following observations:

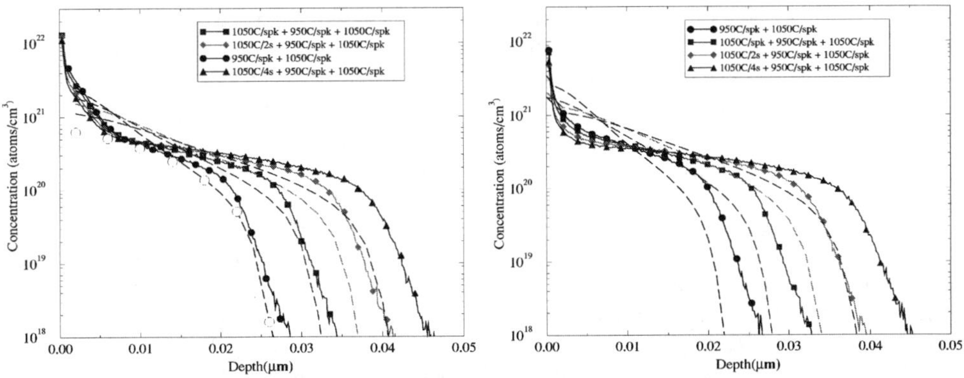

Fig. 2: Comparison of simulation (dashed lines) to experimental (solid lines) data for arsenic profiles annealed for different times/thermal budget. (a) $2 \times 10^{15}\text{cm}^{-2}$ arsenic implanted at 5keV through 50Å oxide, (b) $1.6 \times 10^{15}\text{cm}^{-2}$ arsenic implanted at 2keV through 20Å oxide annealed with varying thermal budgets. The simulations assumed a simple solid solubility model for arsenic. Open circle represents the active concentration needed to match the SIMS profile (950°C + 1050°C).

(a) there is a time dependent enhancement as the anneal proceeds; (b) lower energy arsenic is more enhanced relative to higher energies; and (c) not all mobile arsenic is active.

Solmi *et al.* have suggested formation of arsenic clusters based on experimental data.[5,6] As shown above, our data also indicates the presence of a large number of small arsenic clusters. Further, atomistic calculations predict formation of stable arsenic vacancy complexes. *Ab initio* calculations have shown the formation of stable As_4V complexes.[7,8] Recent *ab-initio* (Density Functional Theory) calculations have shown that the critical process is a rearrangement of arsenic atoms via diffusion leading to interstitial ejection.[9] This can be modeled by a single reaction:

$$\mathrm{As_4Si} \Leftrightarrow \mathrm{As_4V} + \mathrm{I}, \tag{2}$$

where, As_4Si is the concentration of arsenic in a As_4 cluster and is given as:[9]

$$C_{\mathrm{As_4Si}} = \left(\frac{C_{\mathrm{As}}}{C_{\mathrm{Si}}}\right)^3 C_{\mathrm{As}} \left[1 - \left(\frac{C_{\mathrm{As}}}{C_{\mathrm{Si}}}\right)^4\right], \tag{3}$$

Using the approximation $\left(\frac{C_{\mathrm{As}}}{C_{\mathrm{Si}}}\right)^4 \sim 0$, we can incorporate this reaction as a single reaction in continuum process simulators like T-SUPREM4.[3] At very high concentrations, Solmi *et al.* have observed arsenic precipitates at the surface likely to be monoclinic SiAs.[5] In our model, we assume the formation of a precipitate phase beyond a concentration of $1 \times 10^{21}\text{cm}^{-3}$. The precipitate is modeled as a large cluster with no point defect interaction.

RESULTS and DISCUSSIONS

In our continuum simulations, we include all the physical processes described in the previous sections: coupled diffusion, implant damage, clustering and precipitation. For modeling spike anneals exact Temperature-time (T-t) profiles are used. Ramps are modeled as piece-wise linear to match T-t profiles. As implanted SIMS was used as the initial arsenic concentration. The arsenic clustering model has been calibrated over a range of experimental conditions.

Figs. 3(a) and (b) show comparison to different thermal budgets (same experimental data as Fig. 2). Using this model we can satisfactorily explain the anomalous arsenic diffusion reported in the previous section. During diffusion, there is a continous formation of new As_4V clusters due to dissolution of the monoclinic phase. The interstitials ejected by the deactivating arsenic lead to further enhancement of the diffusing arsenic. This enhancement increases with a decrease in energy. This is because more As_4V clusters form at lower energies driven by the higher arsenic concentration. Also note the active concentration (open symbol) is much lower than the total chemical concentration. Fig. 4 shows comparison to experimental data for different implant doses over a range of implant energies. In Fig. 5, we plot the dose dependence of a 2keV arsenic implant after final anneal. Note the active concentrations (symbols). As the implant dose is increased more of the arsenic clusters. However, the active concentrations also increase with increase in dose due to small size of the clusters. Fig. 6 shows comparison to 700°C/2hr anneal. It is worth noting here that the arsenic diffusivity is considerably enhanced.

In Fig. 7, we explore the effect of post-processing on the final junction. After the standard activation anneal one of the wafer sees an additional 700°C/40min anneal. The chemical concentration as seen from SIMS shows a large tail in the arsenic profile. We see our model does a good job of capturing the phenemenon. In our model, at low temperatures interstitials are rapidly ejected by the deactivating arsenic. These interstitials cause enhancement of the arsenic leading to formation of the tail. It should be noted that a simple solubility model cannot capture such a behavior. If a simple solubility model were used, the 700°C/40min would show no junction movement contrary to experimental observation. Finally, in Fig. 8 we calculate the sheet resistance values predicted from our model. As can be seen it compares very favorably with experimental data.

SUMMARY

In summary, we find it is possible to model the extent of arsenic diffusion during front-end and back-end processes that define the final junction depth. The key features of the model can be summarised as: (a) Interstitials from implant damage play a diminished role as implant energies are scaled; (b) As_4V formation and precipitation at high concentrations is critical to accurate modeling of ultra-shallow arsenic junctions. These models when used with device simulations help optimize transistor performance/tradeoffs.

We would like to thank Pavel Fastenko and Scott T. Dunham (University of Washington) for details and discussion regarding their modeling results.

[1]A. Agarwal, H.-J. Gossmann, D. J. Eaglesham, L. Pelaz, D. C. Jacobson, T. E. Haynes, Y. Erokhin, in *Appl. Phys. Lett.* **71**(21), 3141 (1997)

[2]R. Kasnavi, P. B. Griffin, and J. D. Plummer, in Sim of Semicond. Proc. and Dev. Tech. Dig. (SISPAD '98),48, 1998.

[3]T-SUPREM4 User's Manual, Avant! Corp. 2001.

[4]S. T. Dunham, S. Chakravarthi, and Alp H. Gencer. in IEDM, Technical Digest, 1998.

[5]D. Nobili, S. Solmi, A. Parisini, M. Derdour, A. Armigliato and L. Moro, in *Phys. Rev. B*49 (4) 2477 (1994).

[6]D. Nobili and S. Solmi in *J. Appl. Phys.***83** (5) 2481 (1998).

[7]M. Berding, A. Sher, in *Phys. Rev. B* , **58** (7) 3853 (1998).

[8]M. Ramamoorthy and S. Pantelides, in *Phys. Rev. Lett.* **76** (25) 4753 (1996).

[9]S.T. Dunham, P. Fastenko, Z. Qin and G. Henkelman, in *Proc. Of Int. Conf. On Modeling and Simulation of Microsystems (MSM-2001)* (2001); P. Fastenko, S.T. Dunham and G. Henkelman in *MRS Proceedings* **669**, 2001.

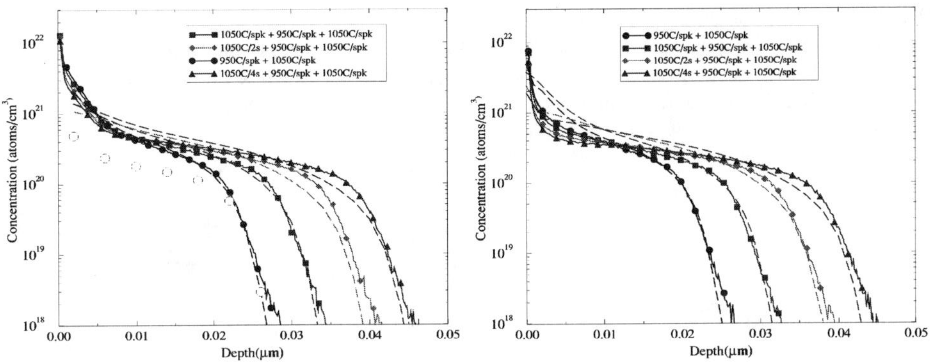

Fig. 3: Comparison of simulation (dashes) and experiment (solid) to arsenic profiles annealed for different times/thermal budget. (a) 2×10^{15}cm^{-2} arsenic implanted at 5keV through 50Å oxide, (b) 1.6×10^{15}cm^{-2} arsenic implanted at 2keV through 20Å oxide annealed with varying thermal budgets. Open circle represents the active concentration after a 950°C + 1050°C anneal.

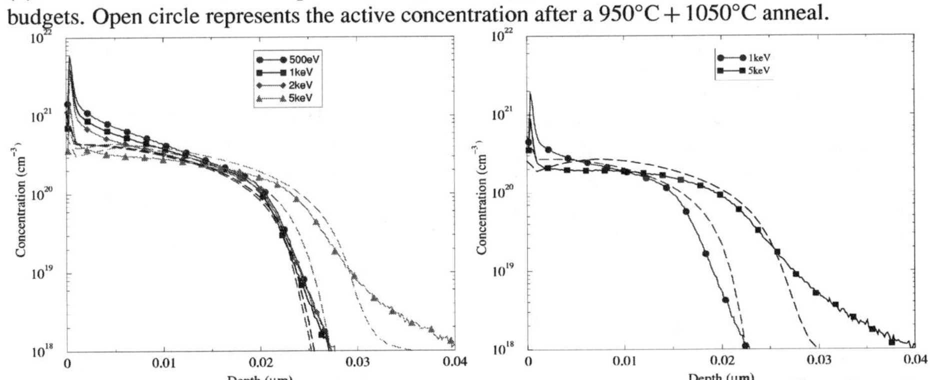

Fig. 4: Comparison of simulation (dashes) and experiment (solid) for (a) 1×10^{15}cm^{-2} arsenic implants at 0.5–5keV, (b) 6×10^{14}cm^{-2} arsenic implants at 1–5keV, after a 950°C + 1050°C spike anneal process.

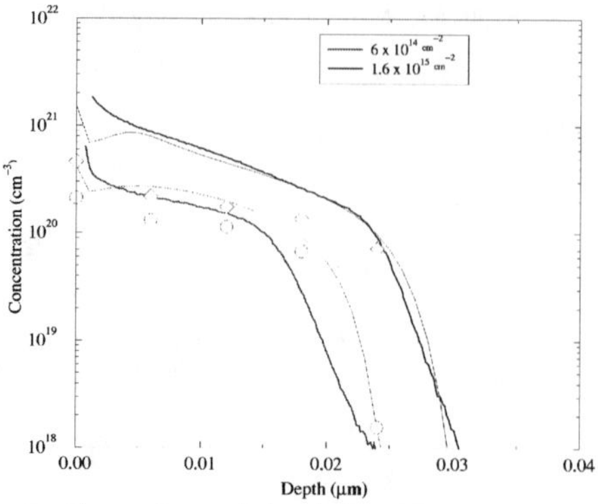

Fig. 5: Comparison of simulation (dashes) and experiment (solid) for varying arsenic implant doses at 2keV. Shown here is comparison to 6×10^{14}–$1.6 \times 10^{15}\text{cm}^{-2}$ arsenic implanted at 2keV after a 950°C + 1050°C activation spike. Open circles represent the active concentration.

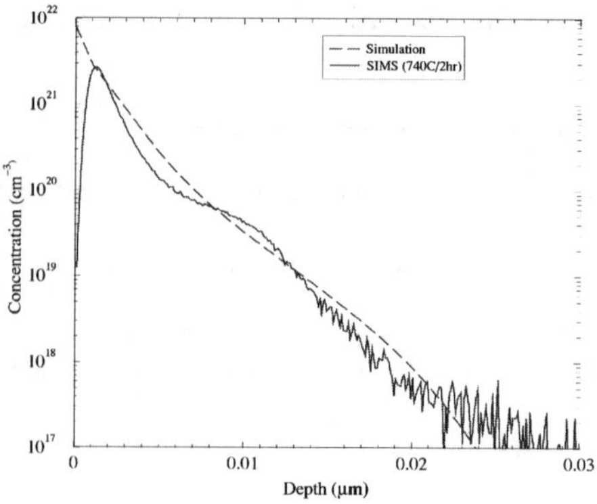

Fig. 6: Comparison of simulation (dashes) and experiment (solid) to $1 \times 10^{15}\text{cm}^{-2}$ arsenic implanted at 1keV through 20Å and annealed for 700°C/2hr.

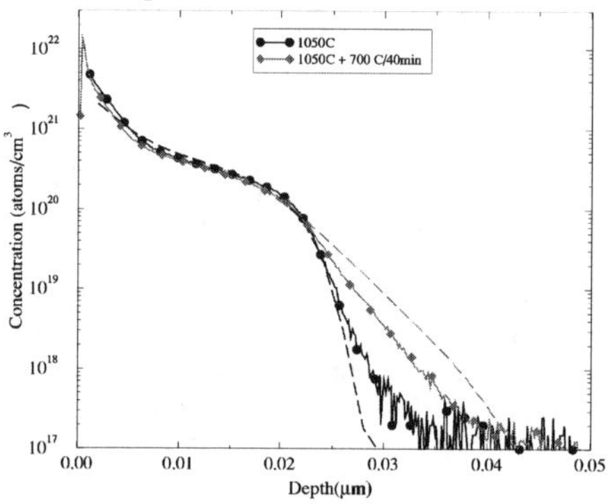

Fig. 7: Comparison of simulation (dashes) and experiment (solid) to $2 \times 10^{15}\text{cm}^{-2}$ arsenic implanted at 5keV through 50Å and annealed to activate at 1050°C. One of the wafer sees an additional 700°C/40min anneal leading to the tail in the arsenic profile.

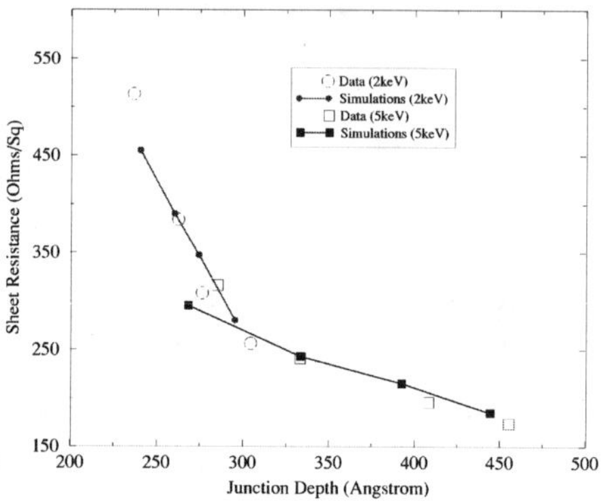

Fig. 8: Comparison of simulation (solid) and experiment (symbols) to sheet resistance and Junction depth data over a range of conditions. Junction depth is measured at a concentration of $1 \times 10^{18}\text{cm}^{-3}$.

Mat. Res. Soc. Symp. Proc. Vol. 717 © 2002 Materials Research Society C3.8

Modeling Fermi Level Effects in Atomistic Simulations

Zudian Qin and Scott T. Dunham
Department of Electrical Engineering, University of Washington,
Seattle, WA 98195

ABSTRACT

In this work, variations in electron potential are incorporated into a Kinetic Lattice Monte Carlo (KLMC) simulator and applied to dopant diffusion in silicon. To account for the effect of dopants, the charge redistribution induced by an external point charge immersed in an electron (hole) sea is solved numerically using the quantum perturbation method. The local carrier concentrations are then determined by summing contributions from all ionized dopant atoms and charged point defects, from which the Fermi level of the system is derived by the Boltzmann equation. KLMC simulations with incorporated Fermi level effects are demonstrated for charged point defect concentration as a function of Fermi level, coupled diffusion phenomenon and field effect on doping fluctuations.

INTRODUCTION

Kinetic Lattice Monte Carlo (KLMC) simulations study diffusion/clustering of defects in silicon at a microscopic level [1,2]. Simulations are performed on a silicon (diamond) lattice structure with impurities and point defects mapped to lattice sites. The system evolves through transitions from one atomic configuration to the next, by virtue of point defect migration/reaction. The rates of these transitions are determined by the migration barriers combined with changes in system energy associated with transitions:

$$\nu = \nu_0 \exp\left(\frac{-E_m}{k_B T}\right) \exp\left(\frac{E_i - E_f}{2k_B T}\right), \tag{1}$$

where E_m is the unbiased migration barrier, E_i and E_f are the system energies before and after the transition, and T is the system temperature. The system energies are calculated based on the atomic scale arrangement of impurities/defects, with parameters from *ab-initio* calculations and/or experimental observations. At each simulation step, one transition is chosen from the possible set based on the relative rates, and the system time is advanced by the inverse of the sum of the rates [1]. By only considering transitions (and not lattice vibrations) associated with defects and impurities present in the system, the KLMC method overcomes the time-scale limits associated with molecular dynamics to consider macroscopic systems and processing time scales.

The Fermi level is of critical importance in modeling dopant diffusion in silicon, but typically has not been included in atomistic simulations. Its importance manifests in two ways [3]: (i) charged point defect concentrations vary spatially with potential, and (ii) ionized dopant atoms as well as charged point defects experience electric fields due to spatial variations in the potential. Both effects are properly modeled in continuous simulations through the quantity of local carrier concentration, which is normally calculated from dopant profiles using the charge neutrality assumption. Unfortunately, dopant concentration is no longer a valid concept at the atomic scale, where dopant atoms are considered to be discrete and point-like. To obtain local carrier concentrations, it is necessary to solve the Poisson equation in the presence of discrete point charges. In the following sections, we first present such a solution and show how a simple analytic form can model it. Then we demonstrate examples of KLMC simulations including potential variations.

THEORY

To clearly define the problem, we consider a system with a defined uniform background carrier (electron or hole) concentration. To maintain charge neutrality in the region, an equal, but opposite dopant (donor or acceptor) concentration is also assumed. A pair of charges is then introduced into the system with the positive (negative) charge held at the origin (point-like) and the other mobile, negative (positive) charge released in the system. The challenge is then to solve the redistribution of the mobile charges in the neighborhood of the point charge at the origin under consideration of the background carrier screening.

We began with a classical solution to the problem, best known in its potential form as the screened Coulomb potential. The classical approach applies Boltzmann statistics and assumes the validity of linearization of the Boltzmann equation [4]. The solved charge distribution has the same form as the screened Coulomb potential:

$$\rho(\vec{r}) = \frac{-e}{4\pi\, L_D^{\,2}\, r} \exp\left(\frac{-r}{L_D}\right), \tag{2}$$

where L_D is the Debye Length

$$L_D = \sqrt{\frac{k_{Si}\, \varepsilon_0\, (k_B T/e)}{e(n+p)}}\,. \tag{3}$$

Here, n and p are the background carrier concentrations. This solution has a simple analytic form, but diverges as r approaches zero, which is physically incorrect and causes a problem in the carrier concentrations derivation. To overcome this characteristic of the classical solution, a quantum approach is necessary.

In order to solve the problem in a practical way, two assumptions are made. The first assumption regards the system. We assume that the conduction band has a parabolic shape near its minima and conduction electrons are free. The second assumption is made on the approach, where we assume the quantum perturbation method is applicable to this problem.

Based on the two above assumptions, the generalized susceptibility of the system, characterizing the system's response to an external sinusoidal perturbation can be calculated as a function of spatial frequency [5]:

$$\chi(\vec{q}) = -\frac{6m^*}{\pi^3\hbar^2} \int_0^\infty \frac{f(E_{\vec{k}})}{2\vec{k}\cdot\vec{q} + q^2}\, d^3\vec{k}\,, \tag{4}$$

where q represents the spatial frequency of the sinusoidal perturbation, m^* is the effective mass of electrons ($0.33m_e$), and the factor of six accounts for multiple band minima. The background carrier concentrations and system temperature enter as parameters through $f(E_{\vec{k}})$, the Fermi-Dirac distribution function, with the Fermi energy chosen to match the background carrier concentrations.

Given the susceptibility function of the system from Eq. (4), the induced charge distribution is calculated in response to an external point charge, which is represented by its Fourier components [5]:

$$\rho(\vec{r}) = -e\int_0^\infty \left[\frac{k_{Si}}{\varepsilon(\vec{q})} - 1\right] \exp(i\vec{q}\cdot\vec{r})\frac{d^3\vec{q}}{(2\pi)^3}, \quad (5)$$

where $\varepsilon(\vec{q})$ is the dielectric function of the system

$$\varepsilon(\vec{q}) = k_{Si} - \frac{e^2}{\varepsilon_0 q^2}\chi(\vec{q}). \quad (6)$$

RESULTS

Eq. (5) was evaluated numerically and the results are shown in Figs. 1 and 2. Fig. 1 shows results for three distinct background carrier concentrations: 5×10^{18}, 1×10^{19} and 5×10^{19} cm^{-3}. Clearly, the charge distribution drops off more rapidly with an increasing background carrier concentration, indicating a stronger screening effect. Fig. 2 shows results for three different system temperatures of 800°C, 900°C and 1000°C. As seen, there is only a weak dependence on temperature within the process temperature regime.

Upon examination of numerical results for different background carrier concentrations and temperatures, we find that the quantum solution can be modeled by:

$$\rho(r) = \rho(0)\frac{r_0}{\sqrt{r^2 + r_0^2}}\exp(-r/L_D), \quad (7)$$

where $\rho(0)$ is the charge density at the origin, and r_0 is a parameter representing best fitting to the numerical results. In fact, r_0 is only a weak function of the background carrier concentration and temperature, and is approximately 0.8 nm. In Fig. 3, both the classical and quantum solutions are plotted, along with Eq. (7), which shows excellent agreement with the quantum solution over the entire range. The quantum solution matches the classical solution at long range, but deviates from it in the short range. Most significantly, it gives a finite value at the origin, where the classical solution diverges. Calculations are carried out over background carrier concentrations up to 3×10^{20} cm^{-3}. Beyond this concentration limit, Friedel oscillations [5] appear and Eq. (7) no longer accurately predicts the numerical results.

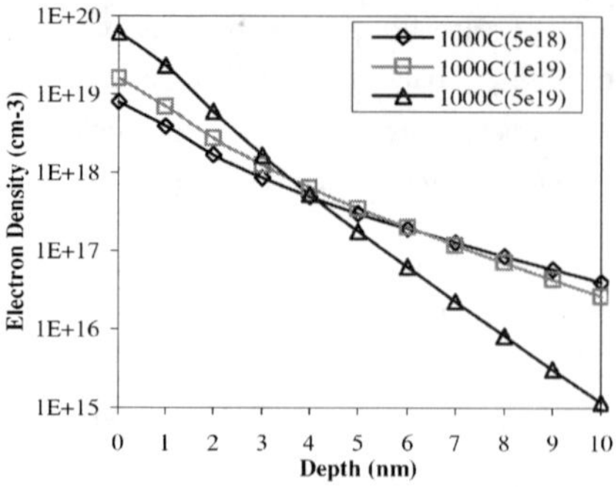

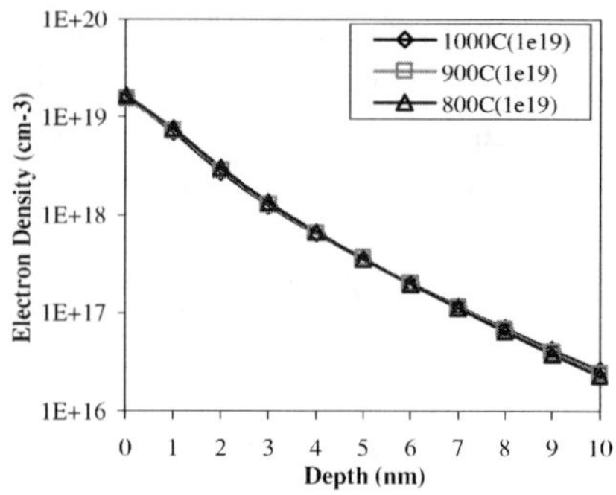

Figure 1. (left side) Electron distribution in the neighborhood of a positive point charge, calculated using quantum perturbation method for three distinct background carrier (electron) concentrations of 5×10^{18}, 1×10^{19} and 5×10^{19} cm^{-3} at temperature of 1000°C. As the background carrier concentration increases, the distribution profile drops off more rapidly due to stronger mobile carriers screening.

Figure 2. (right side) Electron distribution in the neighborhood of a positive point charge, calculated using quantum perturbation method for three different system temperatures of 1000°C, 900°C and 800°C with the background carrier (electron) concentration set to be 1×10^{19} cm^{-3}. Distribution profiles show only a weak dependence on temperature.

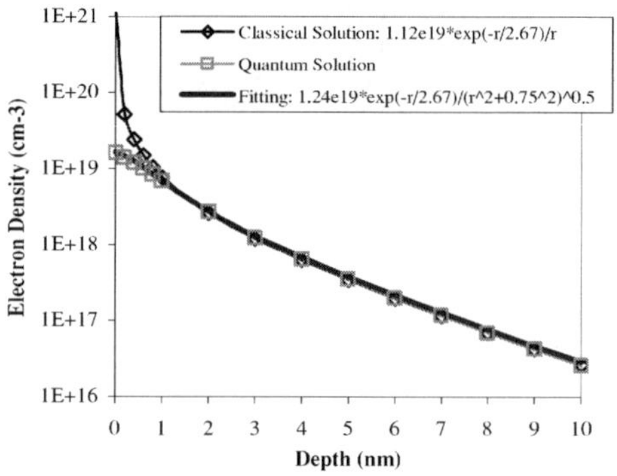

Figure 3. Electron distribution in the neighborhood of a positive point charge from both the classical and quantum solutions for a system with a background carrier (electron) concentration of 1×10^{19} cm^{-3} and a temperature of 1000°C. The classical Debye length in this case is 2.67 nm. The quantum solution matches the classical solution at long range. Also plotted is the fitted curve $1.24\times10^{19}\exp(-r/2.67)/(r^2+0.75^2)^{0.5}$, which shows excellent agreement with the quantum solution over the entire range.

Given the charge distribution in the neighborhood of a single point charge, the local carrier concentrations can be, to first approximation, determined by summing contributions from all ionized dopant atoms and charged defects, with a positive point charge (e.g., an ionized donor) inducing an electron cloud in its neighboring region and a negative charge (e.g., an ionized acceptor) inducing a hole cloud. The Fermi level is then derived from local carrier concentrations by the Boltzmann equation, as is usually done in continuum modeling. In the following section, we show example KLMC simulations with incorporated Fermi level effects.

KLMC SIMULATIONS

As a first test, the charged defect concentration is characterized as a function of the local Fermi level. A negatively charged vacancy, V^-, is put in a box with a non-uniform background Arsenic (As) doping. We then measure the time that the vacancy spends at various locations, the residence time. As seen from Fig. 4, the residence time follows the background carrier concentration curve, as would be predicted by a continuum model.

Multiple dopants diffusing simultaneously interact via field effects. The next simulation illustrates this phenomenon. Initially, there is a uniform background Boron (B) and non-uniform As doping in the system. After the simulation, B is swept out of the junction region where the field gradient is the strongest, as seen in Fig. 5.

Channel dopant fluctuations are a concern for nanoscale devices as they impact transistor characteristics such as threshold voltage. The final simulation shows the effect of fields on dopant location fluctuation. Dopant atoms are first randomly initialized and the Fermi level at various locations is recorded. The diffusion is then performed and the Fermi level again recorded. Fig. 6 compares the results from these two calculations in terms of the Fermi level distribution in the system. The curves clearly show that the system has a more uniform potential after the diffusion, indicating a more uniform dopant distribution as a result of dopant/dopant repulsion, with the standard deviation in electrical potential dropping from 0.014 to 0.012eV.

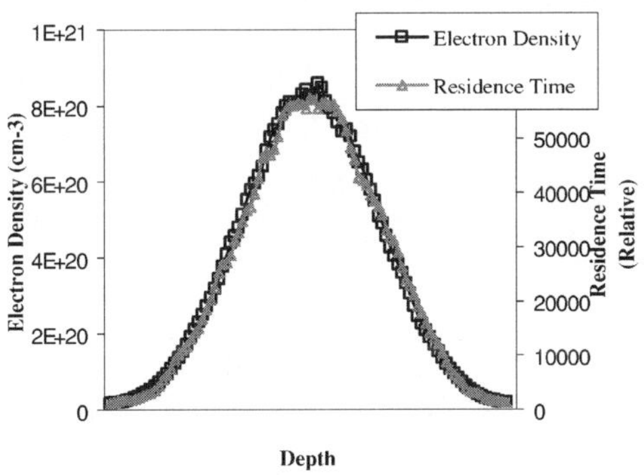

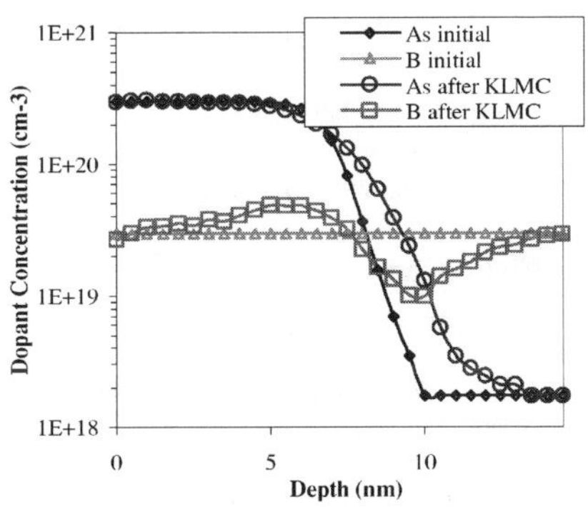

Figure 4. (left side) KLMC simulation shows residence time of a negatively charged vacancy at various locations matches background carrier concentration as predicted by a continuum model. **Figure 5. (right side)** KLMC simulation shows coupled diffusion due to field effects. B is swept out of the junction region by the present strong electric field.

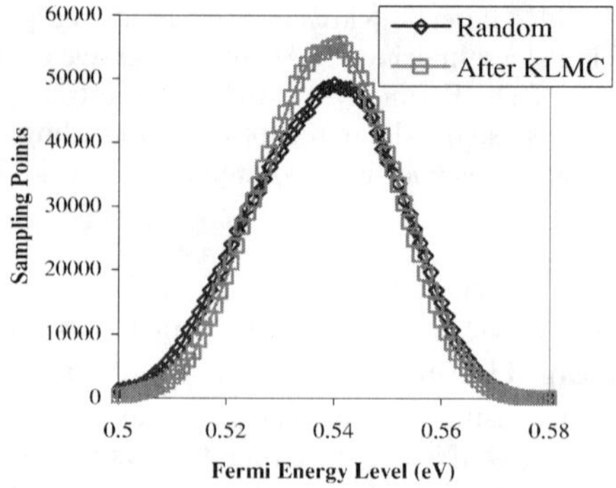

Figure 6. Results of KLMC simulation showing field effect on dopant location fluctuations. Plotted are Fermi level distributions in the system before and after diffusion simulation. The system has a narrowed field distribution after diffusion, indicating effect of dopant/dopant repulsion on dopant redistribution. Calculation gives a standard deviation of 0.12eV for the profile after diffusion, in contrast to 0.14eV for the initial (random) profile.

CONCLUSIONS

We have solved the charge distribution in the neighborhood of a point charge using the quantum perturbation method and have shown that the solution can be expressed as a simple analytic function. KLMC simulations with this model incorporated gave expected results for charged defect concentration as a function of the Fermi level, illustrated coupled diffusion phenomenon and showed reduced channel dopant location fluctuation due to like-dopant repulsion.

ACKNOWLEGEMENT

This work was supported by the Semiconductor Research Corporation.

REFRENCES

1. S. T. Dunham and C. D. Wu, J. Appl. Phys. **78**, 2362 (1995).
2. M. M. Bunea and S. T. Dunham, in *Semiconductor Process and Device Performance Modeling*, edited by S. T. Dunham and J. Nelson, (Mater. Res. Soc. Proc. **490**, Pittsburgh, PA, 1998) pp. 3-8.
3. P. M. Fahey, P. B. Griffin and J. D. Plummer, Rev. Mod. Phys. **61**, 289 (1989).
4. N. W. Ashcroft and N. D. Mermin, *Solid State Physics*, 1st ed. (Holt, Rinehart and Wilson, 1976) pp. 342.
5. J. Chazalviel, *Coulomb Screening by Mobile Charges: Applications to Materials Science, Chemistry, and Biology*, (Birkhäuser, 1998) pp. 27-34.

Epitaxial Techniques, Fluorine, Defect Fundamentals

Mat. Res. Soc. Symp. Proc. Vol. 717 © 2002 Materials Research Society

SELECTIVE SILICON-GERMANIUM SOURCE/DRAIN TECHNOLOGY FOR NANOSCALE CMOS

M. C. Öztürk, N. Pesovic, J. Liu, H. Mo, I. Kang and S. Gannavaram

North Carolina State University
Department of Electrical & Computer Engineering
Centennial Campus, EGRC Building, Campus Box 7920
Raleigh, NC 27695-7920, USA

ABSTRACT

Future CMOS technology nodes bring new challenges to formation of source/drain junctions and their contacts. To avoid MOSFET performance degradation with scaling, series resistance contribution of each junction must be limited to five percent of the device channel resistance. This requires ultra-shallow junctions with extremely low sheet, spreading and contact resistance. In this paper, we present an overview of the SiGe junction technology recently proposed by this laboratory for nanoscale CMOS. The technology is based on selective deposition of boron or phosphorus doped SiGe alloys in source/drain regions isotropically etched to the desired junction depth. Since the dopant atoms naturally occupy the substitutional sites during growth, the need for an activation anneal is completely eliminated limiting the maximum process temperature to 550°C for boron and 750°C for phosphorus. In this temperature range, dopant diffusion is virtually eliminated resulting in extremely abrupt doping profiles. Elimination of the high temperature implant anneal also makes the process compatible with the thermal stability needs of future high-K gate dielectrics. A key advantage of the technology is its potential to reduce the junction contact resistance. SiGe provides a smaller bandgap under the metal contact resulting in a smaller metal-semiconductor barrier height. Since the contact resistivity is an exponential function of the barrier height, the technology provides a significant advantage in reducing the contact resistivity. The results to date indicate that the technology is a promising alternative for nanoscale CMOS source/drain junctions and their contacts.

INTRODUCTION

Future CMOS technology nodes bring new challenges to formation of source/drain junctions and their contacts. Figure 1 shows extension and contact junction depths for from the 2001 edition of International Technology Roadmap for Semiconductors (ITRS)[1]. The roadmap predicts that extension junctions as shallow as 4 nm will be needed as we approach the physical gate length of 9 nm.

Until now, ion-implantation has been quite successful in keeping up with the demands of MOSFET scaling via advances in low-energy implanter design, annealing techniques and significant improvements in understanding of defects and their impact on dopant diffusion. Silicides of Ti and Co were also successfully used to form reliable, low-resistivity contacts. As we approach junction depths on the order of ten nanometers, it is doubtful that the existing integration schemes will continue to provide high performance junctions and contacts. Indeed, the roadmap incorporates elevated source/drain and single drain approaches as potential candidates for future MOSFETs.

In addition to the traditional challenges of forming junctions and contacts, there are new and fundamental challenges emerging as we approach the end of the roadmap. In

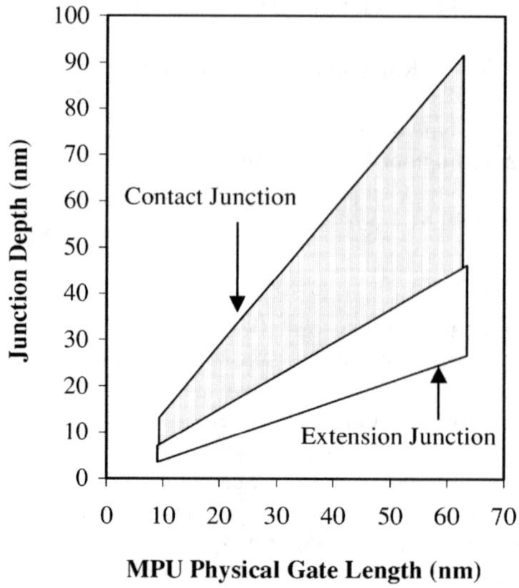

Figure 1 Extension and contact junction depths for different CMOS technology nodes from ITRS2001.

source/drain engineering, a generally accepted goal is to limit the total series resistance of source and drain junctions to less than ten percent of the channel 'on' resistance. Junction specifications in ITRS are also based on this criterion. As we approach the end of the roadmap, contact resistance is emerging as the dominant junction series resistance component. To meet the contact and other series resistance requirements of future junctions, fundamentally new approaches are necessary due to fundamental limitations of silicon as a material.

Another key junction challenge is to form extension junctions with laterally abrupt doping profiles in order to suppress short channel effects. Unfortunately, existing doping processes based on ion-implantation and thermal annealing fall short of meeting the abruptness demands of future junctions.

Finally, it is highly desirable to have a junction and contact formation processes compatible with future high-κ gate dielectrics, which introduce low temperature processing as a new challenge to source/drain formation. The high-κ research community is predicting a temperature limit around 800°C. The alternative is to form the gate stack after junction formation, which involves a fairly complex integration scheme.

Recently, we have proposed a new junction and contact formation technology, which has the potential to address all of the challenges mentioned above[2]. The new technology is based upon selective chemical vapor deposition of in-situ doped SiGe alloys in source/drain regions etched to the desired junction depth X_j. The technology relies heavily on the unique properties of Ge and SiGe. These include improvements in surface chemistry due to the presence of adsorbed Ge atoms, smaller SiGe bandgap, lattice strain and high carrier mobility.

The very first advantage of SiGe over Si is its smaller bandgap. Assuming the contact metal has its Fermi level positioned near the SiGe midgap, the contact barrier height is equal to half the semiconductor bandgap. Since contact resistivity is an exponential function of barrier height, a large improvement in contact resistivity can be achieved by a small reduction in barrier height. Alternatively, a lower active dopant density can yield the same contact resistivity with a smaller barrier height. To fully utilize the bandgap advantage of SiGe, Ge concentration in the alloy must be raised preferably without forming misfit dislocations at the interface, which may lead to leaky junctions. With junction depths approaching a few nanometers, it will be easier to preserve the strain in the alloy.

Another important advantage of SiGe over Si is its ability to provide films with much lower resistivity due to following reasons:

i) Higher dopant activation – The strain in pseudomorphic SiGe serves as a driving force for smaller boron atoms to occupy substitutional sites and compensate the lattice strain. Boron atoms occupying substitutional sites become electrically active. Strain compensation also allows formation of thicker SiGe alloys with high Ge concentrations without misfits.

ii) Higher mobility – Compared to Si, epitaxial SiGe films possess higher electron and hole mobilities.

iii) Higher phosphorus incorporation – It is easier to incorporate high concentrations of phosphorus in SiGe compared to Si primarily due to unique differences in the adsorption/desorption kinetics on the growing film surfaces.

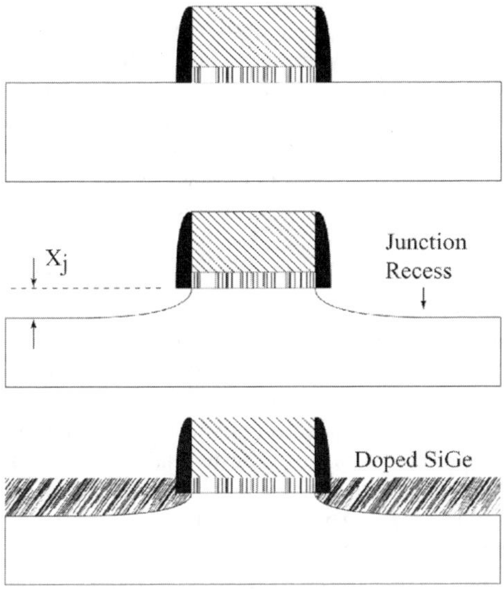

Figure 2 Junction formation by selective deposition of in-situ doped SiGe in source/drain regions etched to the junction depth.

For selective SiGe epitaxy, the proposed junction technology relies on CVD chemistry, which provides the ability to form junctions at very low temperatures. Since dopant atoms are naturally incorporated at substitutional sites during growth, they are electrically active without any subsequent, high temperature annealing. The SiGe layers are deposited at temperatures ranging from 500°C to 800°C where dopant diffusion does not exist - a key requirement for formation of super abrupt junctions.

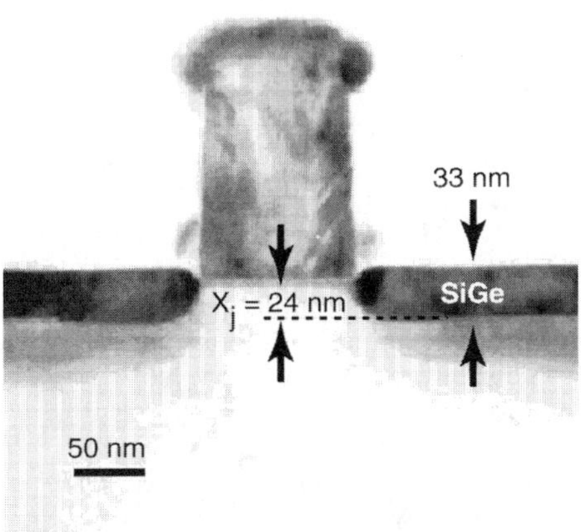

Figure 3 A partially completed MOSFET with selectively deposited SiGe junctions.

Shown in Figure 2 is a possible integration scheme, which combines a thin sidewall spacer with isotropic etching for recess formation. The thickness of the selectively deposited SiGe layer can be made greater than the recess depth providing the ability to fabricate elevated junctions. In ITRS 2001, the spacer thickness for an elevated contact is equal to the extension junction depth. If a thicker sidewall spacer is desired, it can be readily formed after SiGe deposition. This will also allow the first spacer thickness to be less than the extension depth, relaxing the requirements for the etch isotropy. Clearly, it is possible to think of many improvements to these basic integration schemes, which are beyond the scope of this paper. Shown in Figure 3 is the image of a partially completed MOSFET with junctions formed using the above technology. The image was obtained by cross-sectional transmission electron microscopy (XTEM). In this case, selective SiGe deposition took place in recessed junction regions as well as on top of the polysilicon gate electrode. The polysilicon gate could have been protected against SiGe deposition by means of a thin sacrificial layer of oxide or nitride.

SELCETIVE DEPOSITION OF SILICON-GERMANIUM ALLOYS

In this work, SiGe films were deposited in a single-wafer UHV-RTCVD (Ultra-High Vacuum Rapid Thermal Chemical Vapor Deposition) system with a base pressure of 10^{-9} Torr. The system consists of sample entry, intermediate and deposition chambers. The intermediate chamber is cryopumped and it serves as a vacuum buffer between the other two chambers. The deposition chamber is a water-cooled stainless steel chamber with a quartz dome. Two banks of tungsten halogen lamps placed above and around the quartz dome heat the wafer placed on a rotating quartz holder. The system is pumped by oil-free pumps to minimize carbon contamination.

The precursors used for SiGe layers were disilane (Si_2H_6), germane (GeH_4), diborane (B_2H_6), phosphine (PH_3), hydrogen (H_2) and chlorine (Cl_2). Three of the gases, GeH_4, B_2H_6 and PH_3 were diluted in H_2. The total pressure in the deposition chamber was kept under 200 mTorr. Chlorine was added to enhance the selectivity by introducing a mechanism to desorb the ad-atoms from the insulator surface. Boron doped layers were deposited in a temperature range of 450 – 600°C resulting in a growth rate within 10 – 20 nm/min. To obtain similar growth rates, n+ layers were deposited in a higher temperature range (T = 650°C - 810°C) to overcome the surface poisoning effect of phosphorus ad-atoms.

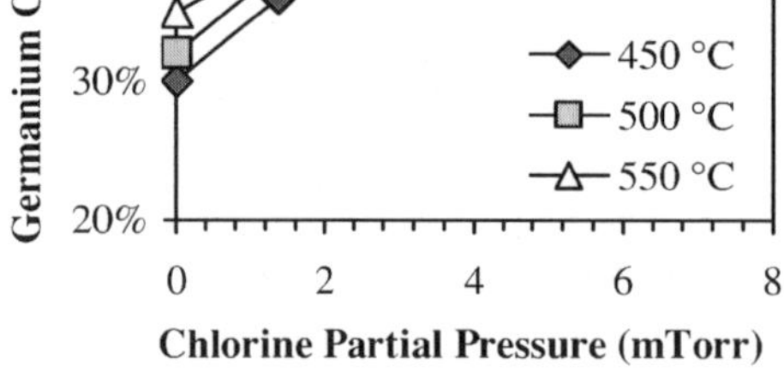

Figure 4 Germanium concentration in boron doped SiGe layers deposited at three different temperatures using different chlorine partial pressures.

Shown in Figure 5 is the germanium concentration measured in boron doped SiGe layers deposited in a temperature range from 450 to 550°C. The data was obtained by X-Ray diffraction analysis (XRD). Figure 4 is representative of the range of Ge concentrations if interest in this study. Ge concentrations above 30% are necessary to reduce the SiGe bandgap. The dependence of Ge concentration on chlorine partial pressure can be tied to chlorine adsorption on the growing layer and competition between Si and Ge ad-atoms for available surface sites. For p+ layers, impact of chlorine on growth rate was found to be negligible at T ≥ 500°C. Chlorine also had a negligible effect on the n+ layer growth rate except for T ≥ 800°C due to etching via $SiCl_2$ desorption.

Shown in Figure 5 is the corresponding surface roughness figures obtained by atomic force microscopy. As shown, very smooth films can be obtained in this temperature range provided the chlorine partial pressure is kept under a certain level. It is interesting to note that three-dimensional growth is suppressed in spite of the fact that Ge concentration is approaching 50%. The smooth surface morphology is evident in the XTEM image shown in Figure 3.

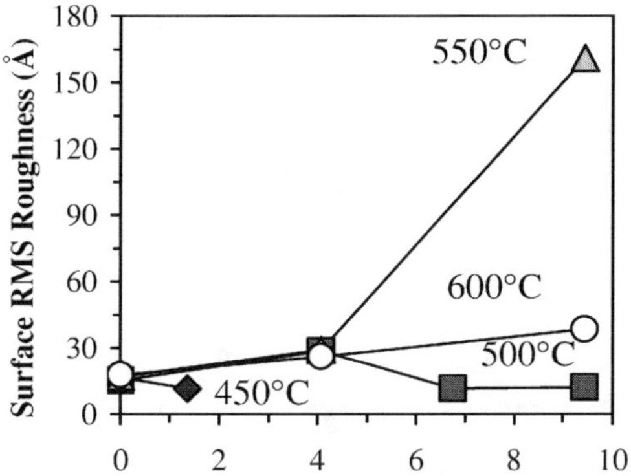

Figure 5 RMS roughness of boron doped SiGe deposited at different temperatures using different chlorine partial pressures.

FORMATION OF ABRUPT DOPING PROFILES

Formation of super abrupt source/drain doping profiles is one of the key challenges posed by future CMOS technologies. Abrupt junction profiles are necessary to suppress the short channel effect and reduce the junction spreading resistance. Shown in Figure 6 are the extension lateral abruptness values predicted in ITRS2001 for different technology nodes.

When junctions are formed by ion-implantation, high temperature annealing is necessary to remove the implantation damage and activate the dopants. The abruptness of the final doping profile is determined by diffusion of dopant atoms influenced by a variety of factors including fundamental properties of silicon, nature and quantity of defects introduced during implantation and temperature profile used for annealing.

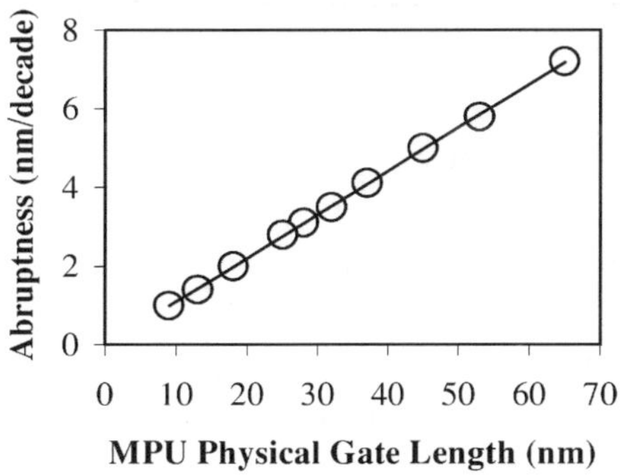

Figure 6 Extension lateral abruptness for different technology nodes from ITRS 2001.

It is well established that junction processes that involve thermal diffusion of dopant atoms cannot achieve the abruptness figures shown in Figure 6. This includes ion-implantation and any other technique based on dopant out-diffusion from solid sources.

One of the key features of the proposed junction technology is the absence of dopant diffusion. Figure 7 shows B^{10} and B^{11} profiles obtained by secondary ion mass spectroscopy (SIMS), which indicates a junction abruptness around ~ 2.18 nm/decade. It should be noted that the measured

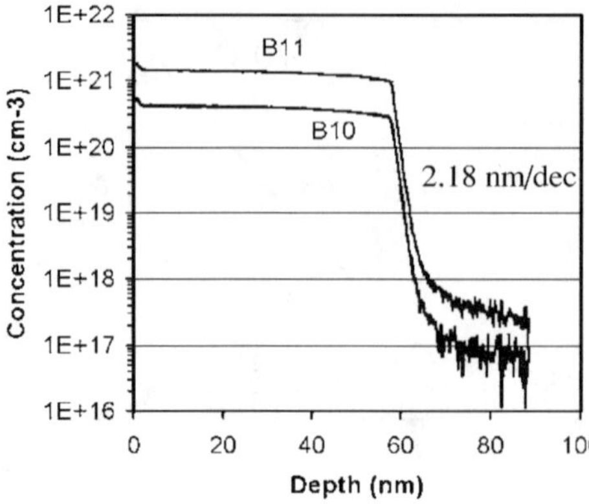

Figure 7 Atomic boron profiles obtained by SIMS for B^{10} and B^{11} isotopes in in-situ doped SiGe on Si.

slope is likely to be greater than the actual slope due to the SIMS cascading effect. Even so, this figure satisfies the ITRS 2001 requirement for the 20 nm node.

JUNCTION SHEET RESISTANCE

The active dopant concentration in junctions formed by traditional diffusion techniques is limited to equilibrium electrical activity limit, set by the maximum junction annealing temperature (< $2x10^{20}$ cm^{-3} @ 1100 °C for boron). In SiGe junction technology, a driving force exists for boron atoms to occupy the substitutional sites since smaller boron atoms strive to compensate the lattice strain in SiGe. Once a boron atom occupies a substitutional site, it becomes electrically active. Consequently, the active boron concentration in SiGe can exceed the equilibrium activation limit in Si by a large amount as verified by Hall Effect measurements on four terminal Van Der Pauw structures. The amount of boron required to fully compensate the lattice strain is in the range of 2-10 % (1-5 x 10^{21} cm^{-3}) for SiGe alloys in the compositional range of 20-80 % Ge. The above-equilibrium dopant activation achieved is stable since the substitutional boron incorporation causes the net strain energy of the system to be minimized. This is in contrast to metastable dopant activation that results from solid phase recrystallization[3] or from alternate annealing techniques such as laser annealing[4,5]. The B^{11} profile shown in Figure 10 demonstrates that it is possible to obtain SiGe layers with boron concentrations exceeding 10^{21} cm^{-3}. With a driving force to occupy the substitutional sites, a large fraction of the incorporated boron is electrically active, resulting in p-type layers with active boron concentrations well above the boron solubility in Si. Plotted in Figure 11 is resistivity as a function of the Cl_2 partial pressure at three different deposition temperatures of 450, 500 and 550°C. The data shows that a SiGe resistivity under $3x10^{-4}$ ohm-cm is possible.

The sheet resistance of a junction formed using the SiGe technology is equal to the ratio of the film resistivity to the depth of the junction recess obtained by etching. Using a resistivity of $3x10^{-4}$ ohm-cm, we have calculated the junction sheet resistance for different technology nodes using the

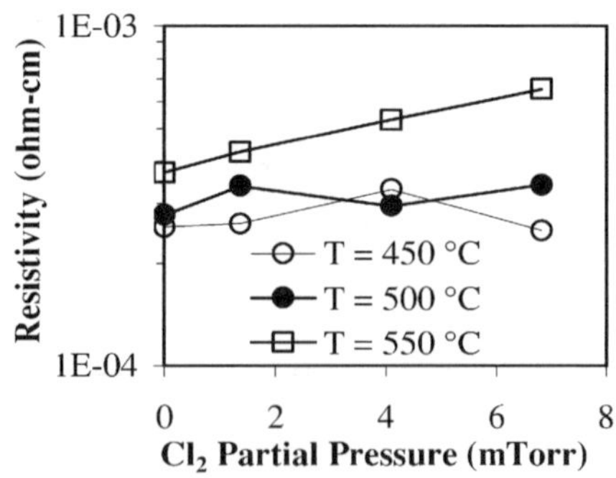

Figure 8 Resistivity of boron doped SiGe grown at different temperatures with different chlorine levels.

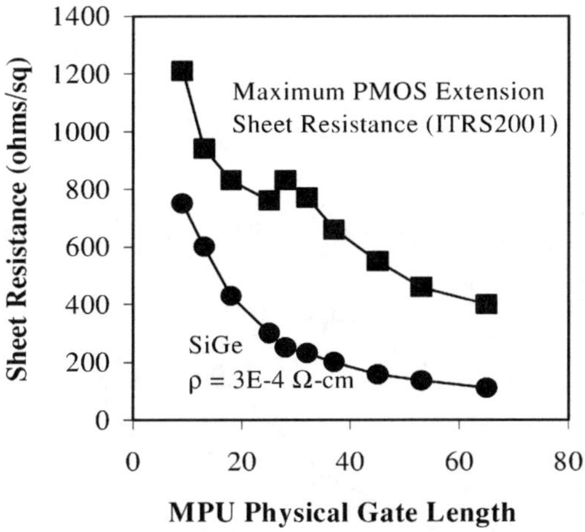

Figure 9 Sheet resistance of SiGe junctions and maximum PMOS extension sheet resistance from ITRS2001.

minimum extension junction depths given in ITRS 2001. The resulting sheet resistance values are plotted in Figure 9. Also shown are the maximum extension sheet resistance values from ITRS 2001, which were obtained by allowing 7% of the total parasitic resistance to the sheet resistance of the extension under the sidewall spacer. Figure 9 shows that boron doped SiGe junctions can easily satisfy the sheet resistance requirements of all technology nodes in ITRS2001.

For n-type doping, there exists an entirely different challenge. Incorporating large concentrations of phosphorus in Si is difficult due to desorption of P_2 dimers from the surface. In SiGe growth, a fraction of the P_2 dimers is replaced by Ge-P dimers, which can also desorb but not as readily as the P_2 dimers. At the same time, there exists an optimum Ge level, above which Ge-P desorption becomes important and starts reducing the phosphorus level in the film. This behavior is observed in Figure 10, which shows the active phosphorus concentration in SiGe for deposition temperatures ranging from 670°C to 810°C. The graph indicates that the phosphorus level in the films is maximized at a Ge fraction of ~ 20%. The data was obtained from Hall Effect measurements.

The best resistivity obtained for n+ layers is ~ 6×10^{-4} ohm-cm corresponding to Ge concentrations within 20-30%. While this resistivity is higher than that of p+ layers, the resulting sheet resistance falls below the maximum limit for all nodes included in Figure 9.

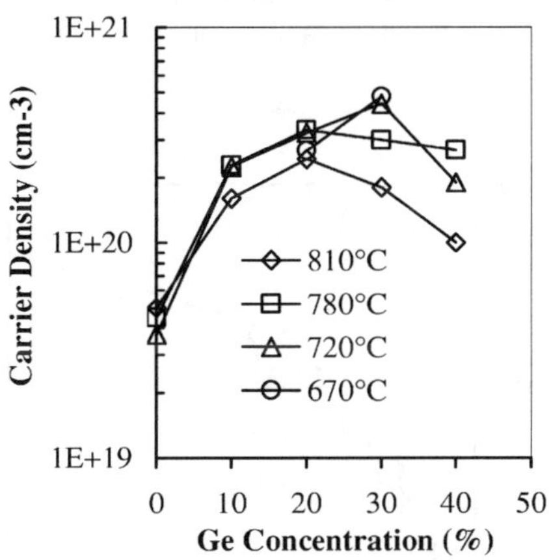

Figure 10 Active phosphorus density in SiGe films grown at different temperatures.

REVERSE BIAS JUNCTION LEAKAGE

Reverse bias junction leakage is a good measure of the defect density within the junction depletion region. In SiGe junction technology dopant diffusion is virtually nonexistent because a high temperature anneal is not employed subsequent to the deposition of the in-situ doped SiGe layer. Thus, the SiGe/Si interface resides within the depletion region, which requires an excellent interface to avoid junction leakage due to carrier generation.

Shown in Figure 11 is the leakage current of boron doped junctions measured at a reverse bias of –1V. The diodes were deposited at different temperatures again using different chlorine levels. The layers deposited at 550°C yield a leakage current of only 20 pA for all chlorine levels considered, which is suggestive of an excellent SiGe/Si interface with a very low defect density. For lower chlorine levels, leakage is even better for layers deposited at 450°C and 500°C.

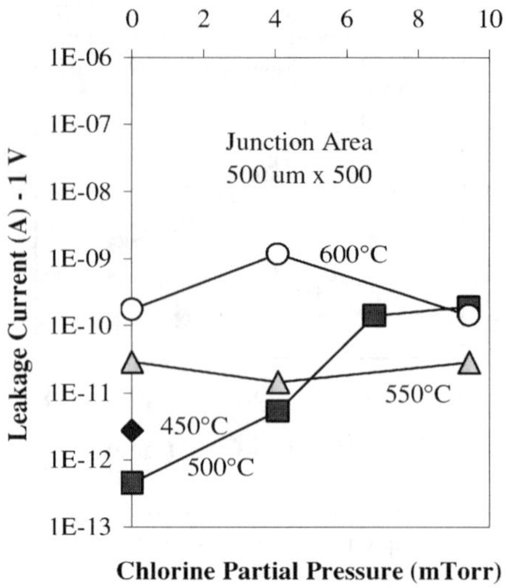

Figure 11 Reverse bias junction leakage for boron doped SiGe junctions deposited at different temperatures as a function of chlorine partial pressure.

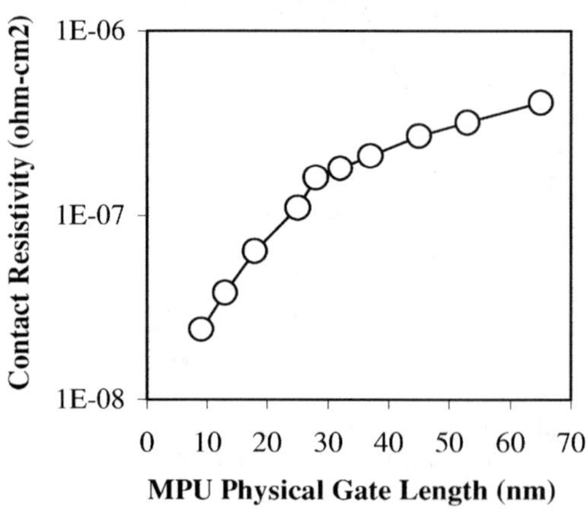

Figure 12 Maximum contact resistivity for different technology nodes from ITRS 2001.

LOW RESISTIVITY CONTACTS

Shown in Figure 12 is the maximum allowable contact resistivity predicted in ITRS 2001 for different technology nodes. The numbers were obtained assuming 100% of the parasitic junction resistance is due to the contact resistance. Since this is not entirely true, the data can only represent an upper bound for contact resistivity.

The resistivity of an ohmic contact is determined by the Schottky barrier height at the metal-semiconductor interface and the doping density under the metal. This is illustrated in

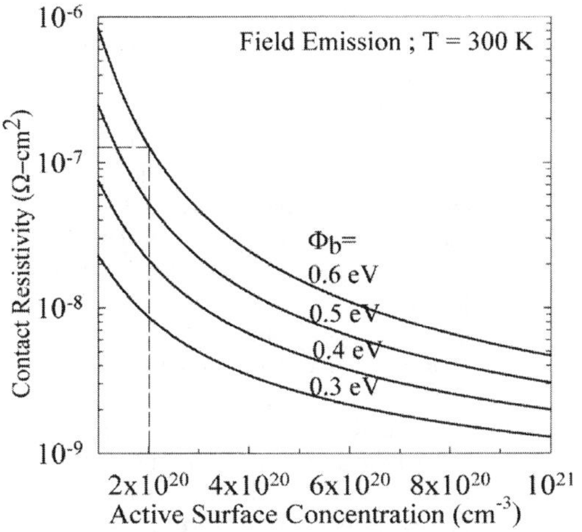

Figure 13 Contact resistivity as a function of doping level for different metal-semiconductor barrier heights.

Figure 13, which shows the contact resistivity plotted as a function of active doping density for different barrier heights. If a single type of metal is to be used for both PMOS and NMOS junctions, it is desirable to choose a metal whose Fermi level lies near the silicon midgap. For Si, this corresponds to a barrier height around 0.6 eV, which is indeed the case with the existing silicide contacts. As shown in Figure 13, with a barrier height of 0.6 eV, a typical doping density of $2x10^{20}$ cm^{-3} yields a contact resistivity of ~ $2x10^{-7}$ $ohm\text{-}cm^2$. According to ITRS requirements shown in Figure 12, this contact resistivity is certainly not sufficient for gate lengths below 30 nm. Therefore, to meet the demands of future devices, we either need to increase the active doping density under the metal contact or reduce the metal-semiconductor barrier height.

If Si is the semiconductor under the metal contact, reducing the barrier height requires two different contact metals with work functions tuned to p+ and n+ Si. Aside from its complex integration scheme, the idea implicitly requires a novel interface to eliminate Fermi level pinning near the Si midgap. An alternative approach would be to keep the barrier height fixed and increase the active doping density under the metal according to Figure 13. However, as mentioned in the previous section, under equilibrium conditions, the active doping level is determined by the solid solubility of the dopant in Si at a given temperature. For boron, the equilibrium limit is about $2x10^{20}$ cm^{-3}, which yields a contact resistivity of $2x10^{-7}$ $\Omega\text{-}cm^2$ as illustrated in Figure 12. Therefore, it is physically impossible to achieve a contact resistivity in the range suggested in Figure 11 for physical gate lengths below 30 nm.

To achieve the desired contact resistivities for future technology nodes, the SiGe junction technology provides a low resistivity semiconductor under the contact metal and a smaller barrier height due to its narrower bandgap. In our laboratory,

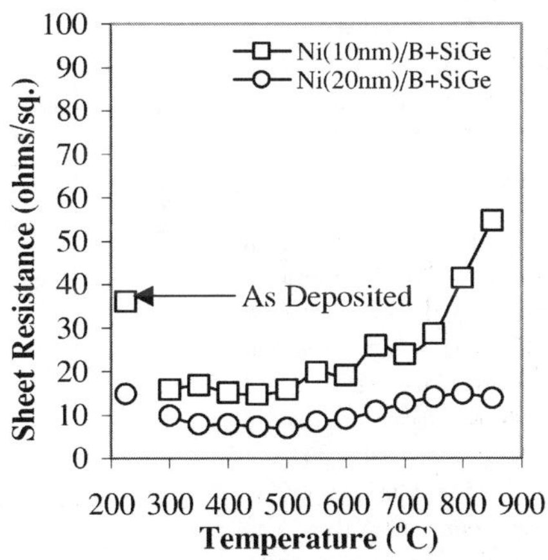

Figure 14 Sheet resistance as a function of annealing temperature for Ni on undoped and heavily boron doped SiGe for 10nm and 20nm thick Ni layers.

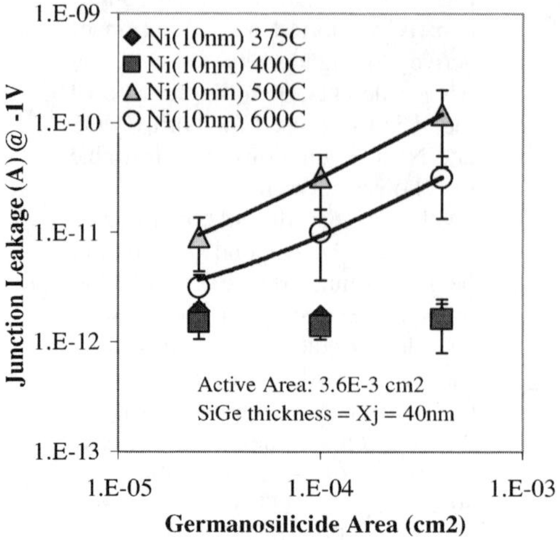

Figure 15 Junction leakage as a function of the area of the germanosilicide contact. The diode active area is the same for all diodes.

we have investigated a variety of metals to form self-aligned contacts to SiGe alloys. Two promising metals for future contacts are Pt and Ni. On Si, both metals form mono silicides and consume much less Si than the disilicides of Ti and Co. This is an important advantage in forming self-aligned contacts to ultra-shallow junctions. Our results indicate that both Ni and Pt can form low resistivity compounds when reacted with SiGe alloys. Shown in Figure 14 is the sheet resistance of nickel germanosilicide ($NiSi_{1-x}Ge_x$) formed at different temperatures after annealing 10 or 20 nm Ni deposited on p+ SiGe for 30 seconds in a rapid thermal annealing (RTA) furnace. The annealing ambient was nitrogen. The thermal stability of the germanosilicide appears to be quite good at least within the temperature range of interest. On the other hand, Ni germanosilicide formed on undoped SiGe is stable only up to ~ 400°C. Improved germanosilicide thermal stability was observed previously for SiGeC alloys where carbon was used for strain compensation, similar to boron in this study[6]. On undoped or lightly doped SiGe, we have determined that a thin interlayer of Pt can improve the thermal stability[7].

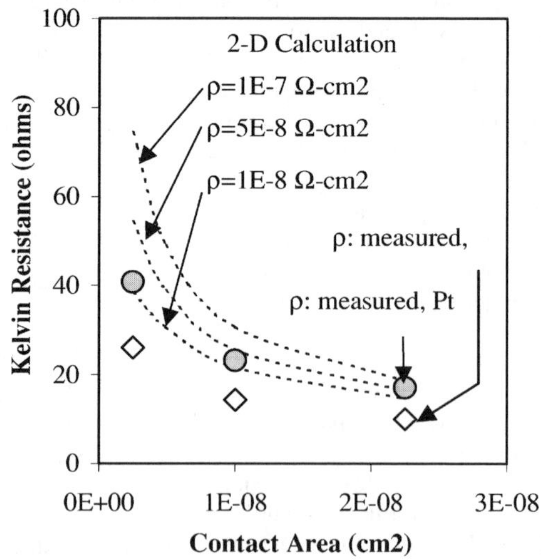

Figure 16 Contact resistance obtained with Ni and Pt contacts on p+ SiGe.

Figure 15 shows the junction leakage of boron-doped junctions with Ni germanosilicide contacts. The SiGe thickness (i.e. junction depth) is 40 nm. The self-aligned contacts were formed in a temperature range of 375 - 600°C. For 500 and 600°C anneals, we can see that the junction leakage increases with the germanosilicide area suggestive of silicide induced leakage. The increase in leakage may be due to agglomeration or a new phase consuming more

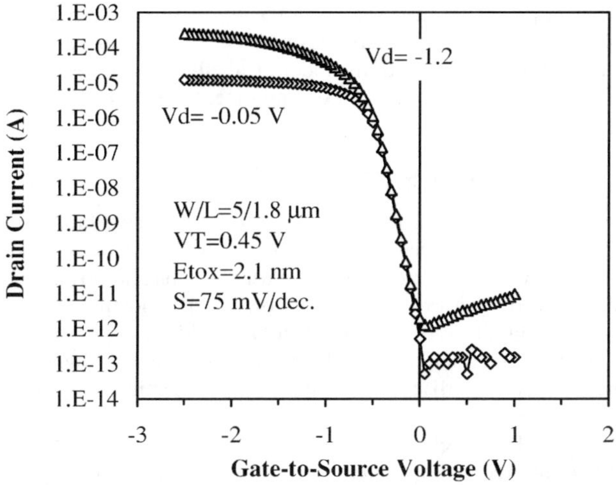

Figure 17 Drain current versus gate-to-source voltage of a p-channel MOSFETs fabricated using the SiGe source/drain junction technology.

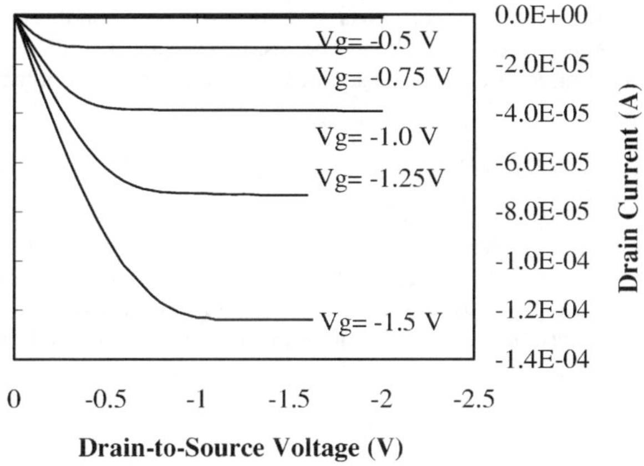

Figure 18 Drain current versus drain-to-source voltage of p-channel MOSFET in Figure 17.

SiGe. However, since XRD analysis does not indicate formation of a new phase at these relatively low temperatures, we suspect that the cause of increased leakage is agglomeration, which is the subject of a current study. On the other hand, when the formation temperature is 400°C or lower, excellent reverse bias behavior is observed independent of the contact area.

Four terminal contact resistance structures were used to measure the contact resistivity. Shown in Figure 16 is the contact resistivity obtained with Ni and Pt germanosilicide contacts on p+ SiGe. Also shown in the figure are three lines obtained using a 2-D analytical model for contact resistivities of 10^{-8}, $5x10^{-8}$ and 10^{-7} ohm-cm^2. Comparison of the measured data points with the theoretical curves indicates that both Pt and Ni can successfully meet the contact resistivity requirements of the technology nodes beyond 30 nm.

MOSFET FABRICATION

Using the junction integration scheme outlined in Figure 2, we have fabricated large area p-channel MOSFETs with ultra-shallow junctions as a first demonstration. The junction recess within 15 – 30 nm was obtained by plasma etching in an SF_6 ambient. The gate dielectric was formed by a process, which combined rapid thermal oxidation and remote plasma CVD silicon nitride. The equivalent gate oxide thickness was 2.1 nm. The sidewalls were formed by oxidation of the polysilicon gate. The self-aligned contacts to p+ SiGe was Ni germanosilicide discussed in the previous section. The starting Ni thickness was 10 nm.

Shown in Figure 17 and 18 are electrical characteristics of a MOSFET fabricated using the SiGe source/drain technology presented in this paper. This is the first demonstration of a working MOSFET with the technology. The device exhibits excellent I-V characteristics with very low off-state leakage indicative of low junction leakage.

CONCLUSIONS

In this paper, we have presented an overview a new source/drain technology based on selective deposition of in-situ doped SiGe alloys. The results indicate that the technology is highly promising in meeting the demands of source/drain junctions and contacts at the end of the roadmap.

ACKNOWLEDGMENTS

This work was supported by the SRC-SEMATECH Front End Processing Research Center. The authors are grateful to Majid Mansoori of Texas Instruments, Bill Taylor of Motorola and Eric Graetz of SEMATECH for their contributions to analytical characterization of the samples. The authors extend their sincere thanks to Joan O'Sullivan and the NCSU microelectronics laboratory personnel for their invaluable help during the course of the study.

REFERENCES

1. International Technology Roadmap for Semiconductors, 2001 Edition, Front End Processes
2. S. Gannavaram, N. Pesovic and M.C. Ozturk, IEDM Technical Digest, 437, (2000).
3. G. L. Olson and J. A. Roth, *Mat. Sci. Reports*, **3**, 1-78 (1988)
4. B. Yu, Y. Wang, H. Wang, Q. Xiang, C. Riccobene, S. Talwar, and M. R. Lin, IEDM Tech. Digest,509-512 (1999)
5. S. Talwar, Y. Wang, and C. Gelatos, Electrochem. Soc. Symp. Proc., 95-105 (2000)
6. A.Eyal,, et al, Appl. Phys. Lett., 69 (1), 1996, 64-66
7. J. Liu, H. Mo & M.C. Ozturk, presented at MRS 2002 Spring Meeting, San Fransisco, CA, (unpublished)

Mat. Res. Soc. Symp. Proc. Vol. 717 © 2002 Materials Research Society C4.3

Thermal Stability and Substitutional Carbon Incorporation far above Solid-Solubility in $Si_{1-x}C_x$ and $Si_{1-x-y}Ge_xC_y$ Layers Grown by Chemical Vapor Deposition using Disilane

M. S. Carroll*, J. C. Sturm, Dept. of Electrical Engineering, Princeton University, Princeton NJ; E. Napolitani, D. De Salvador, and M. Berti INFM and Dept. of Physics, University of Padova, Padova, Italy

*Present address: Agere Systems, Murray Hill NJ

Abstract
Growth conditions for epitaxy of $Si_{1-x-y}Ge_xC_x$ and $Si_{1-x}C_x$ alloy layers on (100) silicon substrates by rapid thermal chemical vapor deposition (RTCVD) with disilane as the silicon source gas are described and the $Si_{1-x}C_x$ conditions are compared to previously reported RTCVD growth conditions for epitaxy of $Si_{1-x}C_x$ using silane as the source gas. The thermal stability of the layers at 850°C in nitrogen is examined using x-ray diffraction as a measure of the average substitutional carbon concentration in the layers after annealing. A characteristic time constant to describe the reduction of average substitutional carbon concentration in the layer is extracted from the XRD measurements. The characteristic time constants are found to agree within a factor of 3 with that observed in previous reports. However, the time constants are found to depend more strongly on the as-grown substitutional carbon concentration than what is predicted by simple precipitation kinetics, assuming carbon diffusion to a constant number of nucleation centers.

Introduction
The incorporation of high concentrations (> 0.1%) of carbon in silicon and SiGe is of wide technological interest for its potential use to engineer dopant diffusion [1], engineer the conduction band in silicon based materials [2, 3], and strain compensation [4]. However, because the solid solubility of substitutional carbon in silicon is $\sim 10^{17}$ cm^{-3} and carbon at concentrations above this can incorporate on interstitial sites or precipitate as β-SiC, special non-equilibrium epitaxial growth conditions are required to grow metastable films with substitutional carbon concentrations greater than 0.1 atomic percent.

It is has been shown that the incorporation of substitutional carbon incorporation increases with reduced growth temperature and higher growth rates [5, 6]. Because the growth rate of silicon and SiGe by rapid thermal chemical vapor deposition is strongly dependent on the choice of silicon source gas and disilane is known to produce higher silicon epitaxial growth rates for similar partial pressures and temperatures as silane and dichlorosilane, disilane is a promising candidate for expanding the available parameter space of growth conditions that produce 100% substitutional carbon alloys with silicon and SiGe, which in turn could allow for increased throughput and an extension of the maximum total carbon concentration that can be incorporated entirely as substitutional carbon grown by RTCVD.

This paper characterizes the growth of $Si_{1-x-y}Ge_xC_x$ and $Si_{1-x}C_x$ layers using a mixture of 10% disilane in hydrogen as the silicon source gas. Growth conditions are described to obtain $Si_{1-x-y}Ge_xC_x$ and $Si_{1-x}C_x$ films with as much as 0.81 and 0.45 atomic percent, respectively, that are 100% substitutional, and the thermal stability of the $Si_{1-x-y}Ge_xC_x$ and $Si_{1-x}C_x$ alloy layers is examined at 850°C for up to 10 and 16 hours of annealing in nitrogen, respectively. X-ray diffraction is used as a measure of the average substitutional carbon concentration in the layers during annealing and characteristic time constants are extracted to describe the carbon concentration dependence on time.

$Si_{1-x-y}Ge_xC_x$ and $Si_{1-x}C_x$ epitaxy using disilane as the silicon source

SiGeC alloy layers were grown by rapid thermal chemical vapor deposition (RTCVD) on (100) silicon substrates. All SiGeC and SiC layers discussed in this work were grown in a homemade reactor described elsewhere [7]. Sources gasses for silicon, germanium and carbon were disilane (10% mixture in hydrogen), germane (0.8% mixture in hydrogen) and methylsilane (1% mixture in hydrogen), respectively. Epitaxy of the alloy layers was done at temperatures ranging from 550-625°C at 10 torr with 3 slpm of hydrogen and 50 sccm of disilane mixture.

The chemical composition of all alloys was determined by secondary ion mass spectroscopy (SIMS) leading to an uncertainty of approximately 15-20% in the total carbon concentrations and 1-2 atomic percent in the germanium concentration. The substitutional carbon concentrations in the alloys were deduced from x-ray diffraction rocking curves around (004) silicon reflection and a linear interpolation between β-SiC and silicon is used to estimate the lattice constant and elastic constants for carbon in silicon [6].

Two $Si_{1-x}C_x$ alloy layers were grown with 20 sccm of methylsilane at 550°C and 625°C resulting in growth rates of 2.5 and 27.7 nm/min and having thicknesses of 100 and 300 nm, respectively. Each of the $Si_{1-x}C_x$ layers were capped with ~ 50 nm of silicon and subjected to a subsequent 5 minute 800°C anneal. Sharp interfaces between the $Si_{1-x}C_x$ and silicon and uniform carbon concentrations in the $Si_{1-x}C_x$ are measured by SIMS, Fig. 1(a). The rocking curve of the $Si_{0.9955}C_{0.0045}$ layer, furthermore, shows a well defined

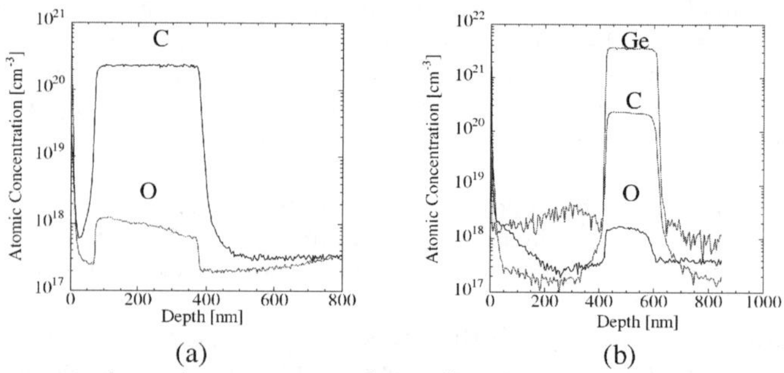

Figure 1. Carbon, germanium and oxygen concentration profiles of the as-grown (a) $Si_{0.996}C_{0.004}$ layer and (b) $Si_{0.9255}Ge_{0.07}C_{0.045}$ layer.

(004) reflection centered at ~ 500 arcseconds from the silicon (004) diffraction peak indicating a tensilely strained single crystal layer, Fig 2 (a). The substitutional carbon concentrations extracted from the XRD measurements of both layers agree well with the

total carbon concentration measured by SIMS demonstrating that 100% substitutional carbon layers may be grown at 625°C with up to at least 0.45% substitutional carbon.

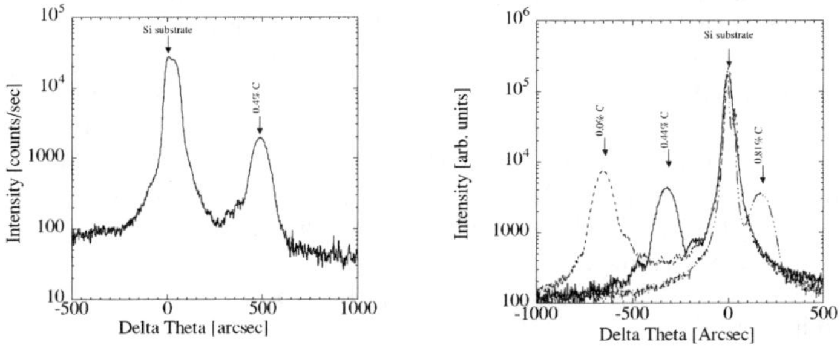

(a) (b)

Figure 2. X-Ray diffraction rocking curves around the (004) Bragg reflection of the (a) $Si_{0.996}C_{0.004}$ layer and (b) and the $Si_{0.93-x}Ge_{0.07}C_x$ layers with x varying from 0 to 0.008.

These results are compared to those reported by Mitchel et al. at 600°C, who studied substitutional carbon incorporation in $Si_{1-x}C_x$ epitaxial layers grown by RTCVD using silane. With the exception of using silane as the silicon source gas their growth conditions were similar to those used in this work (i.e. 12 torr, 2 or 8 slpm of H_2 and methylsilane was the source gas for carbon). It was found in that work that it was necessary to use high silane partial pressures, 300 mtorr, to incorporate all carbon

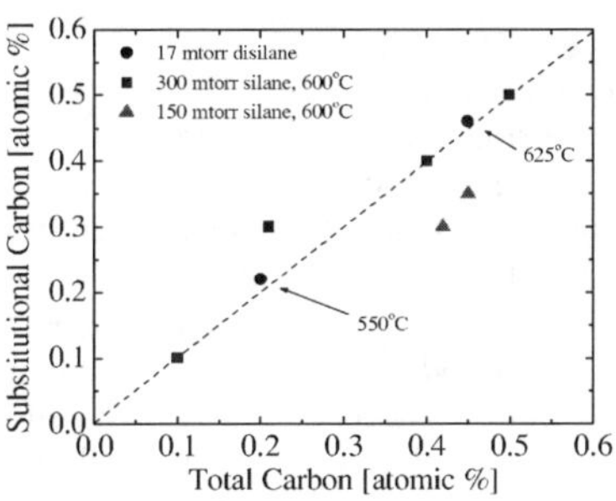

Figure 3. Summary of substitutional carbon incorporation in epitaxial $Si_{1-x}C_x$ for different silane partial pressures and different disilane growth temperatures. The dashed line represents 100% substitutional carbon. Conditions for carbon incorporation using silane were obtained from reference [6].

substitutionally in silicon up to 0.5% atomic percent of carbon at 600°C, Fig. 3. However, using a disilane partial pressure 17 times lower and at a higher temperature, 625°C, a similar amount of carbon was incorporated entirely substitutionally, Fig. 3, which is encouraging as restrictions on both critical growth parameters for 100% substitutional carbon incorporation, low temperature and high partial pressures of the silicon source gas, appear to be relaxed with the use of disilane.

Silicon capped $Si_{0.93-x}Ge_{0.07}C_x$ layers were grown at 625°C with the addition of 25 sccm of germane and either 0, 20 or 50 sccm of methylsilane, while using the same hydrogen and disilane flows and partial pressures. Sharp interfaces and uniform concentrations of carbon and Ge were also obtained for these conditions, Fig 1 (b) (SIMS) and X-ray diffraction rocking curves around the (004) reflection again show well defined reflections from the alloy layers that shifts from the compressively strained lattice constant of the $Si_{0.93}Ge_{0.07}$ to the tensilely strained $Si_{0.922}Ge_{0.07}C_{0.008}$ layer, Fig 2 (b), which further demonstrates that fully strain compensated SiGeC layers up to at least 7 atomic percent germanium may be grown by RTCVD using disilane. The growth rates for all three layers was ~ 30 nm/min and little change in growth rate due to the introduction of carbon was observed. The substitutional carbon concentrations were extracted from the XRD using Ge concentrations obtained from SIMS and found to be 100% substitutional in all of the layers.

Finally, in all layers grown with high carbon concentrations a proportional increase in oxygen concentration is also observed. Oxygen is an undesirable contaminant in the presence of carbon, as it is known to facilitate carbon precipitation [8, 9], however, TEM of all the as-grown $Si_{0.93-x}Ge_{0.07}C_x$ layers showed no evidence of precipitates. The resulting oxygen background is consistently about 200 times less than the carbon concentration at 625°C and the source is as of yet unidentified, Fig 1 (a & b). Because a similar magnitude of increase in oxygen concentration is observed in SiGeC layers grown with dichlorosilane and low oxygen backgrounds are observed in pure silicon grown with disilane, disilane is not suspected as the source of oxygen contamination. Residual oxygenated species in the methylsilane are a potential source of the contamination and different purification schemes for the methylsilane source are currently being examined.

Thermal Stability of Carbon Alloys

Reduction of tensile strain in alloy layers that contain carbon after annealing in nitrogen has been correlated with the formation of precipitates and a decrease in carbon on substitutional sites [10, 11]. Therefore, the thermal stability of the carbon in the SiGeC and SiC layers after annealing at 850°C in a nitrogen ambient for up to 10 and 16 hours, respectively, was monitored by XRD, Fig. 4. To evaluate the contribution of just the Ge to the change of strain over the annealing times the pure $Si_{0.93}Ge_{0.07}$ sample was measured and the strain in the layer was found to be indistinguishable from the as-grown layer within the uncertainty of the Ge concentration due to growth non-uniformity.

However, a rapid increase in compressive strain is observed in the $Si_{0.9219}Ge_{0.07}C_{0.0081}$ layer after 10 hours of annealing indicating a loss of approximately 59% of the average substitutional carbon in the layer. In contrast, the strain in the layers with less carbon are significantly more stable despite differing in concentration by only 2-4 times. The carbon profiles after annealing, measured with SIMS, show out-diffusion of the carbon from the SiC and SiGeC layers affecting the concentration only within the first 10-20 nm of the edges of the carbon layers, however, the carbon concentration in the rest of the layer remains indistinguishable from the as-grown layer. The observed shift in the XRD peak positions, therefore, can not be explained by carbon out-diffusion.

Based on classical theory of phase transitions, the rate of precipitation in supersaturated mixtures can often be described by the semi-empirical Avrami relation,

$$y(t) = \exp[-(t/\tau)] \qquad (1)$$

where y(t) is the number of solute atoms that remain unprecipitated t is the time and τ is the characteristic time constant of the reaction. Loss of substitutional carbon after annealing has indeed been described relatively well by this relationship [12]. Furthermore, the time constant dependence on temperature and carbon concentration have been described previously with some success with the following relationship[10, 12]:

$$\frac{1}{\tau} \sim D\left(\frac{y_o - y^{eq}}{y_{SiC}}\right)^{\frac{1}{3}} \sim D\left(\frac{y_o}{y_{SiC}}\right)^{\frac{1}{3}} \qquad (2)$$

where y_o is the initial carbon concentration, y_{SiC} is the carbon concentration in β-SiC (i.e. y_{SiC} =0.5), D is the diffusivity of the carbon, and y^{eq} is the solid solubility of carbon in silicon, which is negligible compared to the initial carbon concentrations in the layers of this work. This relationship can be derived from the assumption that the precipitation reaction is a diffusion limited process with a fixed number of nucleation sites.

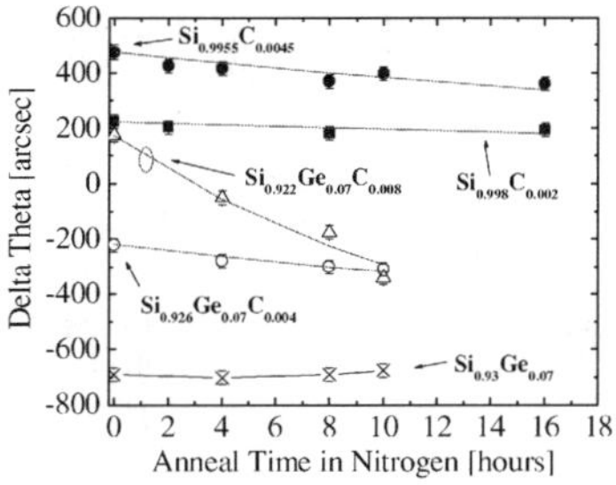

Figure 4. Summary of the positions of the XRD reflections from all alloy layers with respect to the silicon (004) substrate diffraction peak before and after annealing in nitrogen ambient at 850°C. Best fits using a single time constant are shown for all layers except the $Si_{0.93}Ge_{0.07}$ layer.

Characteristic time constants were extracted for the $Si_{1-x-y}Ge_xC_x$ and $Si_{1-x}C_x$ layers annealed in this work and are normalized according to their respective concentrations, eqn. 2, so that they may be compared with one another and previous work, Fig. 5 [12]. The agreement of normalized time constants between this work and that expected from Fischer et al. is within a factor of 3, which is consistent with the observation that the time constant is dominated by a temperature dependent diffusivity with an activation energy similar to that of the diffusivity of carbon in silicon, 3.3 eV.

However, the spread of normalized time constants for different concentrations shows that the time constants are significantly more sensitive to the carbon concentrations than is predicted by this simple classical theory. The presence of 7 atomic percent germanium does not appear to change the time constant more than 20% for layers containing approximately the same carbon concentration, (i.e. the $Si_{0.9955}C_{0.0045}$ and the $Si_{0.9262}Ge_{0.07}C_{0.0038}$), and previous reports indicate a faster rate of carbon loss in $Si_{0.892}Ge_{0.1}C_{0.008}$ compared to $Si_{0.992}C_{0.008}$ of only about 2 at similar temperatures [13], which is not enough to explain the factor of 5 difference that is observed between the lowest and highest carbon content layers in this work. It is not clear, from this limited set of observations, whether the increased sensitivity to carbon concentration may be caused by the increased oxygen concentration observed in these layers, a known catalyst for precipitation [8], or perhaps because of an intrinsic aspect of the carbon precipitation itself.

Conclusion

Growth conditions for SiC and SiGeC layers using disilane as the silicon source gas are presented and carbon concentrations that are 100% substitutional are incorporated into silicon and $Si_{0.93}Ge_{0.07}$ up to 0.45 and 0.81 atomic percent, respectively. It is found that a similar carbon content film may be grown at a higher temperature and lower partial pressure of disilane than what is reported for growth with silane. The thermal stability of

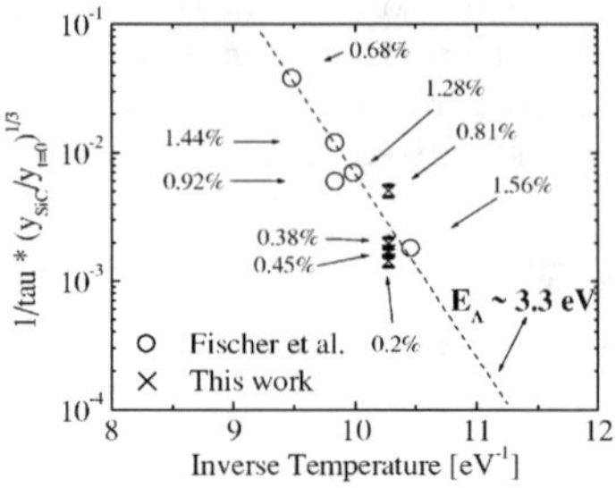

Figure 5. Summary of characteristic time constants normalized according to eqn. 2 so that they may be compared to the work of Fischer et al. [12]. The carbon content in each of the layers is indicated beside the normalized time constant.

all layers grown with disilane, is furthermore, examined and the rate of loss of substitutional carbon from these layer is found to be within a factor of 3 of what would be expected from previous reports. However, the characteristic time constant of the precipitation is found to depend more strongly on the carbon concentration than what is predicted by simple precipitation kinetics assuming carbon diffuses to a fixed number of nucleation centers.

References

[1] H. Ruecker and B. Heinemann, *Solid-State Elec.*, vol. 44, pp. 783, 2000.

[2] R. L. Williams, G. C. Aers, N. L. Rowell, K. L. Brunner, W. Winter, and K. Eberl, *Applied Physics Letters*, vol. 72, pp. 1320, 1998.

[3] K. Brunner, K. Eberl, and W. Winter, *Phys. Rev. Lett.*, vol. 76, pp. 303, 1996.

[4] O. G. Schmidt and K. Eberl, *Phys. Rev. Lett.*, vol. 80, pp. 3396, 1998.

[5] J. Osten, K. Myeongcheol, K. Pressel, and K. Zaumseil, *J. Appl. Phys.*, vol. 80, pp. 6711, 1996.

[6] T. O. Mitchell, J. L. Hoyt, and J. F. Gibbons, *Appl. Phys. Lett.*, vol. 71, pp. 1688, 1997.

[7] J. C. Sturm, P. V. Schwartz, E. J. Prinz, and H. Manoharan, *J. Vac. Sci. Tech. B*, vol. 9, pp. 2011, 1991.

[8] W. J. Taylor, W. Y. Tan, and U. Goesele, *Appl. Phys. Lett.*, vol. 62, pp. 2336, 1995.

[9] F. Shimura, R. S. Hocket, D. A. Reed, and D. H. Wayne, *Appl. Phys. Lett.*, vol. 47, pp. 794, 1985.
[10] P. Warren, J. Mi, F. Overney, and M. Dutoit, *J. of Crystal Growth*, vol. 157, pp. 414-419, 1995.
[11] J. W. Strane, H. J. Stein, S. R. Lee, S. T. Picraux, J. K. Watanabe, and J. W. Mayer, *J. Appl. Phys.*, vol. 76, pp. 3656, 1994.
[12] G. G. Fischer, P. Zaumseil, E. Bugeil, and J. Osten, *J. Appl. Phys.*, vol. 77, pp. 1934, 1995.
[13] L. V. Kulik, D. A. Hits, M. W. Dashiell, J. Kolodzey, *Appl. Phys. Lett.*, vol. 72, pp. 1972, 1998

Mat. Res. Soc. Symp. Proc. Vol. 717 © 2002 Materials Research Society C4.4

Structural and optical properties of the multilayer structures formed by Ge sub-critical insertions in a Si matrix

George E. Cirlin [1,3] , Nikolai D. Zakharov [1], Peter Werner [1], Alexander G. Makarov [2], Andrei F. Tsatsul'nikov [2], Victor M. Ustinov [2], and Nikolai N. Ledentsov [2], Vyatcheslav A. Egorov [3], and Ulrich Gösele [1]
[1] Max-Planck-Institute for Microstructure Physics, Halle D-06120, Germany
[2] A.F.Ioffe Phisico-Technical Institute RAS, 194021 St.Petersburg, Russia
[3] Institute for Analytical Instrumentation RAS, 194021 St.Petersburg, Russia

ABSTRACT

Si/Ge multilayer structures formed by the deposition of relatively small amounts of Ge layers (less then the critical thickness for 3D islands formation) are obtained by molecular beam epitaxy. Their structural and optical properties are investigated in detail. Appropriate growth parameter of the stacked island structures lead to significant increasing of the luminescence efficiency even at room temperature.

INTRODUCTION

The strained SiGe/Si system is expected to play an important role in monolithic integration of Si-based opto-and microelectronic devices. It is well known that Ge/Si system is a prominent example of Stranski-Krastanow growth mode, where three-dimensional (3D) islands appear at the surface after exceeding of a certain *critical* thickness. There are a lot are a lot of investigations in numerous groups (see, e.g., [1,2]. However, already small amounts of deposited Ge (smaller than one monolayer) can form island-like structures.

In the present work, we report on the observation of a new kind of nanostructures formed by *sub-critical* Ge insertions in a Si matrix and their structural and optical properties. Our approach is based on the following assumption: insertion of Ge submonalyers (SML) into Si may lead to the formation of the ensemble of relatively small islands (with lateral sizes less than the hole Bohr radii). This can result in a partial lifting of the selection rule for radiative recombination and exciton formation (which may be stable up to room temperature) via electron and localized-hole interaction. This situation is possible if the Coulomb attraction energy is high enough to localize electrons near the potential barrier which is produced by Ge SML inclusions in the conduction band. In the case of relatively large sizes of QDs or large widths of quantum wells, this barrier may also lead to spatial separation of the electrons and holes [3]. Furthermore, for the SML in other heteroepitaxial systems [4, 5] the narrow photoluminescence (PL) line leads to the increase of the absorption (gain). The PL intensity will also increase if multiple layers are used with Ge SML separated by Si spacers. Due to relatively small strain accumulation in such a system, it is expected that there will be a low probability of dislocations and formation of structural defects. In fact, despite of the total amount of the Ge (sub)monolayers deposited is lower then the critical thickness we show that they exhibit (quasi) 3D properties when capped with a host (Si) material under certain growth conditions.

EXPERIMENTAL DETAILS

All structures are grown by molecular beam epitaxy (MBE) using a Riber SIVA 45 set-up on Si(100) n-type substrates (conductivity 3 Ohm cm). The substrates are five inches in diameter, manufactured by OKMETIC. After chemical preparation by the method described in [6] the substrates are transferred into the MBE set-up loading chamber. The method of chemical treatment allows us to remove the oxide layer from the silicon surface at 840^0C in the growth chamber by direct radiating heating. During the growth process the rotation of the samples is used, and the temperature field inhomogeneity across the surface is about ~5%. In order to grow Ge SMLs we use the SML epitaxy technique, which we have also used to grow the SML insertions in the A_3B_5 and A_2B_6 systems [7–9].

The structures consist of a Si 100 nm buffer layer, a Ge/Si superlattice (20 pairs) and a silicon 20 nm capping layer. The growth rates for Si and Ge are 0.5 and 0.05 Å/s, respectively. The substrate temperature (T_s) is varied within 600^0C –750^0C. Ge content is varied between 0.07 – 0.5 nm. Spacer thickness is 4 – 11 nm. In some cases, n-type doping of the spacers is used. If doping is applied, central part (2 nm) of the spacers is Ge:Sb grown at decreased T_s with electron concentration ~ 10^{16} cm^{-3}. A scheme of the structures is seen in Fig.1a. The growth rates are controlled by two mass spectrometers with feedback which are set to 28 (Si) and 74 (Ge) masses. Total pressure during growth is better than $5x10^{-10}$ Torr. The surface is monitored *in situ* by reflection high-energy electron diffraction (RHEED).The photoluminescence is excited by an argon laser (λ=514.5 nm, maximal excitation density ~500 W/cm^2), and is detected by a Ge-cooled photo-diode. The samples are also investigated by different electron microscopic (TEM) and selected area electron diffraction (SAED) techniques using a JEM 4010 microscope with acceleration voltage 400 kV and a CM 20 (200kV) microscope equipped with a field-emission gun.

RESULTS AND DISCUSSION

The surface morphology is monitored during the growth *in situ* by reflection high-energy electron diffraction and shows no significant changes in comparison with the initial (2x2) surface reconstruction except of slight broadening of the main (0n) reflections. Thus, even on the upper layers of the superlattice the surface remains quite smooth and well pronounced 3D structural formation due to strain relaxation does not occur. Scanning tunnelling microscopy of *uncapped* samples show quite smooth surfaces.

The cross-section TEM images of grown structures taken in a diffraction contrast mode at relatively low resolution are shown in Fig.1b,c. The periodicity in growth [001] direction is 4.4 nm in both cases. From the image analysis follows that samples with the 1-ML Ge deposition are characterized by sharp layers/interfaces, whereas layers of the samples with the 0.5-ML deposition have fluctuations and a Si/Ge interdiffusion. Corresponding plan-view images of the 0.5-ML samples show small islands of high-density.

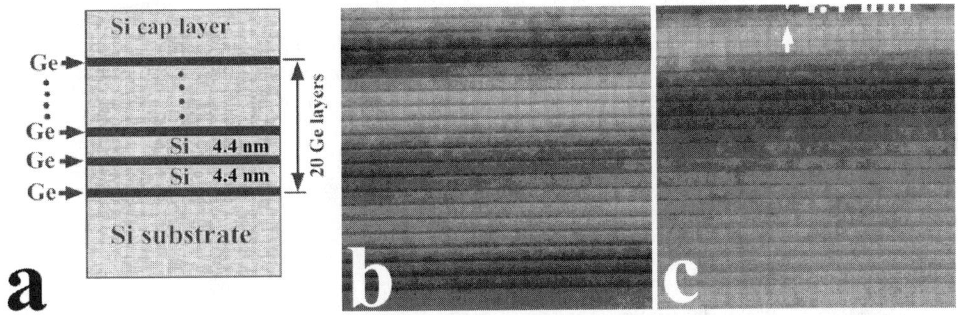

Figure 1. (a) Sequence of the layers in MBE grown structures. Thickness of Ge layers was equal 0.5 ML (0.07 nm) and 1 ML (0.14 nm) in samples A Cross section diffraction contrast images of specimen with 0.5 ML Ge (b) and 1 ML Ge (c). The periodicity of grown structures in [001] direction are equal 4.4 nm in both cases.

In the following the results of the photoluminescence spectroscopy are presented. For *capped* SL we have found that for the growth temperatures within 600 – 700^{0}C and 0.07 – 0.13 nm of Ge deposited lines are appeared in PL spectra in comparison with Si substrate spectrum. These lines are in the ranges (0.98 – 1.01), (1.04 – 1.07) and (1.11-1.13) eV (as an example, in Fig.2 the lines are marked as SL^{TO+0}, SL^{TO} and SL^{NP}, respectively) at 15 K depending on the Ge amount deposited. No such PL lines are observed at lower and higher T_s. For an integer quantity of Ge ML deposited at relatively low T_s (600^{0}C – 700^{0}C) the intensity of Ge-associated PL lines decrease in two orders of magnitude. To clarify the observed features, transmission electron microscopy (TEM) study has been done. In the diffraction pattern taken from the SL formed by 0.5 ML Ge, the number of Fourier harmonic is two times less than in the case of 1 ML one. High-resolution TEM cross-section images manifest the distribution of Ge atoms over tens ML for 0.5 ML Ge sample and very smooth Ge-spread layer for 1 ML case. Compositional spherically shaped non-uniformities of a very high density are clearly resolved in the plane-view TEM image of the 0.5 ML sample (size of the inclusions 3-17 nm, surface density ~ $5x10^{11}$ cm^{-2}) as opposite to the smooth TEM image of the 1 ML structure. Similar results are obtained for the samples with Ge amount higher than 1 ML in sub-critical thickness range. Thus, the appearance of the new PL lines is unambiguously correlates with Ge islands formation.

High temperature growth (T_s = 750^{0}C) leads to the significant changes in the optical properties on Ge thickness dependencies. Deposition of 0.07 nm of Ge results in formation of modulated structure with the composition sinusoidal changes along [-113] and [1-12] directions, distortion of the cubic symmetry and in a relaxation of the misfit stress causing disappearance SL-related PL lines [10]. As opposite to the low temperature growth, islands are formed at 0.1 nm Ge deposited which can be a result of partial re-evaporation of Ge adatoms. In Fig. 3 PL spectra of the sample grown at T_s = 750^{0}C and Ge amount 0.1 nm in each layer taken at different excitation densities are presented. We note that the main trends observed on the spectra are typical for the samples in which TEM shows a formation of Ge-related islands.

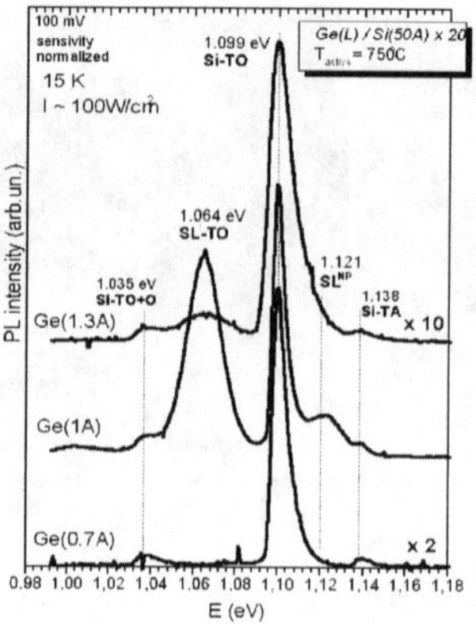

Figure 2. PL spectra for Si/Ge superlattices (20 pairs) grown at different substrate temperatures with 0.5 ML Ge in each layer.

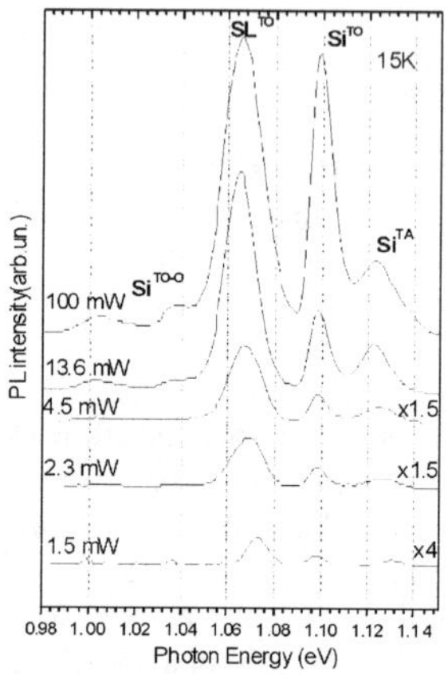

Figure 3. Excitation density dependence of PL spectra taken from the sample grown at $T_s = 750^0$C and Ge deposited 0.1 nm.

It is found that there is a steady increase of the integral intensity ratio of Si–TO and SL–TO lines with the increase of excitation density, and even at very high excitation levels the integral intensity ratios of these lines are comparable. At low excitation levels the SL–TO line dominates the spectra. The full width at half maximum is about 15 meV. The SL–TO line shifts towards long wavelengths with the increase of the excitation density (up to 10 mW) whereas Si-TO replica does not changes an energetic position. This situation is opposite to typical for the systems with type II alignment [5,13] where increasing of the excitation density leads to *blue-shift* of PL line position. For type-I-band structure, as it was detected for e.g. InAs/GaAs QD system the QD PL line does not shift with excitation density changing. A possible explanation of the observed effect can be given as follow. At low excitation densities there is a lack of the electrons in the conductivity band, which are localized in the wells formed by the stress accumulation at the Si/Ge interface. When the states in the wells are filled out (with increasing of the excitation density) electrons can energetically overcome the barrier of Si matrix and, thereby, can form a new energy level below the barrier. This is true in a case of the power of Coulomb potential from the localized holes is high enough to attract electrons. Hence, the exciton can be formed and direct electron-hole transition in real space and k-space could be achieved. In the case of small dispersion of the islands size, further increasing of the excitation density should not lead to the shift of the PL line which we actually observe at excitation densities higher then 10 mW (Fig.3). However, with increasing of the thickness of germanium insertion electron wave function corresponding to this level (*s*-state) is pressed out from the barrier, probability of exciton formation decreases and PL intensity drops. The latter situation is typical to the Ge portions 0.13 nm at $T_s = 600^0C - 700^0C$ where Ge adatoms re-evaporation is negligible but Ge forms continuos layer.

In Fig. 4 shows PL spectra for an Sb-doped Si/Ge structure taken at different temperatures. One can see that with the temperature increasing SL-TO PL line originating from Germanium inclusions practically does not shift. At the same time Si-TO line corresponding to emission from silicon matrix shifts towards lower energies. Highest SL-TO line observation temperature for the doped sample is ~ 220K whereas typical values for undoped samples do not exceed 100K showing higher temperature stability of the intentionally doped structures.

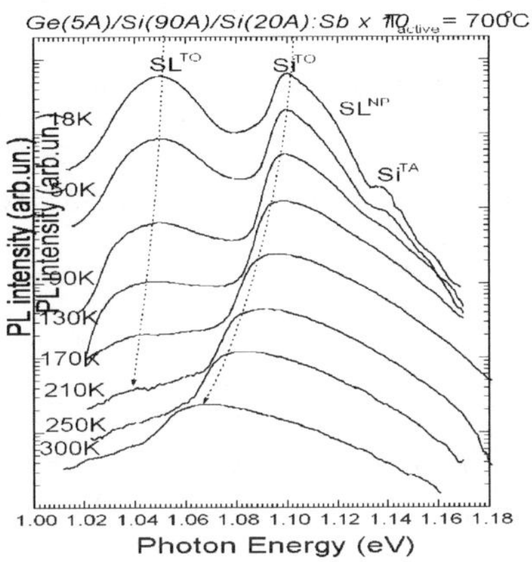

Figure 4. PL spectra of a multi-layer structure consisting of 10 periods of 3 ML Ge within 11nm Si spacers.
Growth temperture: 700°C.
The spectra were taken at different observation temperatures.

CONCLUSIONS

Structural and optical properties of Ge submonolayers embedded in a Si matrix were successfully grown by MBE. It is shown that there is a range of optimal growth conditions in which PL intensity increases due to the formation of Ge nanoscale islands. The islands with appropriate sizes appearance leads to the exciton formation and, at least, partial lifting of the *k*-selection rules. Thereby, the intensity of the luminescence (range of 1.06 eV) could be significantly increased.

ACKNOWLEDGMENTS

The authors thank G.Gerth and A.Frommfeld for their participation during the growth experiments and Dr. V.G.Talalaev for fruitful duscussions. G.E.Cirlin is grateful to the Alexander von Humboldt Foundation. This work was partially supported by the Ministry of Production and Science of the Russian Federation and St.Petersburg Scientific Center.

REFERENCES

1. Ch. Teichert, M.G. Lagally, L.J. Peticolas, J.C. Bean, and J. Tersoff, Phys. Rev. B **53**, 16334 (1996).
2. K. Eberl, O.G. Schmidt, R. Duschl, O. Kienzle, E. Ernst, Y. Rau, Thin Solid Films **369**, 33 (2000)
3. N. N. Ledentsov, J.Böhrer, M. Beer, F.Heinricchsdorff, M. Grundmann, D. Bimberg, S. V. Ivanov, B. Ya. Meltser, S. V. Shaposhnikov, I. N. Yassievich, N. N. Faleev, P. S. Kop'ev and Zh. I. Alferov, *Phys. Rev.* **B 52** 14058 (1995).
4. N. N. Ledentsov, I. L. Krestnikov, M. V. Maximov, S. V. Ivanov, S. L. Sorokin, P. S. Kop'ev, Zh. I. Alferov, D. Bimberg and C. M. Sotomayor Torres, *Appl. Phys. Lett.* **69** 1343 (1996).
5. M. V. Belousov, N. N. Ledentsov, M. V. Maximov, P. D. Wang, I. N. Yassievich, N. N. Faleev, I. A. Kozin, V. M. Ustinov, P. S. Kop'ev and Zh. I. Alferov, *Phys. Rev.* **B 51** 14 346 (1995).
6. G. E.Cirlin, P. Werner, U. Gösele, B.V. Volovik, V. M. Ustinov and N. N. Ledentsov, *Tech. Phys. Lett.* **27,** 14 (2001).
7. P. D. Wang, N. N. Ledentsov, C. M. Sotomayor Torres, P. S. Kop'ev and V. M. Ustinov, *Appl. Phys. Lett.* **64**, 1526 (1994).
8. G. E. Tsyrlin, A. O. Golubok, S. Ya. Tipissev and N. N. Ledenstov, *Semiconductors* **29**, 884 (1995).
9. G. E. Cirlin, V. N. Petrov, A. O. Golubok, S. Ya. Tipissev, V. G. Dubrovskii, G. M. Guryanov, N. N. Ledentsov and D. Bimberg, *Surf. Sci.* **377–379**, 895 (1997).
10. N. D. Zakharov, P. Werner, U. Gösele, G. Gerth, G. Cirlin, V. A. Egorov and B. V. Volovik, *Mat.Sci.Engeneering* **B 87,** 92 (2001).
11. R. Heitz, N.N. Ledentsov, D. Bimberg, A. Yu. Egorov, M. V. Maximov, V. M. Ustinov, A. E. Zhukov, Zh. I. Alferov, G. E. Cirlin, I. P. Soshnikov, N. D. Zakharov, P. Werner and U. Gösele, *Appl.Phys.Lett.* **74,** 1701 (1999).
12. B. V. Volovik, D. S. Sizov, A. F. Tsatsul'nikov, Yu. G. Musikhin, N. N. Ledentsov, V. M. Ustinov, V. A. Egorov, V. N. Petrov, N. K. Polyakov and G. E. Tsyrlin, *Semiconductors* **34,** 1316(2000).
13. S. S. Mikhrin, A. E. Zhukov, A. R. Kovsh, N. A. Maleev, V. M. Ustinov, Yu. M. Shernyakov, I. N. Kayander, E. Yu. Kondrat'eva, D. A. Livshits, I. S.Tarasov, M. V. Maksimov, A. F. Tsatsul'nikov, N. N. Ledentsov, P. S. Kop'ev, D. Bimberg ,Semiconductors **34**, 119 (2000).

Mat. Res. Soc. Symp. Proc. Vol. 717 © 2002 Materials Research Society

Ab-initio Calculations to Model Anomalous Fluorine Behavior

Milan Diebel[1] and Scott T. Dunham[2]
[1]Department of Physics, University of Washington, Seattle, WA 98195-1560, USA
[2]Department of Electrical Engineering, University of Washington, Seattle, WA 98195-2500, USA

ABSTRACT

Implanted fluorine has been observed to behave unusually in silicon, manifesting apparent uphill diffusion [1]. We are further motivated to understand the behavior of implanted fluorine in silicon by experiments which suggest that fluorine reduces boron diffusion [2, 3, 4, 5] and enhances boron activation in shallow junctions [2, 3]. In order to investigate fluorine behavior, we calculated the energy of fluorine defect structures in the framework of density functional theory (DFT). Besides identifying the ground-state configuration of a single fluorine atom in silicon, a set of energetically favorable fluorine defect structures were found. The latter strongly suggests a distinct fluorine diffusion mechanism, which was implemented in a continuum diffusion simulation and compared to experimental data.

INTRODUCTION

As ULSI devices enter the nanoscale, ultra-shallow junctions become necessary. Reduction in transient enhanced diffusion (TED) and enhanced dopant activation are desired. Experimentally, fluorine has shown to have beneficial properties on both boron TED reduction [2, 3, 4, 5] as well as boron activation [2, 3]. However, to utilize these benefits effectively, a fundamental understanding of F behavior is necessary, particularly since F shows anomalous diffusion behavior [1]. Figure 3 shows data reported by Jeng *et al.* In the experiment, 10^{13}cm^{-2} F^+ was implanted at 30 keV and annealed for 30 min at various temperatures.

The anomaly consists of two key features. At temperatures below 550°C, no noticeable F diffusion takes place. However, at higher temperatures, rapid F diffusion is reported. Such a behavior indicates the formation of strongly bound F complexes, since ab-initio calculations give a migration barrier of only 0.7-1.3 eV, which indicates that interstitial F (F_i) is a highly mobile species [8, 9]. The second part of the anomalous behavior is the shapes of the annealed profiles. Instead of a broadening, the annealed profiles sharpen and shift toward the surface.

RESULTS

The analysis is split into two sections. First, the ab-initio results are discussed. Then, a continuum model is developed using the first-principle data to compare to the anomalous fluorine behavior reported experimentally.

Ab-initio Calculations

In order to investigate fluorine behavior in silicon, the energies for various configurations were calculated using the DFT code VASP [10] with a PW91 GGA functional and ultra-soft

Structure	E_b last F [eV]	E_b^{tot} [eV]	E_f [eV]
FV	−2.38	−2.38	+1.00
F_2V	−2.25	−4.63	−1.25
F_3V	−1.95	−6.58	−3.20
F_4V	−0.54	−7.12	−3.74
V_2	—	−1.45	+5.31
FV_2	−2.75	−4.20	+2.56
F_2V_2	−2.87	−7.07	−0.31
F_3V_2	−1.97	−9.04	−2.28
F_4V_2	−2.43	−11.47	−4.71
F_5V_2	−1.82	−13.29	−6.53
F_6V_2	−1.80	−15.09	−8.33

Table 1: Binding energies of F_nV_m configurations. The decreasing binding energy of the F_nV structures is attributed to the increasing crowding of the F atoms. This phenomenon is illustrated in Fig. 1. The total binding energies (third column) are calculated with respect to interstitial fluorine (F_i) and single vacancies (V). The binding energies in the second column are the energy change in adding an additional F to the structure. The fourth column lists the total formation energy, which includes the formation energy of the necessary vacancies.

Vanderbilt-type pseudo-potentials [11]. All calculations were performed in a 64 atom supercell with periodic boundary conditions and $2 \times 2 \times 2$ Monkhorst-Pack **k**-point sampling with an energy cut-off of 320 eV. The structures were fully relaxed requiring a maximal force of 0.005 eV/Å per atom. In addition to fluorine structures, interstitial (I), and vacancy (V) ground-state energies were calculated as references. The calculated formation energies of I and V are 3.78 eV and 3.38 eV respectively.

In its ground-state, a single F atom prefers to reside in a bond center interstitial site. This site is preferred by 0.18 eV over the tetrahedral interstitial configuration and by 1.00 eV over the lowest substitutional site. The same reference structure was found by Taguchi *et al.* [9]. DFT calculations show interstitial F to be highly mobile ($E^m_{F_i^{bc}} \leq 1.3$ eV [9], $E^m_{F_i^{tet}} = 0.7$ eV [8]). However, we found F_nV_m structures to have a rather high binding energy, suggesting decoration of vacancies by fluorine. In Table 1, the formation and binding energies for different F_nV_m configurations are listed. Here the reference point for the total binding energy is interstitial fluorine and single vacancies.

For two or more F atoms, F_nV_m structures are favored over the interstitial configuration. For F_nV structures, the binding energy gained by adding an additional fluorine atom decreases. This behavior is due to the increasing space requirement of the decorating fluorine atoms. This becomes particularly apparent by comparing F_2V with F_3V in Fig. 1. Three fluorine atoms in F_3V are pushed away from each other and a distortion of the surrounding silicon lattice is noticeable. The rather large drop in marginal binding energy between F_3V and F_4V from 1.95 eV to 0.54 eV can be explained with the same argument, since F_4V shows an increasing distortion of the lattice and the bond length (see Fig. 1).

We also investigated F_nV_2 structures, which also show a reduction in binding energy with n, but no sharp drop-off, since more space is available to accommodate all F atoms.

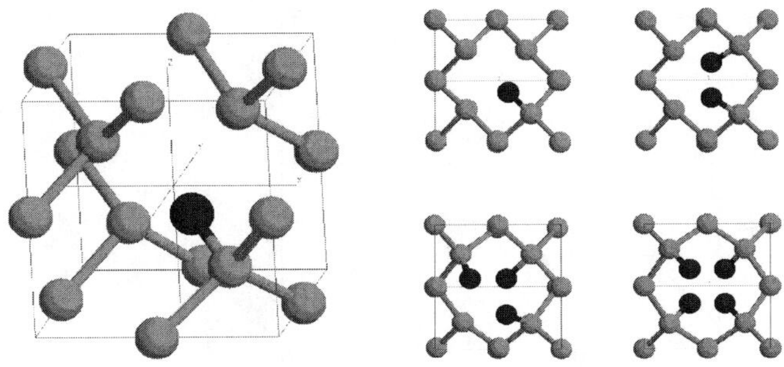

Figure 1: Left: 3D view of a single FV complex. The fluorine atom has moved toward the silicon out of the substitutional silicon site. The Si–F bond length is 1.68 Å (71% of the Si–Si bond length). The silicon atoms are drawn in amber (light), while the fluorine atom is presented in blue (dark). Right: View down the ⟨100⟩ direction (see Fig. 2 left) of the four F_nV structures. An increasing distortion of the surrounding silicon lattice is observed when adding more fluorine atoms. This is due to the repulsion of the F atoms, which becomes particularly apparent by comparing F_2V with F_4V.

The results are also listed in Table 1. In Fig. 2, the equilibrium concentrations at 600°C and 1000°C of F structures versus total F concentration are shown. In the presence of non-equilibrium point defect concentrations, the local equilibrium F_nV_m concentrations need to be multiplied by $(C_V/C_V^*)^m$. Thus, in the presence of excess vacancies during initial stages of implant anneals, almost all fluorine will reside in F_nV_m structures. The evolution of F_nV_m clusters is particularly important during implant anneals due to the high point defect concentrations.

The saturated F_6V_2 structure shows another interesting property: it is stable in the presence of interstitials. Table 2 lists the energy change associated with I reactions with F_nV_m structures. Migration barriers, which might further stabilize F_nV_m structures, are not included. Due to the entanglement of the fluorine atoms with the surrounding silicon lattice, we believe these defects to be immobile.

We also investigated pairing of F_i with I and F_i. DFT predicts interstitial F_2 to be unbound, as the fluorine prefers to remain in an interstitial site rather than forming an F_2 bound-state. F_i binds to I with an energy of 0.46 eV. This number was deduced by comparing FI to F_i and I separated within the same super-cell. This energy difference agrees well with a pure Coulombic model of FI binding as F^-I^+ (separating e^+ and e^- in a silicon environment an equivalent distance leads to an energy difference of 0.50 eV).

Continuum Model

To model fluorine diffusion in the continuum limit, we implemented fluorine decoration of V and V_2. All energetically favorable reactions are treated as diffusion limited and we

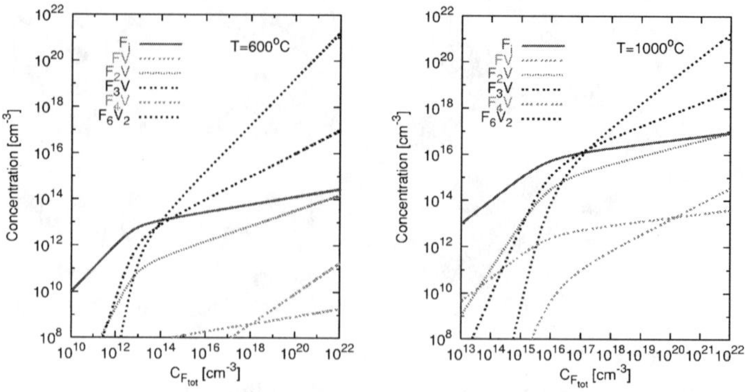

Figure 2: Equilibrium concentration of various F_nV_m structures vs. total F concentration at 600°C and 1000°C. Qualitatively the behavior for both temperatures is similar. For low $C_{F_{tot}}$ the dominant species is F_i due to the entropy of mixing. At high $C_{F_{tot}}$ the major F contribution comes from F_6V_2 clusters. The vacancy formation energy is included in these calculations.

considered only I, V, and F_i to be mobile in the Si phase. The associated formation and dissociation reactions are:

$$F_nV_m + F_i \Leftrightarrow F_{n+1}V_m \quad (1)$$
$$F_i + I \Leftrightarrow FI \quad (2)$$
$$F_nV + I \Leftrightarrow nF_i \quad (3)$$
$$F_nV_2 + I \Leftrightarrow F_4V + (n-4)F_i \quad (4)$$
$$FI + V \Leftrightarrow F_i \quad (5)$$

In addition, extended defect models, including I_n and V_n clusters up to size $n = 10$ plus {311} defects, are used [6]. The extended defect models are crucial for the correct temperature behavior. To match the experimental set up, a thin oxide surface layer of 20 Å is included in the simulation. To limit the number of free parameters, the oxide layer is treated in a very simple way. Only segregation at the interfaces and diffusion of F_i are considered in this phase.

Figure 3 shows the simulation results compared to the experimental data. Qualitatively the correct F behavior is observed, however the simulations show a somewhat stronger temperature dependence than observed experimentally. At 550°C, no significant movement is predicted, while at 850°C, F is entirely removed from silicon. A possible reason for the shortcomings at 550°C might be the fact that the model does not include mobile I_2, which are predicted by ab-initio calculations [7]. We expect I_2 to change the profiles especially in the I-rich tail regions, where mobile I_2 will annihilate V_n clusters, reducing F decoration. The discrepancy at higher T may be due to the fact that we only include V and V_2 decoration. The ab-initio calculations suggest that a complete model need to include also the decoration of larger V_n clusters, since large binding energies are anticipated. Larger more stable F_nV_m clusters can be expected to remain in the V-rich region near the surface. The F signal near the surface might be also due to knock-ons from F segregation to the Si/SiO_2 interface on

$F_nV + I \Leftrightarrow nF_i$	ΔE [eV]
$n = 1$	+4.78
$n = 2$	+2.53
$n = 3$	+0.58
$n = 4$	+0.04
$F_nV_2 + I \Leftrightarrow F_4V + (n-4)F_i$	ΔE [eV]
$n = 4$	+2.81
$n = 5$	+0.99
$n = 6$	−0.81

Table 2: Stability of the F_nV_m cluster in the presence of I. ΔE is defined as $E_{F_nV_m+I} - E_{rhs}$. Our calculations find that the fully saturated F_6V_2 structure is stable even at high I concentrations. This calculation does not include possible migration barriers, which further stabilize F_nV_m structures.

oxide film.

CONCLUSIONS

Ab-initio calculations predict strongly bound F_nV_m clusters. No other comparably stable structures were identified; the binding energy found for FI clusters is significantly smaller. This DFT data was used to identify the diffusion mechanism which was implemented in a continuum model to explain the anomalous fluorine behavior reported by Jeng *et al.* Fast diffusing F_i decorate V_n forming immobile F_nV_m clusters. At higher temperatures, these clusters get annihilated by I. Thus, F decoration of V leads to F dissolving from deeper regions (I excess) and accumulation near surface (V excess). Due to the strong affinity of F for V, for pre-amorphized samples we expect incorporation of F_nV_m clusters during regrowth. The consequences for boron diffusion in such an environment would be TED reduction and activation for amorphizing conditions due to excess V. However, in the case of non-amorphizing implants we expect an increased I concentration due to the formation of F_nV_m clusters, which leads to an enhancement of TED. The predictions for amorphizing conditions are supported by experimental data [2, 4, 5].

ACKNOWLEDGMENTS

This research was funded by the Semiconductor Research Corporation (SRC). The authors like to thank Intel Corporation for donation of a computing cluster used in this work and Pavel Fastenko for providing calibrated extended defect models.

REFERENCES

[1] S.-P. Jeng, T.-P. Ma, R. Canteri, M. Anderle, and G.W. Rubloff, Appl. Phys. Lett. **61**, 1310 (1992).

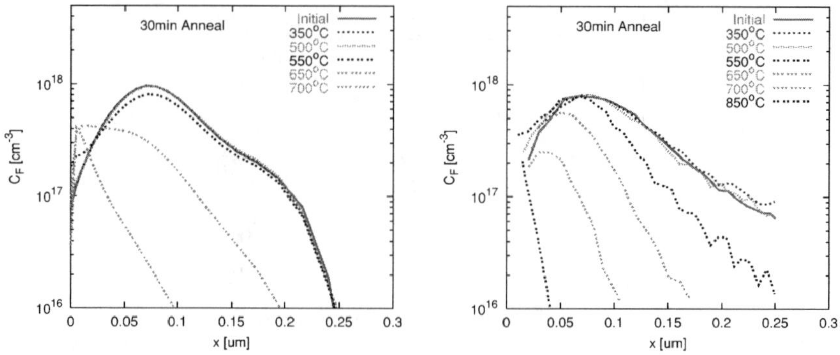

Figure 3: Comparison of simulation (left) with experimental data [1] (right) for 10^{13}cm^{-2} 30 keV F^+ implant annealed for 30 min at various temperatures. The simulation includes F_nV, F_nV_2, and FI clusters in addition to an extended defect model [6]. The initial defect/fluorine profiles were obtained with Monte Carlo implant simulator UT-Marlowe 6.0 and kinetic damage accumulation model.

[2] T.H. Huang and D.L. Kwong, Appl. Phys. Lett. **65**, 1829 (1994).

[3] J. Park and H. Hwang in *Si Front-End Processing: Physics and Technology of Dopant-Defect Interactions*, edited by H.-J. L. Gossmann, T.E. Haynes, M.E. Law, A.N. Larsen, and S. Odanaka, (Mater. Res. Soc. Symp. Proc. **568**, Warrendale, PA, 1999) pp. 71-75.

[4] D.F. Downey, J.W. Chow, E. Ishida, and K.S. Jones, Appl. Phys. Lett. **73**, 1263 (1998).

[5] L.S. Robertson, P.N. Warnes, K.S. Jones, S.K. Earles, M.E. Law, D.F. Downey, S. Falk, and J. Liu in *Si Front-End Processing: Physics and Technology of Dopant-Defect Interactions II*, edited by A. Agarwal, L. Pelaz, H-H. Vuong, P. Packan, M. Kase, (Mater. Res. Soc. Symp. Proc. **610**, Warrendale, PA, 2000) pp. B4.2.1-B4.2.6.

[6] P. Fastenko, S.T. Dunham, and S. Chakravarthi presented at the 2002 MRS Spring Meeting, San Francisco, CA, 2002 (published in this proceeding).

[7] S. K. Estreicher, M. Gharaibeh, P.A. Fedders, and P. Ordejón, Phys. Rev. Lett. **86**, 1247 (2001).

[8] C.G. Van de Walle, F.R. McFeely, and S.T. Pantelides, Phys. Rev. Lett. **61**, 1867 (1988)

[9] A. Taguchi and Y. Hirayama, Solid State Commun. **116**, 595 (2000).

[10] G. Kresse and J. Hafner, Phys. Rev. B **47**, RC558 (1993); G. Kresse and J. Furthmüller, **54**, 11169 (1996).

[11] D. Vanderbilt, Phys. Rev. B **41** 7892 (1990); G. Kresse and J. Hafner, J. Phys. Condens. Matter **6**, 8245 (1994).

Mat. Res. Soc. Symp. Proc. Vol. 717 © 2002 Materials Research Society C4.6

Effect of Fluorine on the Diffusion of Boron in Amorphous Silicon

J. M. Jacques, L. S. Robertson, and K. S. Jones
Department of Materials Science and Engineering, University of Florida, Gainesville, FL 32611

Joe Bennett
International SEMATECH, Austin, TX 78741

Mike Rendon
Motorola, Austin, TX 78741

Abstract

Fluorine and boron co-implantation within amorphous silicon has been studied in order to meet the process challenges regarding p^+ ultra-shallow junction formation. Previous experiments have shown that fluorine can reduce boron TED (Transient Enhanced Diffusion), enhance boron solubility and reduce sheet resistance. In this study, boron diffusion characteristics prior to solid phase epitaxial regrowth (SPER) of the amorphous layer in the presence of fluorine are addressed. Samples were pre-amorphized with Si^+ at a dose of $1x10^{15}$ ions/cm^2 and energy of 70 keV, leading to a deep continuous amorphous surface of approximately 1500 . After pre-amorphization, B^+ was implanted at a dose of $1x10^{15}$ ions/cm^2 and energy of 500 eV, while F^+ was implanted at a dose of $2x10^{15}$ ions/cm^2 and energies ranging from 3 keV to 9 keV. Subsequent furnace anneals for the F^+ implant energy of 6 keV were conducted at 550°C, for times ranging from 5 minutes to 260 minutes. During annealing, the boron in samples co-implanted with fluorine exhibited significant enhanced diffusion within amorphous silicon. After recrystallization, the boron diffusivity was dramatically reduced. Boron in samples with no fluorine did not diffuse during SPER. Prior to annealing, SIMS profiles demonstrated that boron concentration tails broadened with increasing fluorine implant energy. Enhanced dopant motion in as-implanted samples is presumably attributed to implant knock-on or recoil effects.

Introduction

As the semiconductor industry continues to scale silicon based microelectronic devices smaller and smaller, the pressure to achieve shallower junction depths becomes more significant. There has been increasing interest in the co-implantation of boron and fluorine for the formation of ultra-shallow junctions. The Transient Enhanced Diffusion (TED) of boron and other dopant species has been attributed as one of the primary limiting factors surrounding the formation of ultra-shallow, low resistivity junctions within the source and drain extension regions of transistors [1]. Several previous studies have described the unexpected ability of fluorine to significantly inhibit boron TED, while continuously increasing boron solubility limits during high temperature furnace annealing and rapid thermal annealing (RTA) [2-4]. These particular effects have been associated with the ability of fluorine to bind with excess silicon interstitials, thereby reducing the prevalence of boron cluster formation. However, these observations have only been studied within crystalline silicon. During conventional processing, implants are conducted within amorphous silicon, rather than crystalline silicon. Thus, in order to develop a succinct understanding of dopant diffusion processes, studies to determine whether any boron diffusion occurs within amorphous silicon are warranted. This study attempts to determine whether or not boron diffusivity is enhanced within amorphous silicon by the presence of fluorine and to elaborate upon characteristic fluorine effects.

Experimental

Several 200 mm, (100) n-type Czochralski silicon wafers were implanted using an Applied Materials Quantum Implanter for sub 5 keV implants and an Axcelis GSD 200E Implanter for the remaining implants. Wafers were first pre-amorphized with Si^+ at 70 keV and a dose of $1x10^{15}$ atoms/cm^2. The amorphized silicon wafers were then implanted with $^{11}B^+$ at 500 eV and a dose of $1x10^{15}$ atoms/cm^2. Selected wafers were subsequently implanted with F^+ at a dose of $2x10^{15}$ atoms/cm^2 and energies of either 3 keV, 6 keV, or 9 keV. A JEOL 200CX TEM was used to perform cross-sectional transmission electron microscopy (XTEM). XTEM analysis confirmed the generation of a continuous surface amorphous layer of approximately 1500 . The post-implant annealing occurred within a tube furnace in a N_2 ambient at 550°C and 600°C to induce regrowth of the amorphous layer. Samples were annealed at 550°C for times ranging up to 260 minutes, while sample anneal times at 600°C were 2 minutes and 8 minutes. In order to measure amorphous layer depths, measurements were performed upon a Variable Angle Multi-Wavelength Ellipsometer Control Module EC110 (VASE), equipped with a 75W Xenon Light Source LPS-300, at angles of 65°, 70°, and 75°. VASE measurements were verified through XTEM. Secondary Ion Mass Spectroscopy (SIMS) profiles were generated using a Physical Electronics ADEPT-1010 quadruple SIMS instrument. The $^{11}B^+$ counts were obtained using an O_2^+ beam with a beam current of 180 nA and a net impact energy of 600 eV. The beam was continually rastered over a (425 μm x 425 μm) square area, at an angle of 45° under a uniform O_2 ambient of $6x10^{-7}$ torr. The detectable area of the sample was approximately (70 μm x 70 μm) square in dimension.

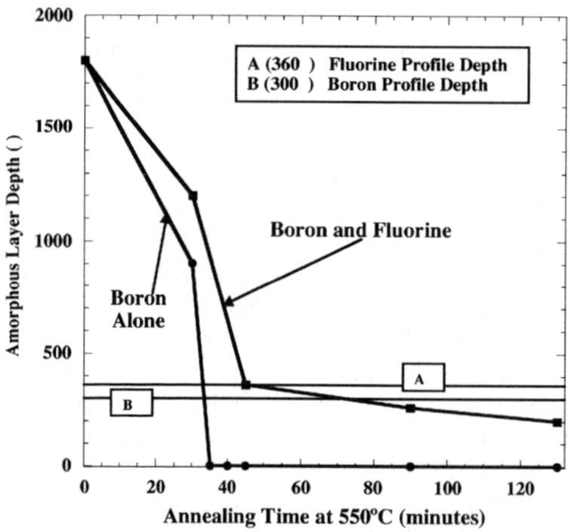

Figure 1: Amorphous Layer Depth as a Function of Annealing Time at 550°C (As-implanted to 130 minutes).

Results and Discussion

In order to study the diffusion characteristics of boron within amorphous silicon, samples were furnace annealed at 550°C. By generating a deep amorphous layer of 1500 , the time required for SPER to commence at this low temperature was greatly increased. Figure 1 shows the SPER characteristics at 550°C. Within this plot, two types of samples are compared, namely those implanted with boron alone and those containing boron co-implanted with fluorine. Figure 1 shows that the samples with boron alone experience an increase in the silicon regrowth velocity once the amorphous-crystalline interface reaches the boron profile. However, the regrowth velocity slows down upon reaching the fluorine profile within the

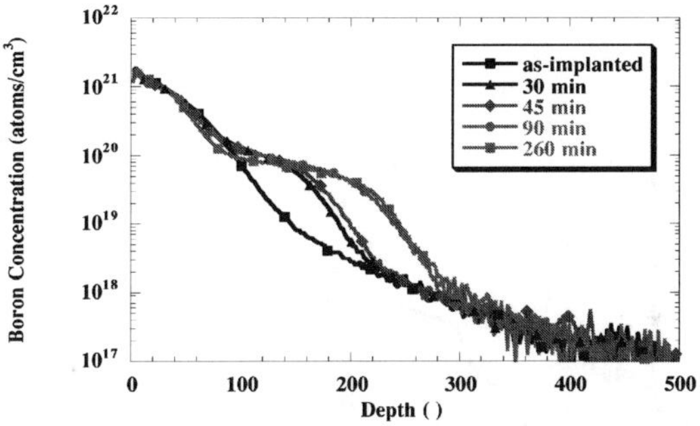

Figure 2:* *Boron Concentration Profiles as a Function of Annealing Time at 550°C (As-implanted to 260 minutes).

samples where boron was co-implanted with fluorine. These observations are consistent with previous studies [5-7]. Based upon this plot we can conclude that SPER ends between 30 minutes and 35 minutes of annealing at 550°C for samples containing boron alone. For samples co-implanted with boron and fluorine, on the other hand, SPER ends after more than 130 minutes of annealing at 550°C. Thus, any boron diffusion observed within the first 30 minutes of annealing at 550°C in the presence of fluorine will be indicative of diffusion within silicon while in the amorphous state.

Figure 2 depicts the boron concentration profiles for samples co-implanted with fluorine. Samples were annealed at 550°C for various times up to 260 minutes. In the presence of fluorine, boron clearly experiences significant enhanced diffusion within the first 30 minutes of annealing at 550°C. At a boron concentration of $2x10^{19}$ atoms/cm^3, the profile shifts by almost 50 , which is nearly a 40% increase in width. In direct contrast, however, no significant boron motion was observed during the130 minute anneal for samples containing boron alone. Figure 2 shows that the fluorine enhanced boron diffusion appears to subside between 45 minutes and 90 minutes. This corresponds well with the data presented in Figure 1. From Figure 1 we ascertain that the amorphous-crystalline interface reaches the boron profile after approximately 70 minutes of annealing, at a depth of about 300 . Based upon these correlations, it is evident that the fluorine enhanced boron diffusion occurs only within amorphous, rather than crystalline silicon.

Figure 3 shows the dopant motion versus annealing time at 550°C for boron at a concentration of $2x10^{19}$ atoms/cm^3. Here, we observe no measurable dopant motion for the sample containing boron alone during the SPER process. However, the samples co-implanted with boron and fluorine exhibit significant boron dopant motion during SPER. During the first 30 minutes of annealing alone, the boron diffused approximately 50 in the presence of fluorine. It is clear that fluorine significantly enhances boron diffusivity within amorphous silicon, with this enhancement substantially subsiding once the amorphous-crystalline interface reaches the boron profile.

Figure 4 depicts the time averaged boron diffusivity at 550°C within amorphous silicon during SPER. (This figure has a shorter time scale of only up to 45 minutes compared to Figure 3 in order

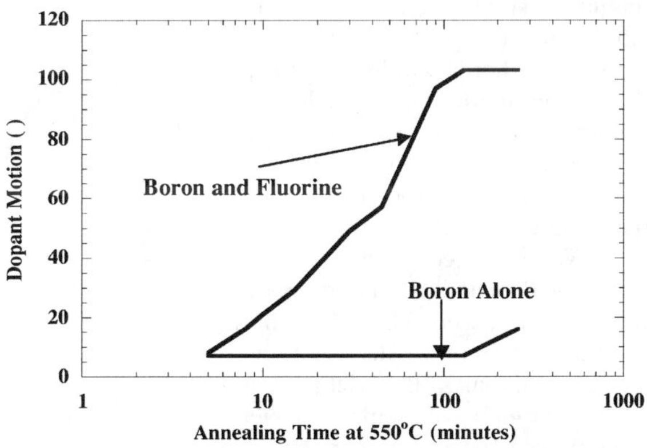

Figure 3:* *Boron Dopant Motion as a Function of Annealing Time at 550°C (As-implanted to 260 minutes).

to isolate diffusion to that within amorphous silicon.) As seen here, the boron diffusivity slightly increases with time, stabilizing to a value of $3x10^{-17}$ cm^2/sec after approximately 15 minutes of annealing. Derived based upon a half Gaussian redistribution of boron dopant at a concentration of $3x10^{19}$ atoms/cm^3, this value is five orders of magnitude larger than the extrapolated equilibrium diffusivity of boron in crystalline silicon [8]. It is interesting to note that the overall boron diffusion profiles exhibited within Figure 5 resemble those of similarly implanted samples after high temperature spike annealing. Based upon these results, it is postulated that fluorine enhanced boron diffusion in amorphous silicon, as part of SPER during RTA ramp-ups, may comprise a significant fraction of the total dopant diffusion after high temperature RTA.

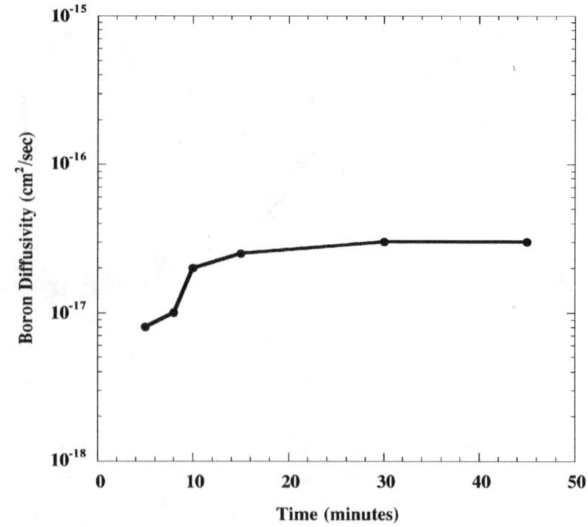

Figure 4: Time-Averaged Boron Diffusivity at 550°C Within Amorphous Silicon During SPER (As-implanted to 45 minutes).

Over the years, the relationship between fluorine and silicon has been extensively investigated. It has been proposed that significant interactions occur between fluorine atoms and dangling or strained silicon bonds, whereby fluorine effectively ties up these bonds [9-11]. Frentrup et al. [9] studied structural changes at the fluorine 1s edge, Ohyu et al. [10] analyzed boron dopant activation levels, and Wong et al. [11] utilized ESR measurements and spin density calculations to collectively investigate the interactions between fluorine and silicon. These studies collaboratively suggest that fluorine has the ability to passivate dangling silicon bonds in amorphous silicon. It is ascertained that by systematically reducing the number of probable interactions between boron atoms and dangling or strained silicon bonds, the fluorine can effectively increase the diffusivity of boron within amorphous silicon. The fluorine is able to decrease the formation energy required for boron diffusion by decreasing the overall dangling or strained silicon bond concentration.

Within the course of these studies, we observed an additional boron motion phenomenon. Figure 5 contains a plot of boron concentration versus depth for as-implanted samples with variable fluorine implant energies. The samples depicted here first received a silicon pre-amorphization implant, followed by a

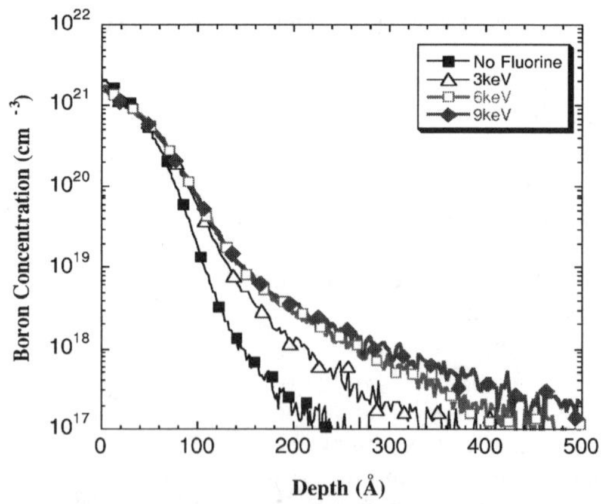

Figure 5: Boron Concentration Profiles for As-Implanted Samples After Variable Energy Fluorine Implants.

boron implant, and in some cases fluorine implants. The fluorine implants were conducted at energies ranging from 3 keV to 9 keV. As seen here, substantial boron motion results after the subsequent fluorine implants, which is depicted by significant broadening of the boron concentration tails. At a boron concentration of $1x10^{18}$ atoms/cm^3, the tails broadened by as much as 100 , denoting a significant boron motion characteristic within as-implanted samples.

Based upon the results presented within Figure 5, it seems logical to also ascertain whether or not the ordering of the implants affects the as-implanted profiles. Figure 6 contains a plot of boron concentrations versus depth for samples with alternating implant orders of 6 keV fluorine. At boron concentrations above $4x10^{18}$ atoms/cm^3, the fluorine + boron sample appears to be shallower than the boron control sample by approximately 15 . This anomaly is attributed to SIMS error commonly encountered at very shallow sample depths. Nonetheless, it is apparent that the boron + fluorine sample exhibits a significantly deeper boron profile than the fluorine + boron sample. The two profiles differ by as much as 50 . Thus, the observed boron motion within as-implanted samples is far less pronounced when the boron is implanted last, after the fluorine.

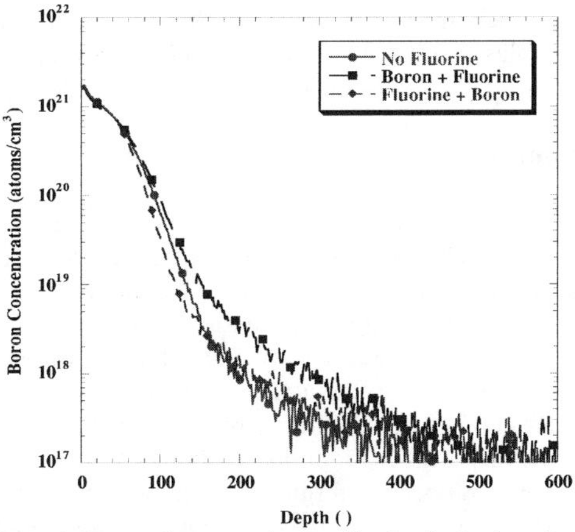

Figure 6: Boron Concentration Profiles for As-Implanted Samples with Alternating Implant Ordering.

Since the samples presented within both Figures 5 and 6 were not annealed, the boron movement is most likely due to silicon recoil or knock-on effects during the fluorine implant process. Based upon SRIM simulations conducted under similar implant conditions, only a small percentage of the total silicon and boron target atoms were affected by recoils during the fluorine implantation process [12]. A significant broadening of the boron concentration profile primarily occurs in the tail region, where only a small percentage of the overall boron profile resides. By implanting the boron before the fluorine, the boron atoms act as target atoms during the fluorine implant and are most susceptible to silicon recoil effects. Thus, it seems plausible that the observed boron dopant motion is indeed the direct result of implant recoil effects.

In order to investigate possible implant ordering effects on boron diffusion, samples were furnace annealed at 600°C. Anneals were conducted at times of 2 minutes and 8 minutes. The boron diffusion characteristics are plotted within Figure 7. As seen here, appreciable boron diffusion occurs within both samples over the anneal

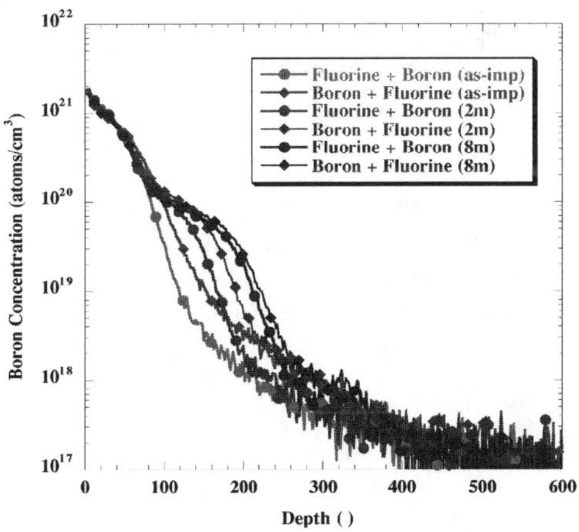

Figure 7: Boron Concentration Profiles as a Function of Annealing Time at 600°C (As-implanted to 8 minutes).

times. In spite of the shallower initial boron profiles, the fluorine + boron profiles begin to approach those of the boron + fluorine samples at longer anneal times. Additional studies of samples with variant fluorine implant energies and anneal temperatures are needed in order to accurately address the effect of implant ordering upon boron diffusion.

Conclusions

Based upon these studies, we have drawn numerous conclusions. The fluorine induced boron enhancement subsides once SPER commences, whereby it has been shown to inhibit subsequent dopant diffusion within crystalline silicon. Rather unexpectedly, fluorine significantly enlarges boron diffusivity within amorphous silicon by 5 orders of magnitude above the extrapolated diffusivity in crystalline silicon at 550°C. It is proposed that through the reduction of dangling and strained bonds within amorphous silicon, fluorine enhances boron diffusion by reducing the formation energy for mobile boron. Knock-on or recoil effects appear to be responsible for enhanced boron motion within as-implanted samples containing co-implanted boron and fluorine. These effects are amplified when boron is implanted before fluorine, as target atoms are recoiled more than incoming ions, and appears to be fluorine implant energy dependent. Clearly, fluorine contributes to numerous boron diffusion and motion characteristics within silicon based microelectronic materials.

References:

1. J.F. Ziegler, ed. Ion Implantation Science and Technology 2000 Ed. New York: Ion Implantation Technology Co., 2000, pg. 283-291.
2. D.F. Downey, J.W. Chow, E. Ishida, & K.S. Jones, Appl. Phys. Lett. **73**, pg. 1263 (1998).
3. K. Ohyu, T. Itoga, and N. Natsuaki, 20th Symp. on Ion Implantation and Submicron Fabrication, RIKEN (1989), p. 149-152.
4. L.S.°Robertson, P.N. Warnes, K.S. Jones, S.K. Earles, M.E. Law, D.F. Downey, S. Falk, and J. Liu. Si Front-End Processing - Physics and Technology of Dopant-Defect Interactions II. Symposium (Materials Research Society Proceedings Vol.610), 2001, Page B4.2.1-6.
5. I. Suni, G. Goltz, M.G. Grimaldi, M.-A. Nicolet, and S.S. Lau, Appl. Phys. Lett. 40 (3), pg. 269 (1982).
6. I. Suni, G. Goltz, M.-A. Nicolet, and S.S. Lau, Thin Solid Films 93, pg. 171 (1982).
7. I. Suni, U. Shreter, M.-A. Nicolet, and J.E. Baker, J. Appl. Phys. 56 (2), pg. 273 (1984).
8. Y. Haddara, B.T. Folmer, M.E. Law, T. Buyuklimanli, *Applied Physics Letters*, **77**(13), pg.1976 (2000).
9. W. Frentrup and A. Mertens, Physica B 208 & 209, pg. 389 (1995).
10. K. Ohyu, T. Itoga, and N. Natsuski, Japanese J. Appl. Phys. Vol. 29, pg. 457 (1990).
11. S. P. Wong, M. C. Poon, H. L. Kwok, and Y. W. Lam, J. Electrochem. Soc.: Solid-State Science and Technology, Vol. 133 (10), pg. 2172 (1986).
12. J.F. Ziegler and J.P. Biersack. The Stopping and Range of Ions in Matter (SRIM-2000.4). Maryland: IMB Co., 1999.

Mat. Res. Soc. Symp. Proc. Vol. 717 © 2002 Materials Research Society C4.7

Ab-Initio Pseudopotential Calculations of Phosphorus Diffusion in Silicon

Xiang-Yang Liu[1], Wolfgang Windl[2], and Michael P. Masquelier[1]
[1]Computational Nanoscience Group, Physical Sciences Research Labs, Motorola, Inc., Los Alamos, NM 87545, U.S.A.
[2]Department of Materials Science and Engineering, Ohio State University, Columbus, OH 43210, U.S.A.

ABSTRACT

In traditional models of P diffusion in Si, vacancy assisted diffusion mechanism has been assumed. More recently, experiments have determined that for intrinsic P diffusion in Si, the interstitial assisted diffusion mechanism dominates. We have performed *ab-initio* pseudopotential calculations to study P diffusion in Si. Special care is taken with regard to structure minimization, charge state effects and corrections. We calculated the defect formation energies and migration barriers for the various competing P-interstitial complex diffusion mechanisms for low concentration P region, as well as the energetics of different charge states P-vacancy complex diffusion. For interstitial mediated diffusion of P-Si_i pair in Si, we find the overall diffusion activation energies calculated are 3.1 eV for neutral case, and 3.4 eV for +1 charge case. This is in agreement with experimental observation that the interstitial mechanism dominates for intrinsic P diffusion in Si. For vacancy mediated diffusion, our calculations are in agreement with previous calculations result in the neutral case. We obtained the lower bounds for diffusion activation energy of 3.8 eV for $(PV)^0$ and 3.4 eV for $(PV)^-$. A further evaluation of the numbers would require a proper treatment of the energy states in the band-gap due to Jahn-Teller relaxations.

INTRODUCTION

Dopant diffusion is an elementary process in ultra-large scale integrated (ULSI) device fabrications. Phosphorus is a common dopant for the fabrication of n or n+ regions in Si devices. Early experimental studies and theoretical analysis on P diffusion in Si associated the P diffusion with P-vacancy (V) complex formation and migration in Si [1]. Later, ample experimental evidence show that P diffusion is mediated by interstitials (I) in intrinsic region [2, 3]. For example, most recent experiments based on the injection of point defects into a P doped Si substrate suggests that the fractional contributions of I mechanism to P diffusion in Si is in the range of 86% - 100% in the worst case analysis [3]. While experiments are able to find the total diffusion coefficient and the influence of point-defects on it, they generally cannot determine the microscopic diffusion mechanisms in the physical processes involved. Several modeling efforts have been made to simulate the transient enhanced diffusions (TEDs) of P diffusion in Si using a combination of I and V mechanisms based on fitting to experimental P diffusion profiles in Si [4]. In this regard, *ab-initio* calculations of the fundamental parameters in the diffusion processes contribute to our knowledge and understanding, and modeling of defects influences on dopant diffusions.

Previous *ab-initio* calculations that are relevant to P diffusion in Si have been done before [5, 6]. The works have been on the neutral species, and on either dopant-vacancy mechanism only [6] or only calculations of the dopant-defect formation energies (because of limited computational power at the time) [5]. In this work, we studied in detail both the P-interstitial and –vacancy mechanisms using *ab-initio* pseudopotential calculations. Special care is taken with regard to structure minimization, charge state effects and corrections.

METHODS

For our calculations, we used the efficient plane-wave DFT code VASP (Vienna *Ab initio* Simulation Package) [7] with ultra-soft pseudopotentials within GGA. Cut-off energy of 173.5 eV and 64-atom supercells were used. The supercell calculations were done with optimized Si lattice constant of 5.4617 Å. k-point sampling was 2x2x2 in Monkhorst-Pack meshes and the atoms were fully relaxed to reach criteria that the magnitude of forces on each atom was less than 0.05 eV/Å. Electronic states were occupied according to a Methfessel-Paxton distribution with a smearing width of 0.05 eV [8]. Furthermore, real-space projection scheme was used for efficient computations.

For the charged state, a neutralizing background charge was imposed automatically in VASP calculations. Special care was taken to ensure a correct treatment of the charged state total energy result. First, we identify the valence band reference energy for the extra/missing electron E_V. This is determined as the band energy of the top of the valence band, which is the Γ point in the Si band structure. The Fermi level E_F is also known to cause a shift to the charged defect in total energy. We define the Fermi level at the top of the valence band to be zero. To compute the energy of a system with charge Q, $E = E_{total} + E_V Q + E_F Q$, where E_{total} is the directly computed total energy of the charged defect, and E is the corrected total energy of the charged defect taking into extra/missing electron and Fermi level effect. We also apply corrections to monopole interaction, and for cases where the electrons are excited to the conduction bands and finite size corrections [9].

For migration calculations, we used the nudged elastic band method (NEBM) implemented into VASP to determine the diffusion paths and migration barriers [10]. The NEBM starts from a chain of geometries interpolating between the initial and final geometries. Then the atomic configurations in the different geometries are iteratively optimized using the ionic forces components that perpendicular to hyper-tangent. The chain optimization steps are repeated and the energy along the path is determined by spline interpolation based on the total energy of all geometries in the path and the tangential projection of the 3N force on each geometry. A systematic approach in migration path finding like NEBM is expected to be more reliable than dragging an atom from minimum to minimum across the estimated or guessed saddle, especially in the cases with multi-atoms migration event. This has been proven to be useful in previous modeling work of B diffusion that involved two of the authors of this paper [9].

RESULTS

P-Interstitial Diffusion

We carried out the P-interstitial defects formation study first. For neutral case, the lowest energy structures of P-I defect found are [110] dumbbell-like interstitial and a structure with ~ C_2 symmetry where the P atom is in an interstitial position making four-fold bonds with nearby Si atoms. These two structures are degenerate in energy. We name these two structures $P_i(X)^0$ [Fig. 1(a)] and $P_i(X2)^0$ [Fig. 1(b)] respectively. The $P_i(X2)^0$ structure was found during the structural relaxation of P atom in tetrahedral *(T)* interstitial site. It could be viewed as a transformation of $P_i(X)^0$ via local structural changes, but has a different symmetry property. If we put P atom in a bond-centered interstitial site ("*B*" interstitial) in Si lattice, the structure relaxes to the similar structure. The formation energy of $P_i(X)^0$ and $P_i(X2)^0$ with respect to reference state of lowest energy P charge state, P_s^+, is 3.5 eV - E_F. This results in binding energy of 1.2 eV relative to dissociation products $Si_i(X)^0$ ([110] dumb-bell like structure of Si interstitial) and P_s^0 (substitutional P) and 1.5 eV relative to dissociation products $Si_i(X)^-$ and P_s^+. The lowest energy structure found in the B-Si_i pair neutral case, B_s-$Si_i(T_{1n})^0$ [9] (T_{Nn} denotes a tetrahedral interstitial in *Nth* neighbor position with respect to B_s [5]), if B atom is substituted with P atom, the structure will be dissociated to Si_i hexagonal *(H)* interstitial and a substitutional P_s in Si, separated by 3.07 Å in distance. For the +1

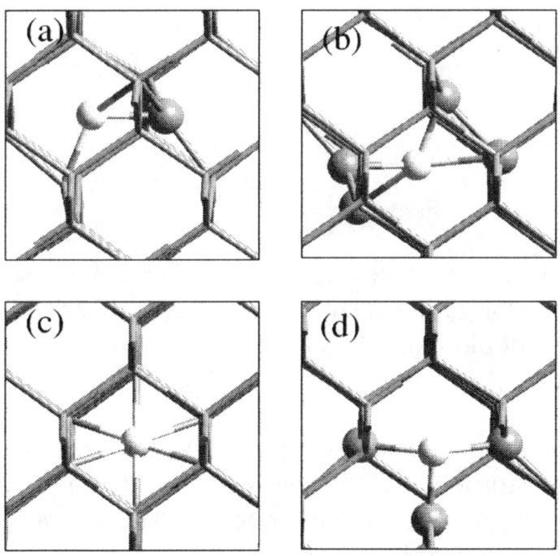

Fig. 1 Lowest energy structures of P-Si_i pair in different charge states: (a) $P_i(X)$, (b) $P_i(X2)$, (c) $P_i(H)$, and $P_i(S)$. The small light gray ball is B atom, the large dark gray balls are Si atoms involved in the cluster. All other Si atoms are shown as stick-only network.

charge system, we identified two lowest energy structures too. These are $P_i(H)^+$ (P interstitial in H interstitial site, Fig. 1(c)) and $P_i(S)^+$ ([100] dumbbell-like shared interstitial between P and Si atoms, Fig. 1(d)) with which structure the B-Si_i pair was found to play a central role in B transient-enhanced diffusion (TED) [9]. The binding energies of $P_i(H)^+$ and $P_i(S)^+$ with respect to dissociation products of $Si_i(X)^0$ and P_s^+ are 0.6 eV and 0.5 eV respectively. For the −1 charge system, we find the lowest energy structure of $P_i(X2)^-$ with binding energy of 2.0 eV with respect to dissociation products of $Si_i(X)^-$ and P_s^0. The reason of such a high binding energy is due to the high energy states of $Si_i(X)^-$ and P_s^0 with which both have one electron in the conduction band in their ground states. The Fermi-level dependence of formation energies of different charge states of P-Si_i pair in their lowest energy structures with respect to reference state P_s^+ is shown in Fig. 2.

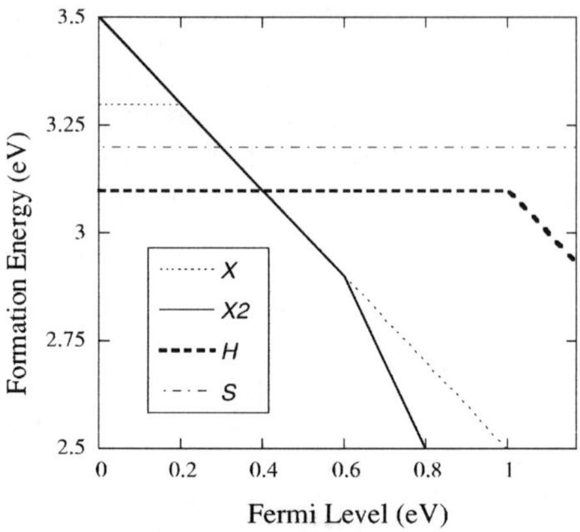

Fig. 2 Formation energies for P-Si_i complexes with respect to reference state P_s^+ as a function of the Fermi level. For charge states where the structure changes (to another structure), the charge states are not plotted.

We studied the diffusion mechanisms next. For the neutral case, since $P_i(X2)^0$ can be viewed as a local structural distortion of $P_i(X)^0$, we start from $P_i(X)^0$ which has the lowest formation energy. We find the P atom to migrate to H site with a migration barrier of 0.6 eV. From this H site to another H site, two transition states of $P_i(X)^0$ are in the migration path. At the second H site, P interstitial is out of the local diffusion range from the original $P_i(X)^0$ position. A long range diffusion is possible with migration from the second $P_i(H)^0$, from which another $P_i(X)^0$ configuration can be accessed without barrier. Another diffusion path we find involves $P_i(S)^0$ configuration although it's not stable in the neutral case by its own. The migration paths are $P_i(X)^0 \rightarrow P_i(S)^0 \rightarrow P_i(X2)^0 \rightarrow P_i(X)^0 \rightarrow P_i(X2)^0 \rightarrow P_i(H)^0$ with maximum barrier height of 0.7 eV. We also noticed that from $P_i(X2)^0 \rightarrow P_i(X)^0$, it is a migration path with very small barrier (equal or less than 0.1 eV). For the charged species, we find $P_i(H)^+ \rightarrow P_i(X^*)^+ \rightarrow P_i(H)^+$, with a barrier 0.3 eV, where $P_i(X^*)^+$ is a symmetric variation of $P_i(X)^+$; and $P_i(X2)^- \rightarrow P_i(H)^- \rightarrow P_i(X2)^-$ with a barrier of 1.4 eV. For the neutral case, if we allow the charge variation of the saddle point (lose electron during migration), the migration barrier will be reduced to 0.2 eV. The overall diffusion activation energies calculated from the sum of formation energy and migration energy for P-Si_i pair are thus 3.1 eV for neutral case, and 3.4 eV for +1 charge case. This compares well with experimentally measured activation energy range of 2.81 (low temperature) to 3.61 eV (high temperature) for the intrinsic diffusivities of P in Si [11].

P-Vacancy Diffusion

The dopant vacancy complex diffusion involves both direct exchange of dopant and vacancy mechanism and vacancy ring migration mechanism in which the vacancy migrates through the lattice sites in the ring to another neighbor site of the dopant. For neutral P-vacancy migration, we obtained 1.3 eV and 0.8 eV for direct change and ring migration respectively. This is in close agreement with a previous calculation result for the total exchange barrier of 1.05 eV [6]. For –1 charge, migration barriers calculated are 1.2 eV and 0.9 eV for direct change and ring migration respectively. In the calculations of formation energies of P-vacancy complexes, we used a larger 216 atoms supercell to allow easier Jahn-Teller relaxations. The isolated vacancy formation energy calculated is 3.69 eV, same as the result in a previous calculation [6]. The formation energies obtained for $(PV)^0$ and $(PV)^-$ with respect to reference state P_s^+ are 2.5 eV (at mid-gap) and 2.2 eV (at mid-gap) respectively. This yields the overall diffusion activation energy of 3.8 eV for $(PV)^0$ and 3.4 eV for $(PV)^-$. The neutral case number obtained is in good agreement with previous calculations [6]. With the much smaller band-gap introduced by LDA/GGA calculations, the energy states in the band-gap due to Jahn-Teller relaxations are lower than they should be. So the activation energies computed for PV complexes diffusion are the lower bounds only. A further evaluations of the numbers would require a proper treatment of the energy states in the band-gap due to Jahn-Teller relaxations.

CONCLUSIONS

We have performed *ab-initio* pesudopotential calculations to study P diffusion in Si. Structural exploration, charge effects, and charge corrections are paid attention in our calculations. For interstitial mediated diffusion of P-Si_i pair in Si, we find the overall diffusion activation energies calculated are 3.1 eV for neutral case, and 3.4 eV for +1 charge case. This is in agreement with experimental observation that the interstitial mechanism dominates for intrinsic P diffusion in Si. For vacancy mediated diffusion, our calculations are in agreement with previous calculations result in the neutral case. We obtained the lower bounds for diffusion activation energy of 3.8 eV for $(PV)^0$ and 3.4 eV for $(PV)^-$. A further evaluation of the numbers would require a proper treatment of the energy states in the band-gap due to Jahn-Teller relaxations.

ACKONLEDGEMENT

We thank Dr. Keith Beardmore at Motorola Computational Nanoscience Group for helpful discussions.

REFERENCES

1. R. B. Fair and J. C. C. Tsai, *J. Electrochem. Soc.* **124**, 1107 (1977).
2. P.Fahey, P. B. Griffin, and J. D. Plummer, *Rev. Mod. Phys.* **61**, 289 (1989), and references therein. See also, H. Bracht, *MRS Bulletin*, June 2000, p. 22-27.
3. A. Ural, P. B. Griffin, and J. D. Plummer, *J. of Appl. Phys.* **85**, 6440 (1999).
4. M. Uematsu, *J. Appl. Phys.* **82**, 2228 (1997); S. T. Dunham, *J. Electrochem. Soc.* **139**, 2628 (1992).
5. C. S. Nichols, C. G. Vande Valle, and S. T. Pantelides, *Phys. Rev. B,* **40**, 5484 (1989).
6. J. S. Nelson, P. A. Schultz, and A. F. Wright, *Appl. Phys. Lett.* **73**, 247 (1998).
7. G. Kresse and J. Hafner, *Phys. Rev. B* **47**, 558 (1993); **49**, 14251 (1994); G. Kresse and J. Furthmüller, *Comput. Mater. Sci.* **6**, 15 (1996); *Phys. Rev. B* **55**, 11169 (1996).
8. M. Methfessel and A. T. Paxton, *Phys. Rev. B* **40**, 3616 (1989).
9. W. Windl, M. M. Bunea, R. Stumpf, S. T. Dunham, and M. P. Masquelier, *Phys. Rev. Lett.* **83**, 4345 (1999).
10. H. Jónsson, G. Mills, and K. W. Jacobsen, in *Classical and Quantum Dynamics in Condensed Phase Simulations*, edited by B. J. Berne, G. Ciccotti, and D. F. Coker (World Scientific, Singapore, 1998), p. 385.
11. Y. M. Haddara, B. T. Folmer, M. E. Law, and T. Buyuklimanli, *Appl. Phys. Lett.* **77**, 1976 (2000).

Mat. Res. Soc. Symp. Proc. Vol. 717 © 2002 Materials Research Society

Modeling of TED and Point Defect Parameter Extraction

Heidi Meyer[1] and Scott T. Dunham[1,2]
[1]Department of Materials Science and Engineering, University of Washington, Seattle, WA 98195, USA
[2]Department of Electrical Engineering, University of Washington, Seattle, WA 98195, USA

ABSTRACT

This work investigates simple transient enhanced diffusion (TED) behavior, which is a reflection of the interstitial behavior in the system. The analysis shows that TED depends mainly on two factors: the initial damage profiles and the $D_I C_I^*$ product. We find that, based on these two inputs, the extent of TED can be accurately calculated. Alternatively, by comparison to experiments, we calculate the interstitial diffusion capacity ($D_I C_I^*$) which is compared to values previously extracted from metal diffusion and silicon self-diffusion experiments.

INTRODUCTION

Ion implantation is an effective and popular method for introducing dopant atoms into silicon [1]. However, this process causes considerable damage due to high energy collisions of the implanted ions with the stationary silicon lattice atoms. A subsequent annealing step is performed in order to repair this implant damage, restoring the lattice back to its crystalline state. During annealing, transient enhanced diffusion (TED) occurs, a burst of diffusion many times faster than equilibrium diffusion [2].

Understanding the damage evolution following ion implantation is crucial for accurately modeling TED. In this work, we use a simple model based on the diffusion capacity and damage profiles to explain TED behavior. We then analyze marker layer experiments to extract $D_I C_I^*$ and compare the values to previous reports.

MODELING

It is possible to define the extent of TED in terms of the equivalent time under equilibrium conditions which produces the same amount of diffusion:

$$t_{\mathrm{TED}}^{\mathrm{eff}} = \int_0^\infty \left(\frac{D}{D^*} - 1\right) dt = \int_0^\infty \left[f_I \left(\frac{C_I}{C_I^*} - 1\right) + f_V \left(\frac{C_V}{C_V^*} - 1\right)\right] dt \cong f_I \int_0^\infty \left(\frac{C_I}{C_I^*} - 1\right) dt, \quad (1)$$

where D is the diffusion coefficient of the dopant under investigation, D^* is the equilibrium diffusivity, f_I and f_V are the fractions of diffusion attributable to diffusion by the interstitial and vacancy mechanisms respectively, C_I is the concentration of the interstitials, and C_I^* represents the equilibrium value of interstitials. Because $C_I \gg C_I^*$ during TED, the interstitial mechanism dominates.

Assuming first a simple delta function approximation for the damage profile along with rapid interface regrowth (Fig. 1), a simple relationship between excess interstitial dose (Q) and interstitial supersaturation results:

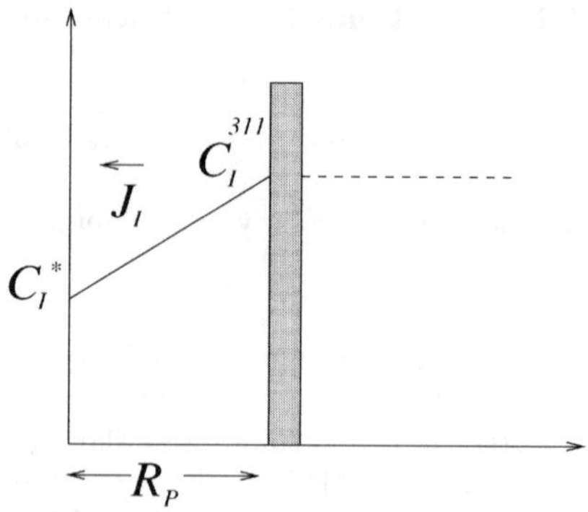

Figure 1: A simple picture of TED, where the excess interstitials from implant damage are approximated using a delta function.

$$Q = \int_0^\infty J_I(t)dt = \int_0^\infty \frac{D_I\left[C_I^{311}(t) - C_I^*\right]}{R_P}dt = \frac{D_I C_I^*}{R_P}\int_0^\infty \left(\frac{C_I(t)}{C_I^*} - 1\right)dt = \frac{D_I C_I^*}{R_P}\left(\frac{t_{\text{TED}}^{\text{eff}}}{f_I}\right) \quad (2)$$

where R_P represents the projected range of the implant, J_I is the interstitial flux, and C_I^{311} is the time-dependent concentration of interstitials that are in equilibrium with the {311} clusters. This leads to the simple expression $t_{\text{TED}}^{\text{eff}} = (f_I Q R_P)/(D_I C_I^*)$, which is independent of $C_I^{311}(t)$, and thus the interstitial clustering model.

By assuming fast interstitial/vacancy recombination, we can generalize this result for an arbitrary damage profile by replacing QR_P with the first moment of the net damage distribution (see Fig. 2):

$$QR_P \rightarrow \int_0^\infty x\left(C_I^{\text{dam}} - C_V^{\text{dam}}\right)dx. \quad (3)$$

Thus,

$$t_{\text{TED}}^{\text{eff}} = f_I \frac{\int_0^\infty x\left(C_I^{\text{dam}} - C_V^{\text{dam}}\right)}{D_I C_I^*}dx. \quad (4)$$

This expression highlights that TED depends on the net damage (due to ion implantation) and the interstitial diffusion capacity ($D_I C_I^*$). To verify that Eq. (4) accurately predicts TED, the results were compared to a full TED simulation which incorporated a moment-based {311} model with excellent agreement (Fig. 3(a)).

The analysis presented above was for marker layers below the implant damage. It is of interest to note how the effective TED time varies for shallower depths. Assuming $D_V^{eff} \gg D_I^{eff}$ (Fig. 2(b)):

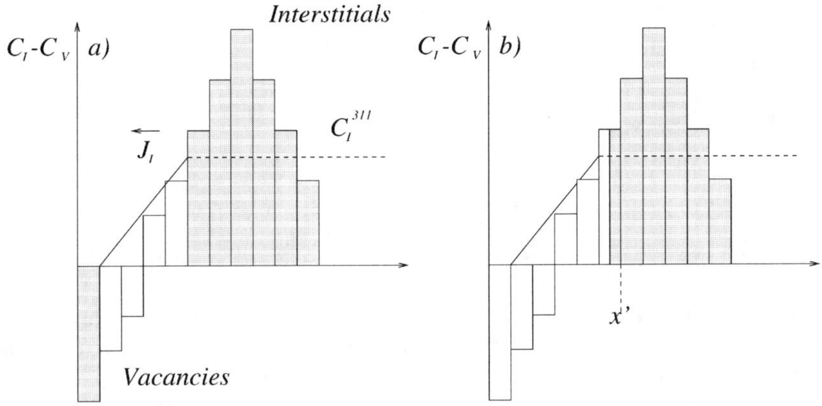

Figure 2: Schematics of net damage and fluxes during annealing. Assuming $D_I^{eff} \gg D_V^{eff}$ (a), the flux is proportional to the distance between the edges of vacancy-rich and interstitial-rich regions. For $D_V^{eff} \gg D_I^{eff}$ (b), fast vacancy diffusion leads to annihilation of interstitial clusters up to x'. In both cases, the final TED below the damage layer is as given by Eq. (4).

$$t_{\mathrm{TED}}^{\mathrm{eff}}\left(x_0\right) = \int_0^{\max\left(x_0,x'\right)} \frac{x\left(C_I^{\mathrm{dam}} - C_V^{\mathrm{dam}}\right)}{D_I C_I^*} dx + \int_{\max\left(x_0,x'\right)}^{\infty} \frac{x_0\left(C_I^{\mathrm{dam}} - C_V^{\mathrm{dam}}\right)}{D_I C_I^*} dx \qquad (5)$$

where x' (the shallow edge of the interstitial excess distribution after vacancies are gone) is:

$$\int_0^{x'} x\left(C_I^{\mathrm{dam}} - C_V^{\mathrm{dam}}\right) dx = 0. \qquad (6)$$

Figs. 3(b) and 4 illustrate the accuracy of Eq. (5) compared to a full TED simulation. Eq. (4) allows us to define a +N factor (damage approximated by $C_I^{\mathrm{dam}} = NC_{\mathrm{ion}}$) directly from the implant profiles:

$$+\mathrm{N} = \frac{\int_0^{\infty} \mathrm{x}\left(\mathrm{C_I^{dam}} - \mathrm{C_V^{dam}}\right) \mathrm{dx}}{\int_0^{\infty} \mathrm{x}\left(\mathrm{C_{ion}}\right) \mathrm{dx}} \qquad (7)$$

+N factors significantly greater than 1 occur for heavy ions for which the net damage profile is significantly deeper than the implanted profile. Based on this relation, +N factors range from near 1 for B to about 5 for As. Fig. 4 shows the results from Eq. (7), which compares the aforementioned analytical expression and full TED analyses based on the +N and +1 models. As seen, the +N calculation closely matches the TED simulation.

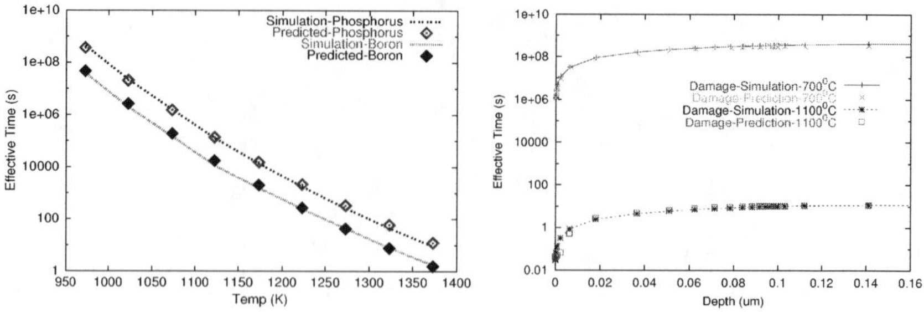

Figure 3: Comparison of analytic predictions to full TED simulation for (a) temperature dependence of $t_{\text{TED}}^{\text{eff}}$ following $1 \times 10^{14}\text{cm}^{-2}$ 35keV boron and phosphorus implants and (b) depth dependence of $t_{\text{TED}}^{\text{eff}}$ for $1 \times 10^{14}\text{cm}^{-2}$ 35keV phosphorus implant.

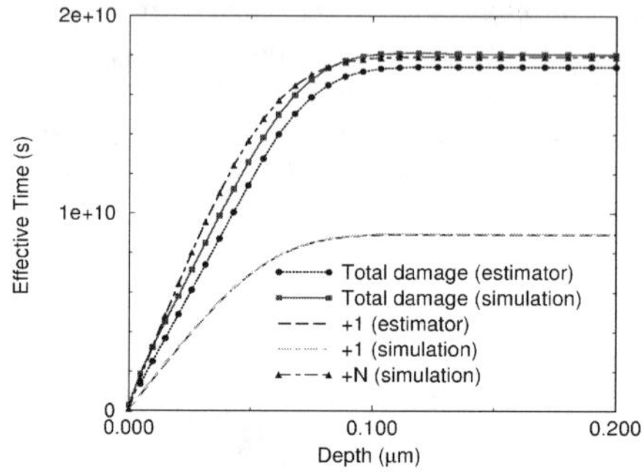

Figure 4: Comparison of $t_{\text{TED}}^{\text{eff}}$ versus depth for 30keV $5 \times 10^{13}\text{cm}^{-2}$ As implant annealed at 600^0C based on Eq. (5) versus full TED simulation using net damage, +1 model, and +N model from Eq. (6).

EXTRACTION OF INTERSTITIAL DIFFUSION CAPACITY FROM TED

Traditionally, there have been two main methods used to extract $D_I C_I^*$. The first is metal diffusion experiments. Stolwijk *et al.* [3] and Morehead *et al.* [5] conducted in-depth studies via Au diffusion. Both Grunebaum *et al.* [4] and Bracht *et al.* [6] have done metal diffusion studies for Zn. The challenges with metal diffusion are to separate the diffusion of vacancies and interstitials from that of the metal. The second method is silicon self-diffusion (isotope) experiments [7]. Challenges with this technique include separating the vacancy and interstitial mechanisms and the unknown correlation factor (ϕ_I) for tracer diffusion, $D_{Si}^{Tracer} = \phi_I D_I C_I^* + \phi_V D_V C_V^*$. An advantage of using TED to determine $D_I C_I^*$ is that it depends only on interstitial diffusion due to the large interstitial supersaturation. This analysis depends only on the interstitial diffusion in the system and the net damage profiles from Monte Carlo simulations (UT-Marlowe was used). Data from marker layer experiments

Author	Prefactor	Activation Energy [eV]
This work	5.00×10^{25}	4.86
Bracht *et al.*[6]	1.50×10^{25}	4.95
Morehead *et al.*[5]	2.00×10^{25}	5.00
Grunebaum *et al.*[4]	3.00×10^{25}	4.79
Stolwijk *et al.*[3]	4.57×10^{25}	4.84
Ural *et al.*[7]	7.45×10^{25}	4.68

Table 1: Point defect parameter comparison for metal diffusion and silicon self-diffusion experiments and TED analysis.

was used to to extract $D_I C_I^*$. These experiments consist of lightly doped marker layers which are epitaxially grown or implanted. An additional implant (usually Si) follows. The sample is annealed and the profile broadens. The diffusivity enhancement is proportional to the transient interstitial supersaturation and can be determined readily via SIMS.

The use of marker layers also decouples the damage itself from the effects of the damage on a nearby region. To avoid complications due to clustering or Fermi level effects, experiments with low dose markers were chosen: boron marker layer experiments from Agarwal *et al.* [8], Chao *et al.* [9], Cowern *et al.* [10], Huang *et al.* [11], and Packan *et al.* [12], and a phosphorus marker layer experiment from Park *et al.* [13]. In all cases, a value of $f_I = 1$ was assumed. Fig. 5(a) shows the extracted fit of $D_I C_I^*$ to the TED sources used. The temperature dependence of each coefficient was assumed to be of Arrhenius form $D_I C_I^* = A_0 \exp(-E_a/kT)$. Table 1 shows the extracted values. Our results are intermediate between values from metal diffusion results, and silicon self-diffusion experiments at high temperature; thus, supporting previous calculated results, (Fig. 5(b)).

CONCLUSIONS

This work has demonstrated that TED can be accurately predicted by a simple analytical expression, which depends only on the net damage and the $D_I C_I^*$ product. +N factors can also be calculated based on the net damage profiles from Monte Carlo simulations. $D_I C_I^*$ products were extracted using marker layer experiments for TED data. The resulting values fall in-between values extracted from metal diffusion and silicon self-diffusion experiments.

ACKNOWLEDGMENTS

The authors would like to thank the Semiconductor Research Corporation (SRC) for funding this work.

REFERENCES

[1] E. Chason, S.T. Picraux, J.M. Poete, J.O. Borland, M.I. Current, T. Diaz de la Rubia, D.J. Eaglesham, O.W. Holland, M.E. Law, C.W. Magee, J.W. Mayer, J. Melngailis, and A.F. Tash, J. Appl. Phys. **81**, 6513 (1997).

[2] J.D. Plummer, M.D. Deal, and P.B. Griffin. *Silicon VLSI Technology : Fundamentals, Practice and Modeling*, edited by C. Sodini (Prentice Hall, Upper Saddle River, New

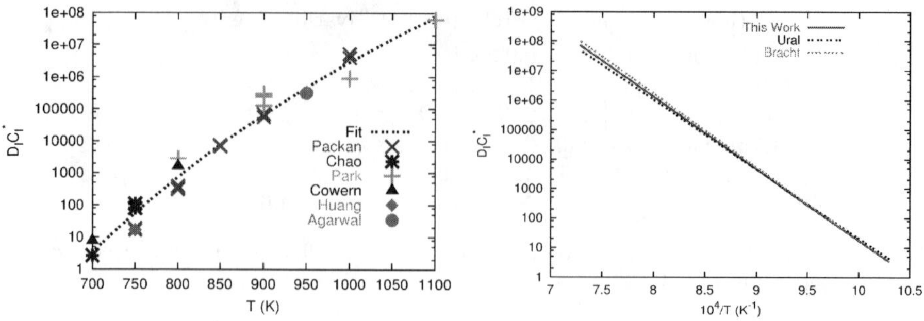

Figure 5: Values of $D_IC_I^*$ extracted at various temperatures from TED data (a) along with the best fit. In (b), extracted values of $D_IC_I^*$ are compared to results of Bracht *et al.* (metal diffusion) and Ural *et al.* (silicon self-diffusion).

Jersey, 2000) pp. 486-487.

[3] N.A. Stolwijk, B. Schuster, J. Holxl, H. Mehrer, and W. Frank, Physica B&C **116B**, 335 (1983).

[4] D. Grunebaum, Th. Czekalla, N.A. Stolwijk, H. Mehrer, I. Yonenage, and K. Sumino, Appl. Phys. A**53**, 65 (1991).

[5] F.F. Morehead, in *Defects in Electronic Materials*, edited by M. Stavola, S.J. Pearton, G. Davies, (Mater. Res. Soc. Symp. Proc. **104**, Pittsburgh, PA, 1988) pp. 99-104.

[6] H. Bracht, N.A. Stolwijk, and H. Mehrer, Phys. Rev. B **53**, 16542 (1995).

[7] A. Ural, Ph.D. thesis, Stanford University (2001).

[8] A. Agarwal, H.-J. Gossman, D.C. Jacobson, D.J. Eaglesham, M. Sosnowski, J.M. Poate, I. Yamada, J. Matsuo, and T.E. Haynes, Appl. Phys. Lett. **73**, 2015 (1998).

[9] H.S. Chao, P.B. Griffin, J.D. Plummer, and C.S. Rafferty, Appl. Phys. Lett. **69**, 2113 (1996).

[10] N.E.B. Cowern, G. Mannino, P.A. Stolk, F. Roozeboom, H.G.A. Huizing, J.G.M. van Berkum, F. Cristiano, and A. Claverie, Physical Review Letters **82**, 4460, (1999).

[11] R. Huang, Ph.D. thesis, Stanford University (1994).

[12] P.A. Packan, Ph.D. thesis, Stanford University (1991).

[13] H. Park and M.E. Law, Appl. Phys. Lett. **58**, 732, (1991).

Defect and Diffusion Models

Mat. Res. Soc. Symp. Proc. Vol. 717 © 2002 Materials Research Society

Simulation of Transient Enhanced Diffusion in Silicon Taking into Account Ostwald Ripening of Defects

Masashi Uematsu
NTT Basic Research Laboratories, 3-1 Morinosato-Wakamiya, Atsugi, 243-0198, Japan

ABSTRACT

The transient enhanced diffusion (TED) of high-dose implanted P is simulated taking into account Ostwald ripening of end-of-range (EOR) defects. First, we integrated a basic diffusion model based on the simulation of in-diffusion, where no implanted damages are involved. Second, from low-dose implantation, we developed a model for TED due to {311} self-interstitial (I) clusters involving Ostwald ripening and the dissolution of {311} clusters. Third, from medium-dose implantation, we showed that P-I clusters should be taken into account, and during the diffusion, the clusters are dissolved to emit self-interstitials that also contribute to TED. Finally, from high-dose implantation, EOR defects are modeled and we derived a formula to describe the time-dependence for Ostwald ripening of EOR defects, which is more significant at higher temperatures and longer annealing times. The simulation satisfactorily predicts the TED for annealing conditions, where the calculations overestimate the diffusion without taking Ostwald ripening into account.

INTRODUCTION

The transient enhanced diffusion (TED) of dopants in silicon is a central issue in silicon device processing because TED is a limiting factor in the scaling down of device size [1]. End-of-range (EOR) defects form at the amorphous/crystalline (a/c) interface following amorphizing implantation [2]. EOR defects act as both a sink for and source of self-interstitials (I), depending on temperature and annealing time, and hence affect TED. In addition, Ostwald ripening of EOR defects reduces their efficiency as a source of self-interstitials, which further complicates TED. In this study, TED of high-dose implanted P is simulated taking into account Ostwald ripening of EOR defects.

Our simulation is integrated step-by-step so that the simulation is done in a unified manner with a reduced number of variables and parameters, because diffusion of high-concentration P in Si during post-implantation annealing is a complex phenomena and the simulation tends to include many variables and parameters. The present simulation is started with integrating a basic diffusion model based on in-diffusion, where no implantation damages are involved. Then, diffusion after low-dose implantation is simulated and we develop a model for {311} self-interstitial clusters with Ostwald ripening and the dissolution of the clusters. In addition, we show that electrically inactive and immobile P-I clusters should be taken into account for a consistent simulation for medium-dose implanted P. Finally, from high-dose implantation, a model for EOR defects with Ostwald ripening is developed. We show that the simulation satisfactorily predicts the diffusion of high-dose implanted P and describe the effect of Ostwald ripening of EOR defects on TED.

INTEGRATED DIFFUSION MODEL

In order to simulate impurity diffusion in a unified manner with the essential parameters, we propose a basic diffusion model [3] that consists of the substitutional-interstitial (S-I) mechanism [the kick-out and the Frank-Turnbull (F-T) mechanism] and the vacancy mechanism. These mechanisms are described by the following reactions:

$$A_i \Leftrightarrow A_s + I \quad \text{(the kick-out mechanism)} \quad (1.1)$$

$$A_i + V \Leftrightarrow A_s \quad \text{(the F-T mechanism)} \quad (1.2)$$

$$(A_s\text{–}V) \Leftrightarrow A_s + V \quad \text{(the vacancy mechanism),} \quad (1.3)$$

where A_i and A_s denotes interstitial and substitutional impurities, and I and V are self-interstitials and vacancies. In reaction (1.3), (A_s–V) denotes the nearest-neighbor pair configuration in which A_s becomes mobile through site-exchange with vacancies. In this paper, v^{2-} and v^- are used to denote the configurations with doubly and singly negatively charged vacancies [V^{2-} and V^-].

This diffusion process can be represented in analogy with an electrical circuit, as shown in Figure 1. Impurity flux Λ in this model is analogous to the current flowing through this circuit. In circuit 1 for the S-I mechanism, the contributions of self-interstitials (Λ_I) and vacancies (Λ_V) are in parallel, and hence the *faster* one governs subcircuit 12, and its contribution, Λ_{12}, is determined by the larger one of Λ_I and Λ_V. When two parallel contributions are comparable, their sum determines the subcircuit. In addition, subcircuit 11 of the impurity interstitial contribution (Λ_i) and subcircuit 12 are in series, and the *slower* one, *i.e.*, the smaller one of Λ_i and Λ_{12}, determines the S-I mechanism (Λ_1). Furthermore, the contribution of the S-I mechanism (Λ_1) and of the vacancy mechanism (Λ_v) are in parallel, and therefore, the *faster* one, *i.e.*, the larger one of Λ_1 and Λ_v, (or their sum) finally governs the diffusion.

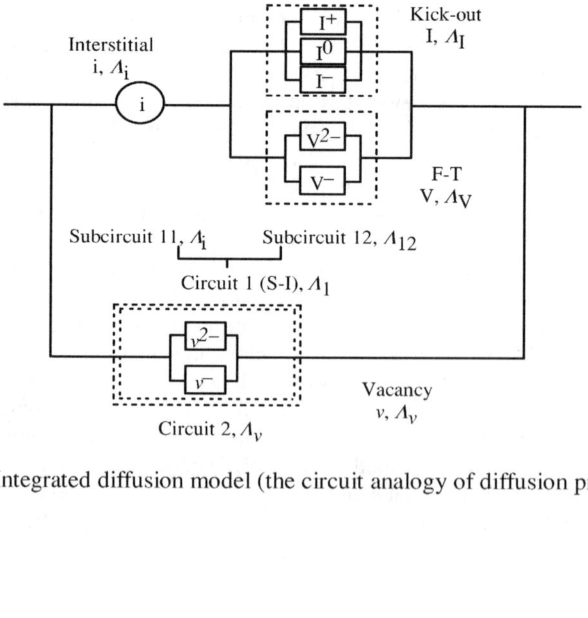

Figure 1. Integrated diffusion model (the circuit analogy of diffusion processes).

Next, we extract the essential parameters for P in-diffusion based on the following reported knowledge: (A) Intrinsic P diffusion is known to be retarded by nitridation, which induces vacancy supersaturation. This excludes the contribution of the vacancy mechanism to P diffusion in the intrinsic region. (B) In contrast, for a heavily doped P sample, the nitridation retardation effect is much less significant than in the intrinsic cases. This indicates that the vacancy mechanism starts to contribute to P diffusion when P concentration is very high. (C) The diffusivity of P diffusion is proportional to the square of the P concentration in high P concentration region. Therefore, the vacancy mechanism *via* V^{2-} (denoted as v^{2-}) should be the dominant P diffusion mechanism in this region. In high P concentration region, the contributions of V^{-} and V^{0} are quite minor because the *n* concentration is so high that V^{2-} concentration is tremendously increased due to the Fermi-level effect. (D) A phosphorus isoconcentration study showed that the P diffusivity is almost proportional to the P concentration in medium and low P concentration region. This linear dependence means that the charge difference between P_i and P_s^{+} is one. Therefore, we assume only P_i^{0} primarily contribute to P diffusion, and even if P_i^{+} and P_i^{-} exist, their contributions to P diffusion are quite minor. From (A)-(D), therefore, we conclude that P diffusion is primarily governed by P_i^{0}, I, and v^{2-}. Based on the model, we construct diffusion equations that take into account these essential diffusion species and the corresponding reactions (1). These diffusion equations (simultaneous partial differential equations) are solved numerically and we simulate P in-diffusion profiles [4]. Figure 2 shows the simulation results of P in-diffusion profiles at 900 °C, and the profiles at 1000 and 1100 °C can also be simulated by the model. The effective diffusivities of P_i^{0}, I, and v^{2-} used in the simulation are consistent with those from experiments, and this confirms the validity of our simulation.

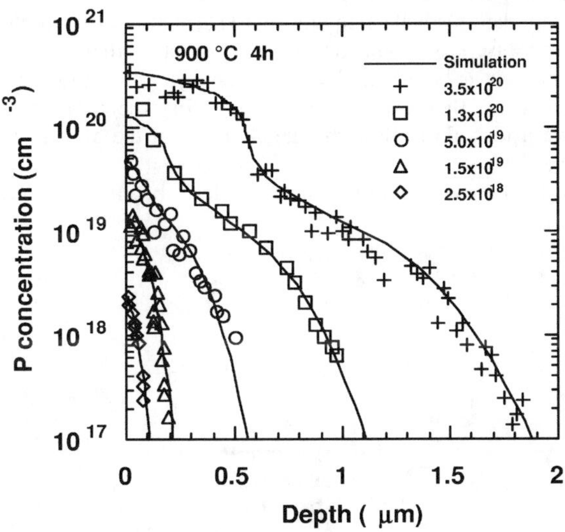

Figure 2. Simulated P in-diffusion profiles at 900 °C. Experimental data are from Ref. 4.

Based on the model shown in Figure 1 and the simulation results mentioned above, we discuss the mechanism of P diffusion and clarify the mechanism for the appearance of the kink-and-tail profiles. In the plateau region, *i.e.*, from the surface to a depth of about 0.5 μm ($C_S \sim 1\times10^{20}$ cm^{-3}), the vacancy mechanism *via* V^{2-} (v^{2-}) governs the diffusion: Λ_v^{2-} determines the process because Λ_v^{2-} is larger than Λ_I, Λ_V^{2-}, or Λ_V^{-}. On the other hand, the kick-out mechanism dominates in the region deeper than about 0.5 μm, where self-interstitials govern the diffusion in the kink region, namely from about 0.5 to 0.8 μm ($C_S \sim 2\times10^{19}$ cm^{-3}) and P interstitials dominate in the tail region. In the kink region, Λ_I determines the diffusion because Λ_I is larger than Λ_v^{2-}, Λ_V^{2-}, or Λ_V^{-} but smaller than Λ_i. In the tail region, Λ_i determines the diffusion because Λ_i is larger than Λ_v^{2-} but smaller than Λ_I. This changeover from Λ_v^{2-} *via* Λ_I to Λ_i is the mechanism by which the kink-and-tail profiles appear when surface P concentration is high. In contrast, when surface P concentration is low, the contribution of Λ_v^{2-} becomes low and the kick-out mechanism almost totally governs the diffusion in the whole region, and therefore no kink-and-tail profiles appear.

{311} SELF-INTERSTITIAL CLUSTERS

{311} self-interstitial clusters produced by ion implantation damage are the source of the supersaturated self-interstitials that cause TED [1]. Although a kinetic equation has been proposed to describe the time evolution of the cluster [5], it is difficult to correctly predict the cluster evolution because of the need to take into account the decrease in the evolution rate with elapsed time as small clusters grow into larger ones (Ostwald ripening). Ostwald ripening of {311} self-interstitial clusters is schematically shown in Figure 3. In the present study, we simulate Ostwald ripening of {311} self-interstitial clusters based on an analytic model [6], because it is a simpler approach with a smaller number of parameters than atomistic simulation, but is based on physical meanings, enabling us to do quantitative simulations. We analyze the experimental data for the cluster evolution and semi-empirically derive a formula to describe the decrease in the evolution rate with time due to Ostwald ripening.

The evolution of {311} self-interstitial clusters is described by the reaction,

$$C \underset{k_b}{\overset{k_f C_c}{\leftrightarrows}} I \xrightarrow{D_I} I \text{ diffusion}, \qquad (2)$$

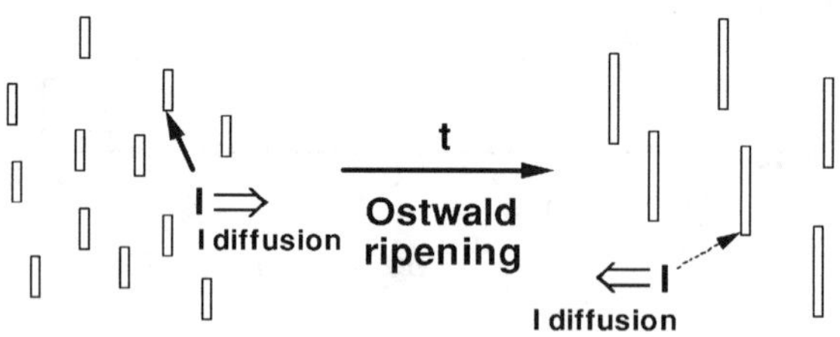

Figure 3. Ostwald ripening of {311} self-interstitial clusters is schematically shown.

and the kinetic equation of $k_f C_c C_I - k_b C_c$ is incorporated in the diffusion equations for C_I and C_c, where D_I and C_I are the diffusivity and the concentration of free self-interstitials, C_c means the concentration of self-interstitials trapped in the clusters (not that of the clusters), and k_f and k_b denote the rate constants of the cluster growth and dissolution, respectively.

The evolution of {311} self-interstitial clusters consists of two time stages determined by the smaller one of either the cluster growth, $k_f C_c C_I$, or the self-interstitial diffusion, $D_I(d^2 C_I/dx^2)$. At the first stage, where $k_f C_c C_I > D_I(d^2 C_I/dx^2)$ holds, the self-interstitial diffusion away from the Si implanted region is slower than the rate at which the cluster can trap the self-interstitials. Therefore, the self-interstitial diffusion is the rate-determining step for this stage and the C_I is maintained at a constant value C_I^*. With the elapse of time during the first stage, cluster growth rate k_f will slow down due to Ostwald ripening, as described below. And finally, when $k_f C_c C_I < D_I(d^2 C_I/dx^2)$ holds, the cluster cannot maintain the C_I^* value anymore, and the self-interstitial concentration decreases abruptly. Then Ostwald ripening will cease, and hence k_f becomes constant. This is the second stage of the evolution, where the cluster dissolution becomes the rate-determining step. At this stage $dC_c/dt = -k_b C_c$ holds, and C_c and C_I show an exponential decay with time according to $\exp(-k_b t)$.

The decrease in the k_f with time due to Ostwald ripening is described as follows: As the clusters grow larger due to Ostwald ripening, the number of self-interstitials per cluster, N_I, increases and the cluster density decreases (Figure 3). Then free self-interstitials may be less likely to be trapped by the clusters, which means the decrease in k_f with time. Dividing the total self-interstitial concentration in the cluster by the cluster density (data from Ref. 1), the time dependence of N_I during Ostwald ripening can be fitted by $1/(-at+b)$. Therefore, we use $k_f = -at+b$ because k_f may be inversely proportional to N_I, as mentioned above. This relation is used to fit the B TED data shown in Figure 4.

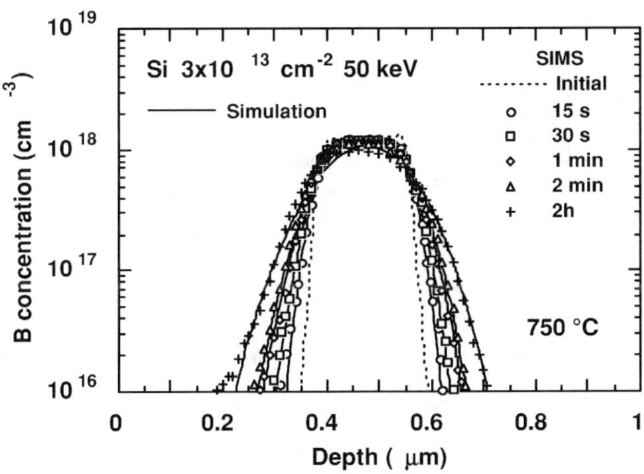

Figure 4. Simulated (solid lines) and SIMS (symbols) B profiles with 50 keV, 3×10^{13} cm^{-2} Si implantation and annealing at 750 °C. SIMS data are from Ref. 7.

Parameters used for the simulation are C_I^* to determine the B diffusion enhancement at the first stage, a and b to determine the duration time of the first stage, and k_b to govern the second stage. At the first stage, $dC_I/dt \approx 0$ and $dC_c/dt \approx D_I(d^2C_I/dx^2)$ hold, and the C_I is maintained at C_I^* even though C_c actually decreases with time. Therefore, C_I^* is termed the pseudo-equilibrium concentration, and is determined by the balance between the cluster growth and dissolution. According to reaction (2), C_I^* is described as k_b/k_f, and hence the enhancement in diffusivity, C_I^*/C_I^{eq}, is almost constant during the first stage, where C_I^{eq} denotes the thermal equilibrium concentration of self-interstitials. These four parameters, C_I^*, a, b, and k_b are determined by the simulation of low-concentration B TED induced by low-dose Si implantation, where the effects of {311} clusters can be extracted. For the initial profile of C_c, a "+1" model is used. We satisfactorily fitted the B TED profiles at various temperatures (*e.g.*, Figure 4), which indicates that the above model is reasonable.

P-I CLUSTERS

As implantation dose becomes higher, some impurities, such as B [1], may form clusters with supersaturated self-interstitials caused by the implantation even at concentrations far below the solubility limit. These clusters are electrically inactive and immobile and therefore greatly affect the diffusion profiles. In this section, we show that electrically inactive and immobile P-I clusters should be taken into account to achieve a consistent simulation for implanted P [8]. Because little is known about P-I clusters [8,9], we use a model for P-I clusters that is analogous to that of B-I clusters [10] to simulate the diffusion profiles of medium-dose implanted P [11] (Figure 5).

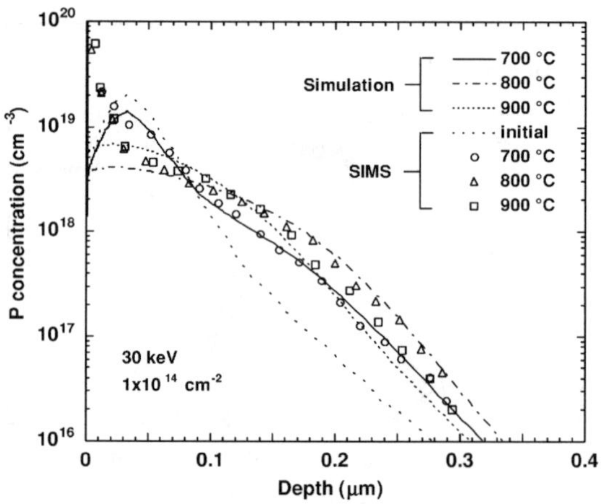

Figure 5. Simulated (lines) and SIMS (symbols) total P profiles with 30 keV, 1×10^{14} cm^{-2} P implantation and annealing at 700 °C (15 min), 800 °C (5 min), and 900 °C (5 min). SIMS data are from Ref. 11.

Our model for B-I and P-I clusters is based on an analytic approach, similarly to that for {311} clusters, because it enables us to reduce a number of parameters compared with atomistic simulation. Assuming that P-I clustering takes place during the very early stage of annealing [12], the initial P-I cluster profile for the simulation is given by

$$C_{\mathrm{Pcl}} = AQ_{\mathrm{d}}C_{\mathrm{s}}(n/n_{\mathrm{i}}), \tag{3}$$

where Q_d is the implantation dose, C_s is the P concentration, n is the carrier concentration, and A is a temperature dependent parameter. This formula has been originally given based on the experimental observations that B-I clustering depends on dose, B concentration, and background doping level. In addition, the C_c for {311} self-interstitial clusters is calculated by subtracting the amount of self-interstitials trapped by P-I clusters from the "+1" self-interstitials. We calculate the C_c in the same way as for B-I clustering, where trapping of two self-interstitials per B atom led to the best fittings [10], which seems to suggest that BI_2 is the precursor of B-I clustering [12]. Furthermore, similarly to B-I clusters, P-I clusters are dissolved (with the rate constant k_d) during the diffusion to emit self-interstitials that also contribute to TED. The two parameters for P-I clusters, A and k_d, are determined from the simulation shown in Fig. 5. In the simulation, P pile-up to the surface oxide is also taken into account [8]. As will be described in the next section, neglecting P-I clustering leads to an overestimation of the P diffusion at the temperature lower than about 800 °C and this is the reason we originally propose P-I clustering from the simulation. In contrast, at higher temperatures, P-I clusters dissolve so fast that the clustering is hardly observed.

EOR DEFECTS

Following amorphizing implantation, EOR defects form at the a/c interface and affect TED. In addition, Ostwald ripening of EOR defects reduces their efficiency as a source of self-interstitials, which is more significant at higher temperatures and longer annealing times [13]. First, therefore, we simulate low-concentration B TED profiles at relatively low temperatures induced by high-dose Si implantation [14] (Figure 6). The reaction between EOR defects and self-interstitials is described as $-k_{\mathrm{EOR}}C_{\mathrm{EOR}}(C_{\mathrm{I}}-C_{\mathrm{I}}^{\mathrm{EOR}})$ and this equation is incorporated in the diffusion equations for C_{I} and C_{EOR} [15]. Here, C_{EOR} is the concentration of self-interstitials trapped in EOR defects (not that of the defects), $C_{\mathrm{I}}^{\mathrm{EOR}}$ is the self-interstitial concentrations maintained by EOR defects, and k_{EOR} is the self-interstitial capturing efficiency. This reaction term means that EOR defects act as a sink for self-interstitials when $C_{\mathrm{I}} > C_{\mathrm{I}}^{\mathrm{EOR}}$, but as a source when $C_{\mathrm{I}} < C_{\mathrm{I}}^{\mathrm{EOR}}$. $C_{\mathrm{I}}^{\mathrm{EOR}}$ values are a few orders of magnitude smaller than C_{I}^{*} values, the pseudo-equilibrium self-interstitial concentration maintained by {311} clusters. When EOR defects act as a sink for self-interstitials, $k_{\mathrm{EOR}}C_{\mathrm{EOR}}$ becomes an essential parameter for determining the efficiency with which self-interstitials released from {311} clusters are trapped. After C_{I} enhanced by {311} clusters becomes smaller than $C_{\mathrm{I}}^{\mathrm{EOR}}$, EOR defects become a source of self-interstitials and C_{I} is maintained at a constant value, $C_{\mathrm{I}}^{\mathrm{EOR}}$. Therefore, the parameters for EOR defects are $C_{\mathrm{I}}^{\mathrm{EOR}}$ and $k_{\mathrm{EOR}}C_{\mathrm{EOR}}$, and they are determined from the fittings of TED profiles (*e.g.*, Figure 6). The depth profiles of C_{I} deduced from the simulation are shown in Figure 7. The solid phase epitaxial (SPE) regrowth of the amorphous region, which occurs at low temperatures, or during the rump-up stage of annealing, almost totally eliminates the damage in this region; only a tail of damage beyond the a/c interface remains and becomes a source of TED [14]. In the simulation, therefore, the initial profiles of self-interstitials in {311} clusters are defined as the implanted ion profiles beneath the a/c interface, which are shown as "+1" self-interstitials (shaded area) in Figure 7.

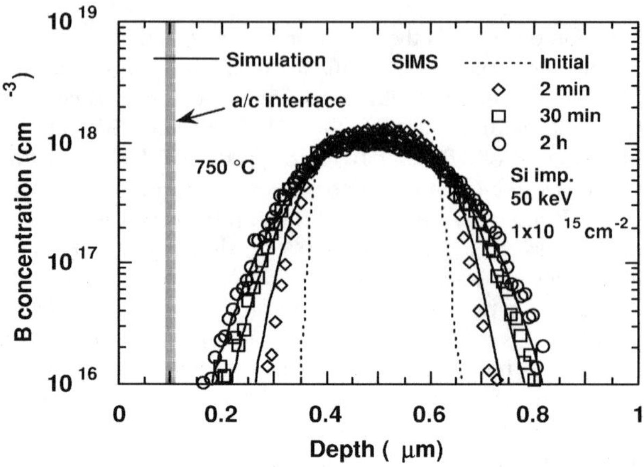

Figure 6. Simulated (solid lines) and SIMS (symbols) B profiles with 50 keV, $1x10^{15}$ cm^{-2} Si implantation and annealing at 750 °C. SIMS data are from Ref. 14.

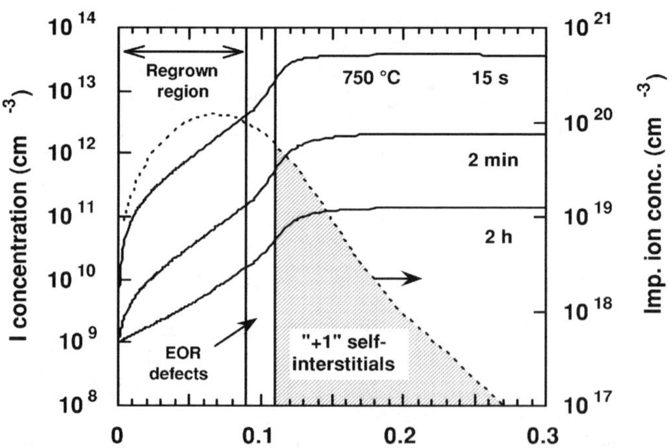

Figure 7. The depth profiles of C_I deduced from the simulation in Figure 7. The dotted line shows the as-implanted ion profile, and the shaded area indicates "+1" self-interstitials beneath the a/c interface.

Next, we take into account Ostwald ripening of EOR defects, by which the supersaturation of self-interstitials induced by the defects decreases with time. Ostwald ripening of EOR defects has been investigated and modeled by Claverie *et al.*, and the model correctly predicts the increase of the mean radius of dislocation loops and the decrease of C_I that governs TED [13]. We derive a formula to describe the time-dependent C_I^{EOR} based on their model, and apply this time-dependent C_I^{EOR} value to our model mentioned above (*e.g.*, Figure 6), where a constant C_I^{EOR} was used. The concentration of free self-interstitials in equilibrium with a dislocation loop of mean radius r_d is described as

$$C_I^{EOR}/C_I^{eq} = \exp(B/r_d), \qquad r_d^2 = Kt/2 + r_0^2, \tag{4}$$

where K is the ripening rate, r_0 is the initial mean radius, and B is a constant that only depends on temperature [16]. Equation (4) leads to

$$C_I^{EOR}/C_I^{eq} = \exp[B/(Kt/2 + r_0^2)^{1/2}], \tag{5}$$

where r_d is approximated by the square root of r_d^2. Concerning parameter values, K and r_0 values from Ref. 13 are used. The value of B is determined so that Eq. (5) at t =0 leads to the C_I^{EOR} values obtained under the conditions where Ostwald ripening can be neglected. Figure 8 shows the simulation results for high-dose implanted P (experimental data are from Refs. 11 and 17). Similarly to the initial profiles of {311} clusters, P-I clustering is taken into account only beneath the a/c interface [18] because the clustering is unlikely to occur in the regrown regions.

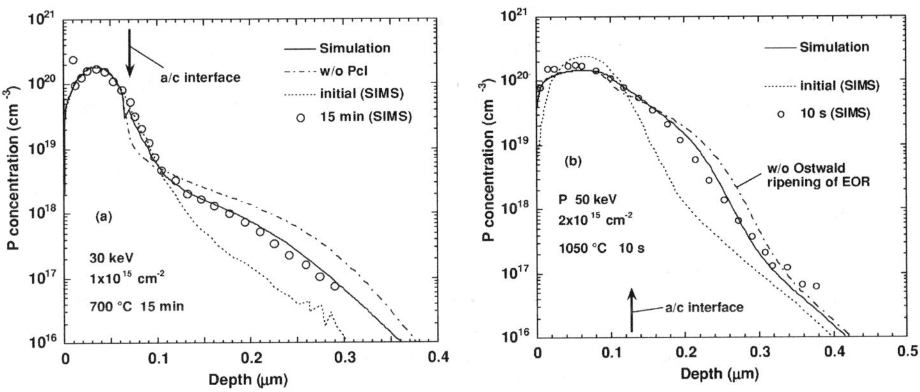

Figure 8. Simulated (solid lines) and experimental (symbols) total P profiles with (a) 30 keV, 1×10^{15} cm^{-2} P implantation and annealing at 750 °C for 15 min and (b) 50 keV, 2×10^{15} cm^{-2} P implantation and annealing at 1050 °C for 10 s. Experimental data are from Refs. 11 and 17. At lower temperatures, the effect of P-I clusters is significant. In contrast, at higher temperatures, the effect of Ostwald ripening of EOR defects is significant.

At 700 °C, P-I clusters still exist after 1h even though their dissolution is much faster than that of B, and therefore neglecting P-I clustering leads to an overestimation of the P diffusion. On the other hand, the effect of Ostwald ripening of EOR defects can be neglected at 700 °C. In contrast, at 1050 °C, C_I^{EOR} is reduced by Ostwald ripening, and hence taking Ostwald ripening into account is indispensable for the simulation. The influence of Ostwald ripening is significant for annealing times longer than about 10 and 1 s at 1000 and 1050 °C, respectively. However, P-I clusters dissolve so fast (within 10 and 0.1 s at 900 and 1050 °C, respectively) that the contribution of P-I clusters is minor at 1050 °C. We should mention that P precipitation is not included in this study because the P concentration covered is lower than the P solubility limits, C_{sol}. However, the present simulation can be extended to P concentrations higher than C_{sol} if the term, $-k_{ppt}C_{ppt}(C_S-C_{sol})$, is introduced. Here, C_S and C_{ppt} denote the concentration of active P atoms and P atoms in P precipitates, and k_{ppt} is the growth rate of the precipitates. This reaction term is incorporated in the diffusion equations for C_S and C_{ppt}, and the diffusion simulation is done in the same way as mentioned above.

CONCLUSIONS

The TED of high-dose implanted P has been simulated taking into account Ostwald ripening of EOR defects. Starting with the integration of a basic diffusion model from in-diffusion, where no implanted damages are involved, our simulation is integrated step-by-step with increasing implantation doses for {311} self-interstitial clusters, P-I clusters, and EOR defects. This leads to a unified simulation with a reduced number of variables and essential parameters with clear physical meanings. The simulation satisfactorily predicts the diffusion of high-dose implanted P, and the effect of Ostwald ripening of EOR defects on TED was described. In addition, the present simulation for P diffusion can also be applied to B, BF_2, and As implantation [19]. Therefore, our simulation is helpful in developing more widely applicable diffusion simulators, especially two- or three-dimensional simulators.

REFERENCES

1. P. A. Stolk, H.-J. Gossmann, D. J. Eaglesham, D. C. Jacobson, C. S. Rafferty, G. H. Gilmer, M. Jaraiz, J. M. Poate, H. S. Luftman, and T. E. Haynes, J. Appl. Phys. **81**, 6031 (1997).
2. K. S. Jones, S. Prussin, and E. R. Weber, Appl. Phys. A **45**, 1 (1988).
3. M. Uematsu, J. Appl. Phys. **82**, 2228 (1997).
4. M. Yoshida, E. Arai, H. Nakamura, and Y. Terunuma, J. Appl. Phys. **45**, 1498 (1974).
5. C. S. Rafferty, G. H. Gilmer, M. Jaraiz, D. J. Eaglesham, and H.-J. Gossmann, Appl. Phys. Lett. **68**, 2395 (1996).
6. M. Uematsu, Jpn. J. Appl. Phys. **36**, L982 (1997); J. Appl. Phys. **83**, 120 (1998).
7. H. S. Chao, P. B. Griffin, J. D. Plummer, and C. S. Rafferty, Appl. Phys. Lett. **69**, 2113 (1996).

8. E. Schroer and M. Uematsu, Jpn. J. Appl. Phys. **38**, 7 (1999); M. Uematsu, Jpn. J. Appl. Phys. **38**, 6188 (1999).
9. P. H. Keys, K. S. Jones, M. E. Law, M. Puga-Lambers, and S. M. Cea, MRS Spring Meeting 2001, J5.5.
10. M. Uematsu, J. Appl. Phys. **84**, 4781 (1998).
11. P. S. Choi, T. Su, R. D. Chang, P. K. Chu, and D. L. Kwong, *Process Physics and Modeling in Semiconductor Technology* (1996) Electrochem. Soc. Proc. vol. **96-4**, p. 149.
12. L. Pelaz, G. H. Gilmer, H.-J. Gossmann, C. S. Rafferty, M. Jaraiz, and J. Barbolla, Appl. Phys. Lett. **74**, 3657 (1999).
13. C. Bonafos, D. Mathiot, and A. Claverie, J. Appl. Phys. **83**, 3008 (1998); E. Lampin, V. Senez, and A. Claverie, J. Appl. Phys. **85**, 8137 (1999).
14. H. S. Chao, S. W. Crowder, P. B. Griffin, and J. D. Plummer, J. Appl. Phys. **79**, 2352 (1996).
15. M. Uematsu, Jpn. J. Appl. Phys. **37**, 5866 (1998).
16. M. Uematsu, Jpn. J. Appl. Phys. **38**, L1213 (1999).
17. G. S. Oehrlein, R. Ghez, J. D. Fehribach, E. F. Gorey, T. O. Sedgwick, S. A. Cohen, and V. R. Deline, *Proc. 3th Int. Conf. Defects in Semicond.*, eds. L. C. Kimerling and J. M. Parsey, Jr. (Metallurgical Society of AIME, Warrenda, PA, 1984) p. 539.
18. M. Uematsu, Jpn. J. Appl. Phys. **38**, 6188 (1999).
19. M. Uematsu, Jpn. J. Appl. Phys. **38**, 3433 (1999); **39**, 1006 (2000); **39**, 1608 (2000).

A MODEL FOR BORON T.E.D. IN SILICON : FULL COUPLINGS OF DOPANT WITH FREE AND CLUSTERED INTERSTITIALS.

F. Boucard[1,2,3], D. Mathiot[1], E. Guichard[2], and P. Rivallin[3]
[1] Laboratoire PHASE-CNRS, 23 rue du Loess, F-67037 Strasbourg Cedex 2, France
[2] SILVACO DATA SYSTEMS , 55, rue Blaise Pascal, F-38330 Montbonnot, France
[3] CEA-LETI, 17 Av. des Martyrs F-38054 Grenoble cedex 9, France

ABSTRACT

In this contribution we present a model for transient enhanced diffusion of boron in silicon. This model is based on the usual pair diffusion mechanism including non-equilibrium reactions between the dopant and the free point defects, taking into account their various charge states. In addition to, and fully coupled with the dopant diffusion we model the growth and dissolution of the interstitials and boron interstitials clusters associated with the anneal of the self-interstitial supersaturation created by the implantation step. It is thus possible to simulate a rather large set of experimental conditions, from conventional predeposition steps, to RTA after low energy implantation.

INTRODUCTION

One of the most important challenges in developing ULSI technology today is to shrink device sizes to their limit. Each generation requires a large effort in research and development, where technological computer aided design (TCAD) can play a key role.

Ion implantation is the common technique used for doping advanced silicon devices in microelectronics. However this technique induces a huge supersaturation of point defects in the Si crystal which leads to an anomalous broadening of the dopant profile during the high temperature activation anneal. This phenomenon, known as transient enhanced diffusion (TED) is particularly noticeable and embarrassing for boron diffusion, where it is a serious issue for the formation of the ultra shallow junctions needed for next generation of devices. It is also now well established that TED is strongly correlated with the evolution of the self-interstitial supersaturation governed by the nucleation and evolution, during the high temperature anneal, of a variety of extended defects structures like boron interstitial clusters (BIC) [1] or interstitials clusters (IC) [2].

Thus, predictive process modeling, needed for deep submicron MOSFET technologies, requires the development of accurate diffusion models taking into account the full set of interactions between the dopant and the point or extended defects (clusters). The purpose of this contribution is to show that it is possible to extend the usual dopant pair diffusion model to take into account these interstitial-related clusters. Special attention is paid on the fact that the "new" model is fully consistent with the equilibrium "normal" model.

BASIC DIFFUSION MECHANISM

As a starting point of our modeling effort we used an extension of the basic dopant pair diffusion model proposed several years ago by one of us [3]. Briefly, the basic idea of this model is that isolated substitutional boron atoms (B_S) are immobile. The dopant diffusion occurs only *via* the diffusion of boron / self-interstitial (BI) and (to a less extend) of boron /

vacancy (BV) pairs. However, in the present work, we do not assume local equilibrium between the pairs and their components. As previously, we considered all the possible charge states of the free defects and of the pairs, the relative concentrations of which depend on the local Fermi level position. Since we want to be able to handle experimental cases where, as after ion implantation, the defect concentrations are of the same order of magnitude or even higher than the dopant concentration, the concentrations of all the charged species (dopant, free point defects, pairs and clusters) are accounted for in the neutrality equation used to compute the electron concentration. Nevertheless, since electronic exchanges are extremely faster than atomic reactions, we assume that the various charge states of a given species are always in local equilibrium, and thus, in order to minimize the number of equations, we explicitly describe only one reaction path for the formation of each pair, the other charge states being consistently computed through the position of the defect related deep level. The pairing reactions between the dopant and point-defects explicitly considered in the present work are the following :

$$B_s^- + I^o \underset{k_a^r}{\overset{k_a^f}{\rightleftarrows}} BI^-$$

$$B_s^- + V^+ \underset{k_b^r}{\overset{k_b^f}{\rightleftarrows}} BV^o$$

$$BI^- + V^o \underset{k_c^r}{\overset{k_c^f}{\rightleftarrows}} B_s^-$$

$$BV^o + I^o \underset{k_d^r}{\overset{k_d^f}{\rightleftarrows}} B_s^- + h^+$$

The evolution of the boron and free defect concentrations are then given by the following set of continuity equations :

$$\frac{\partial[B_S]}{\partial t} = -\frac{\partial J_B^{tot}}{\partial x} - \Sigma\left(\frac{\partial[BI^i]}{\partial t} + \frac{\partial[BV^i]}{\partial t}\right) + (G-R)_{BIC}$$

$$\frac{\partial[I]}{\partial t} = -\frac{\partial J_I}{\partial x} - \Sigma\left(\frac{\partial[BI^i]}{\partial t}\right) - R_{I/V} + (G-R)_{In} + (G-R)_{BIC}$$

$$\frac{\partial[V]}{\partial t} = -\frac{\partial J_V}{\partial x} - \Sigma\left(\frac{\partial[BV^i]}{\partial t}\right) - R_{I/V}$$

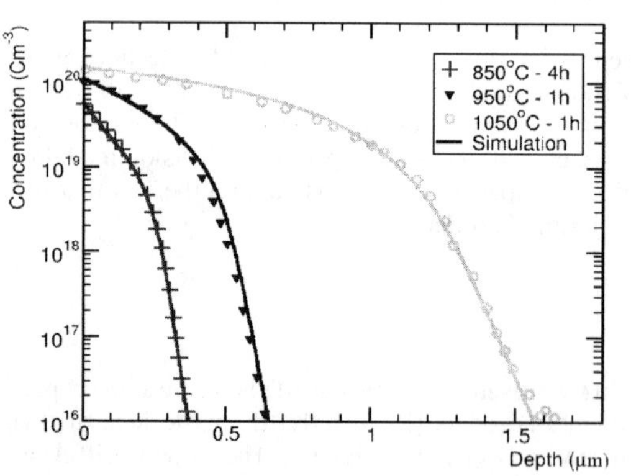

Fig.1 : Comparison between calculated and experimental boron predeposition profiles. Experimental data are from ref. [4]

where the various fluxes (J_X) include built-in field drift effects on charged species and $R_{I/V}$ account for the bimolecular recombination between self-interstitials (I) and vacancies (V) [3]. The additional $(G\text{-}R)_{In}$ and $(G\text{-}R)_{BIC}$ terms account for the generation or consumption of the various species due to the interstitial-related clusters formation and dissolution as described in the following sections.

From these continuity equations solved by PROMIS [5], boron profiles can be theoretically calculated in any experimental situation if appropriate initial conditions are assumed for the point defects. As a first indication of the ability of our model to handle the complex couplings between boron and the free point defects, we show on Fig.1 the results of the simulation of conventional predeposition steps (diffusion with a constant surface concentration), where the initial concentrations of the defects are assumed to have their equilibrium values. For this specific case the (G-R) terms linked to the interstitial-related clusters evolution vanish, and such simulations can thus be used to fit the basic parameters describing the dopant interactions with the free point defects. As shown on Fig.1, the agreement between the experimental profiles [4] and the simulated ones, with parameters corresponding to a I contribution to the total B diffusion of the order of 85 %, is remarkably good.

INTERSTITIAL CLUSTERS

In order to be able to simulate the T.E.D. of boron, the basic pair diffusion model is fully coupled, through generation and recombination terms, to the equations describing the evolution of the point defects in connection with the growth and dissolution of interstitial clusters.

For this purpose we describe the kinetics of the IC's using the model we developed recently for the growth of {311} defects [6]. In this model, a cluster containing n interstitials evolve to a cluster of size n+1 by interaction with a free interstitial in agreement to the following reactions with the corresponding relations describing the time evolution of the various concentrations :

$$\begin{matrix} I^{free} + I^{free} \underset{k_2^r}{\overset{k_2^f}{\rightleftarrows}} I_2 \\ I_2 + I^{free} \underset{k_3^r}{\overset{k_3^f}{\rightleftarrows}} I_3 \\ \vdots \\ I_{n-1} + I^{free} \underset{k_n^r}{\overset{k_n^f}{\rightleftarrows}} I_n \end{matrix} \quad \Leftrightarrow \quad \begin{cases} \dfrac{\partial[I^{free}]}{\partial t} = 2\left(k_2^f\left[I^{free}\right]^2 - k_2^r[I_2]\right) + \sum\limits_{j=3}^{N_{max}} k_j^f\left[I^{free}\right]\left[I_{j-1}\right] - k_j^r\left[I_j\right] \\ \dfrac{\partial[I_2]}{\partial t} = k_2^f\left[I^{free}\right]^2 - k_2^r[I_2] - k_3^f\left[I^{free}\right][I_2] + k_3^r[I_3] \\ \vdots \\ \dfrac{\partial[I_j]}{\partial t} = k_j^f\left[I^{free}\right][I_{j-1}] - k_j^r\left[I_j\right] - k_{j+1}^f\left[I^{free}\right]\left[I_j\right] + k_{j+1}^r\left[I_{j+1}\right] \end{cases}$$

According to [6], the various rate constants are given by

$$\begin{cases} k_n^f = 4\pi r_n^{eff} D_I.\exp\left(-\dfrac{\Delta g_n}{kT}\right) \\ \Delta g_n = \max\left(-kT\ln\left(\dfrac{[I^{free}]}{[I]^{eq}}\right) + E_f(n); 0\right) \\ k_n^r = 6\dfrac{D_I}{\lambda^2}\theta_n.\exp\left(-\dfrac{E^f(n) - E_f^I}{kT}\right) \end{cases}$$

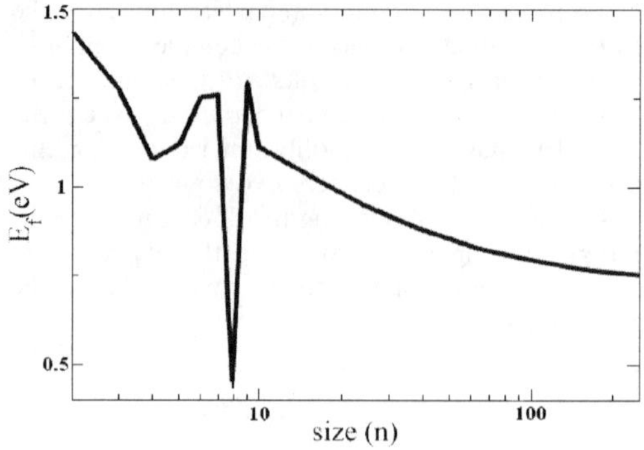

Fig.2 : Fitted formation energy per atom included in a cluster of size n

where r_n^{eff} is the effective capture radius of the defect of size n, Δg_n is the effective energy barrier that an interstitial must overcome, $E_f(n)$ is the energetic cost for adding one Si atom to a cluster of size n, i.e. the formation energy per I atom in the cluster of size n, E_f^I is the formation energy for an isolated self-interstitial, θ_n is the number of dissociation sites, and λ is the lattice distance [6].

The validity of the interstitial cluster growth model is tested by comparing its predictions with the experimental data obtained by Cowern et al. [7]. Briefly, this experiment consists in measuring the enhanced diffusivity of two buried boron marker layers induced by a superficial silicon implantation at 40 keV with a dose of 2.10^{13} at/cm^2. To start the calculation, we used here a simple +1 model to describe the initial self-interstitial concentration induced by the Si implant. In order to obtain the best fit, the values of the various formation energies have to be slightly changed as compared to the initial values proposed by Cowern et al. We think that this is due to the fact that the way we estimate the formation rate constant k_n is somewhat different. In our model, this rate constant depends on the current free self-interstitial supersaturation, in such a way that the formation can be reaction limited for small sizes when the I supersaturation has significantly decreased [6]. However, in agreement with Cowern et al., we found that good fits is possible only by considering some stable "magic" sizes, i.e. cluster containing 4 and 8 atoms. For clusters of size larger than 10 we used {311} formation energies tending asymptotically toward 0.67 eV for large values of n, as depicted on Fig.2.

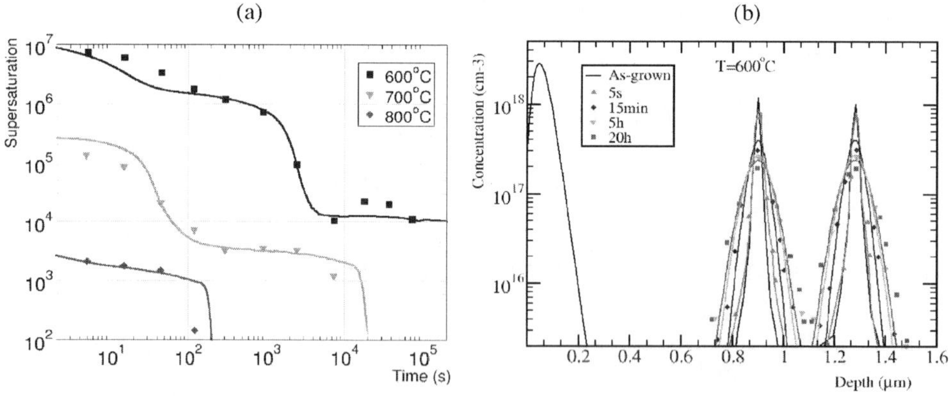

Fig.3 : Comparison between experimental and calculated I supersaturation (a), and corresponding calculated B profiles corresponding to ref.[7] (b)

The corresponding calculated curves are given on Fig.3. On Fig.3(a) we show the excellent agreement between the calculated evolution of the self-interstitial supersaturation as compared to the values extracted from the experimental profile broadening. It is emphasized that our approach allows a particularly nice description of the time evolution of the I supersaturation, especially for low thermal budgets. Since our I evolution model is coupled with the B diffusion equations, it is then possible to simulate the corresponding B broadening. The calculated B profiles are shown on Fig.3(b), and exhibit all the features experimentally observed in [7], i.e. the exponential low concentration tail for low thermal budgets evolving to symmetrically broadened profiles for longer times.

BORON INTERSTITIAL CLUSTER FORMATION

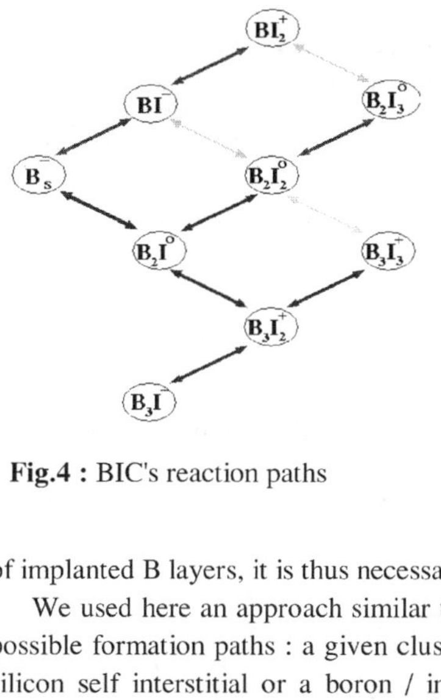

Fig.4 : BIC's reaction paths

The simulation of Cowern and coworker's experiment presented above is possible without taking into account the boron interstitial complexes formation, because the boron concentration is low enough. However, when a high concentration of free self-interstitials co-exists with a significant boron concentration, even below the solubility limit, as for example when crystalline silicon is implanted with medium doses of boron, the presence of these immobile BIC's are usually invoked to account for the experimental observations, as for example the loss of electrical activity of the dopant. In order to obtain meaningful simulations of the redistribution and activation of implanted B layers, it is thus necessary to model the BIC's formation.

We used here an approach similar to the one used for the IC's, but considering the various possible formation paths : a given cluster can grow or dissolve by the addition or release of a silicon self interstitial or a boron / interstitial pair. In order to reduce the total number of equations to handle, we restricted ourselves to a limited number of BIC's. In absence of any direct experimental data concerning these clusters, we chose the BIC's structures and charge states obtained from a recent ab-initio theoretical calculation [8]. The corresponding formation paths are given on Fig.5. The various kinetics parameters were also computed with the help of the formation and binding energies given in [8].

The model parameters were eventually tuned by fitting on the experimental profiles of Pelaz et.al., which correspond to the redistribution of a buried B layer, with a peak concentration in the 10^{19} cm^{-3} range, following defect creation by a 2.10^{13} cm^{-2} Si implant, annealed at 800°C for 35 min. Here again we used a simple +1 model to initialize the I concentration. Fig.6 shows the comparison between the calculated and experimental B profiles, and evidences the quality of the simulation.

As a final test of the capability of our fully coupled model to handle all the interactions existing between boron and the free and clustered self-interstitials, we show on Fig.7 the good

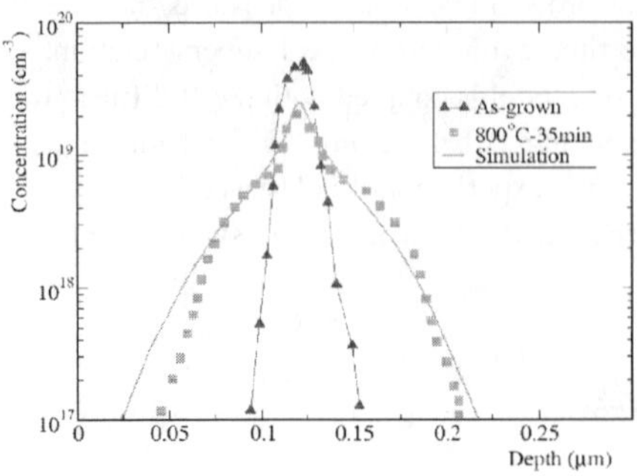

Fig.5 : Comparison between experimental and simulated B profiles evidencing the influence of the BIC's. Data are from ref.[9]

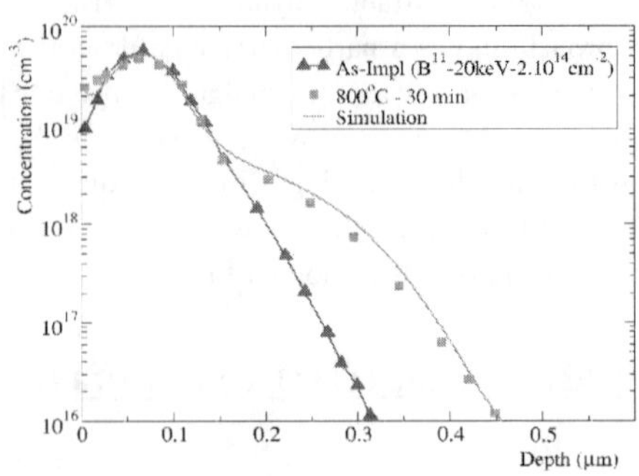

Fig.6 : Comparison between experimental and simulated B profiles implanted at 20 keV / 2.10^{14} cm^{-2}, annealed at 800°C for 30 min.

agreement we obtained for the simulation of the redistribution at 800°C, 30 min of a boron profile implanted at 20 keV / 2.10^{14} cm^{-2} (Fig 7).

CONCLUSION

In this contribution we have shown that, by fully coupling the dopant pair-diffusion model with the kinetics description of the growth and dissolution of the various I-related clusters, it is possible to simulate boron diffusion in a very large range of experimental conditions, from conventional predeposition steps to RTA after low energy implantation.

REFERENCES

1. N.E.B. Cowern, K.T.F. Jansen and H.F.F. Jos, J. Appl. Phys. **68**, 6191 (1990).
2. F. Cristiano, J. Grisiola, B. Colombeau, M. Omri, B. de Mauduit, and A. Claverie, J. Appl. Phys. **87**, 842 (2000).
3. D. Mathiot and J.C. Pfister, J. Appl. Phys. **55**, 3518 (1984).
4. Orr Arienzo, R.Glang, R.F. Lever and R.K. Lewis, J. Appl. Phys., **63**(1),117 (1988).
5. P. Pichler, W. Jungling, S. Selberherr, E. Guerrero, and H. W. Pötzl, IEEE Trans. Computer-Aided Design **4**, 384 (1985).
6. C. Ortiz and D. Mathiot, M.R.S. Symp. Proc. **669**, J5.6, (2001).
7. N.E.B. Cowern, G.Mannino, P.A. Stolk, F. Roozeboom, H.G.A. Huizing, J.G.M. van Berkum, F. Christiano, A. Claverie, M. Jaraìz, Phys. Rev. Lett. **82,** 4460 (1999).
8. J. Lenosky, B.Sadigh, S.K. Theiss, M.-J. Caturla and T.Diaz de la Rubia, Appl. Phys. Lett **77**, 1834 (2000).
9. L. Pelaz, M. Jaraiz, G.H. Gilmer, H.-J. Gossmann, C.S. Rafferty, D.J. Eaglesham and J.M. Poate, Appl. Phys. Lett. **70**,2285 (1997).

Mat. Res. Soc. Symp. Proc. Vol. 717 © 2002 Materials Research Society C5.4

Complete Suppression of the Transient Enhanced Diffusion of B Implanted in Preamorphized Si by Interstitial Trapping in a Spatially Separated C-Rich Layer

E. Napolitani[1], A. Coati[1], D. De Salvador[1], A. Carnera[1], S. Mirabella[2], S. Scalese[2], F. Priolo[2]
[1]INFM and Dipartimento di Fisica, via Marzolo 8, 35131 Padova, ITALY.
[2]INFM and Dipartimento di Fisica e Astronomia, Corso Italia 57, 95129 Catania, ITALY.

ABSTRACT

A method for completely suppressing the transient enhanced diffusion (TED) of boron implanted in preamorphized silicon is demonstrated. Boron is implanted in a molecular beam epitaxy (MBE) grown silicon sample that has been previously amorphized by silicon implantation. The sample is then annealed in order to epitaxially regrow the amorphous layer and electrically activate the dopant. The back-flow of silicon interstitials released by the preamorphization end-of-range (EOR) damage is completely trapped by a carbon-rich silicon layer interposed by MBE between the damage and the implanted boron. No appreciable TED is observed in the samples up to complete dissolution of the EOR damage, and complete electrical activation is obtained. The method might be considered for the realization of ultra shallow junctions for the far future complementary metal-oxide semiconductor technology nodes.

INTRODUCTION

The realization of ultra shallow junctions for the future complementary metal-oxide semiconductor (CMOS) technology nodes [1] is extremely challenging for the ion implantation community, as well as for researchers involved in point defect-dopant interactions in silicon. In fact, even by implanting the dopant extremely close to the surface, i.e. by reducing the implant energy well below 1 keV, the broadening, produced during the post-implantation activation annealing by the transient enhanced diffusion (TED) [2,3], still severely hinders the application of the above processes to the most aggressive CMOS device scaling rules.

Several attempts have been made to control and reduce the TED of boron in silicon. Implantation in preamorphized silicon is one of the most studied and promising methods, due to the advantage of suppressing dopant channeling and the capability of the end-of-range (EOR) defects (mainly dislocation loops) to trap the excess interstitials left by the preamorphization implant. However, since the dislocation size and density critically depend on the implant and annealing conditions, interstitial trapping and, consequently, TED suppression not always occur [4, 5], and in most cases an increase of enhanced diffusion is even observed due to the release of interstitials from EOR defects dissolution [4].

Substitutional C incorporation in Si is another promising method for reducing TED, due to its high efficiency in trapping the silicon I, as reported by Stolk et al. [6]. However, there are several studies reporting the detrimental effect on the electrical properties of the material due to the presence of carbon [7,8]. Therefore, even if C is effective in reducing TED, a method for TED suppression involving C must avoid, for electrical considerations, the presence if C in the active part of the device, and in particular in the doped region.

In this paper we demonstrate a method for suppressing the TED of B implanted in preamorphized Si, in which substitutional C is used to trap the excess I. The back-flow of silicon I released by the preamorphization EOR damage is suppressed by a C-rich silicon layer interposed between the damage and the implanted B. Carbon is therefore incorporated in a region spatially well separated from the shallow implanted dopant, and complete suppression of TED is achieved with good electrical activation.

EXPERIMENT

Fig. 1a shows a schematic description of the experiment. Molecular beam epitaxy (MBE) was used to grow on Si a 100 nm thick silicon layer enriched in carbon with a C concentration of $2.4x10^{20}/cm^3$, corresponding to a total C dose of $2.4x10^{15}/cm^2$. The layer was covered by a pure 250 nm thick Si cap. The temperature of the substrate was kept at 500 °C during growth. A second similar sample was grown without C, as a reference. Both samples were then amorphized from the surface to a depth of about 550 nm by a 250 keV Si implantation at a dose of $3x10^{15}/cm^2$ at liquid nitrogen temperature and subsequently implanted with 10 keV B at a dose of $1x10^{14}/cm^2$. After solid phase epitaxy (SPE), obtained with a 450 °C, 30 min plus 700 °C, 30 min furnace annealing under N_2 flux [9], we recovered the original structure with a good crystalline quality, as verified by Rutherford backscattering spectromery-channeling. Finally, on these regrown samples, we performed isochronal rapid thermal annealing (RTA) at 900 and 1100 °C for 30 s.

The chemical concentration depth profiles of C and B were then obtained by secondary ion mass spectrometry (SIMS), using a CAMECA IMS-4*f* instrument, with a 3 keV O_2^+ or 14.5 keV Cs^+ analyzing beam while collecting B^+ or C^- secondary ions respectively. The substitutionality of C was measured by high resolution X-ray diffraction (HRXRD) (004) rocking curves, taken by means of a Philips MRD diffractometer in standard setup (Bartels monochromator in Ge (220) setting, Cu tube at 40kV and 40mA). The dopant activation of the samples was evaluated by measuring the sheet resistance with a four-point probe apparatus. The extended defects produced by the preamorphization implant were characterized by transmission electron microscopy (TEM).

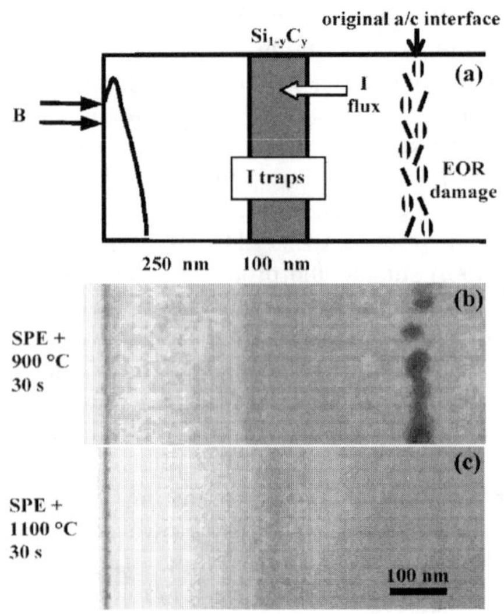

***Figure 1** (a) Schematic description of the experiment. Cross sectional TEM images of the sample with C after annealing at 900°C (b) and 1100°C (c).*

RESULTS

The excess I left beyond the original a/c interface after the SPE cycle will evolve during the following RTA in complex agglomerates, such as the {311}'s and dislocation loops. The formation and dissolution of the above defects will sustain for a certain time a flux of I towards the surface, producing TED of the B implanted close to surface. In our experiment, the substitutional C-rich layer is located just between the EOR damage and the B, so as to trap the I's flowing towards the surface and, consequently,

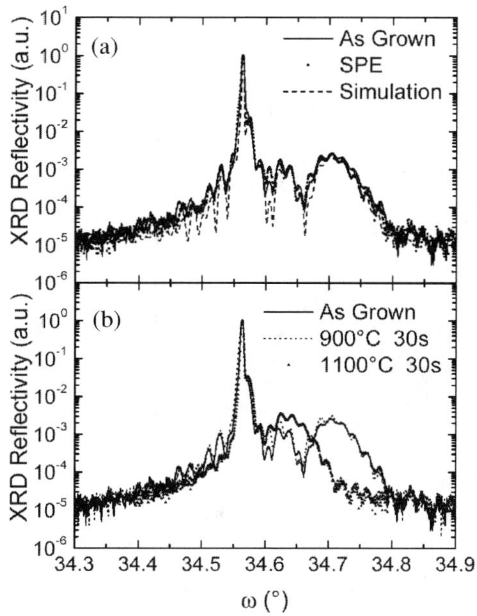

Figure 2 *(a) HRXRD curves of the as grown sample (continuous), the sample after SPE (dotted), and a simulation assuming C 100% substitutional (dashed). (b) The as grown curve is compared to the curves obtained after SPE plus 30s RTA at 900 °C (dashed) and 1100°C (triangles).*

suppress TED (see Fig. 1a). Fig. 1b and 1c show the cross sectional TEM images of the sample with C after annealing at 900°C and 1100°C respectively. After 30 s at 900 °C we obtained the complete dissolution of {311} defects (Fig. 1b), which are the most significant source of interstitials that induces TED. However, a large amount of interstitial type dislocation loops are still observed, which may act as a reservoir for a lot of I's. These defects are completely dissolved after 30s at 1100 °C, where no EOR defects are anymore visible (Fig. 1c). These defects are known to be highly stable and able to store the excess I's without releasing them, even after aggressive thermal processes at 1000°C [10], so their presence wouldn't be a problem for the CMOS planar technology. The 1100 °C RTA was performed in order to check the ability of the C-rich layer to stop the interstitials until complete dissolution of the EOR defects.

In order to be effective as an interstitial's trap, the C must be in a substitutional position [11]. Fig. 2 shows the HRXRD curves taken at different steps of the process. Fig. 2a reports the curves relative to the as grown sample and the sample after SPE, compared to the dynamical theory simulation obtained by assuming the chemical C profile, as obtained by SIMS measurements, to be fully substitutional. A broad peak is present, centered at an angle of 34.7°, due to the tensile strain induced in the lattice by the substitutional C. The three curves are almost identical, demonstrating that C is 100% substitutional either before and after SPE. Fig. 2b shows the HRXRD curves after the RTA processes at 900°C and 1100°C, compared to the as grown one. A very small shift of the C peak is observed in the 900 °C curve with respect to the as grown sample, as a result of the interaction of the interstitial flow with the C-rich layer. However, the large part of C is still substitutional after the 900°C RTA, and therefore still efficient as a I trap. After the 1100 °C the shift of the C peak is instead significant, indicating a reduction of about a factor of two of the substitutional C. This is due to the strong interaction of the I flux with the C-rich layer, due to the complete dissolution of the EOR damage. However, even after such an aggressive thermal process, a significant amount of C is still substitutional.

Fig. 3 shows the SIMS concentration profiles of boron implanted at 10 keV, $1x10^{14}/cm^2$ in preamorphized Si before and after SPE for the sample with C (identical profiles were obtained for the reference sample), and after SPE plus 900°C or 1100°C RTA for 30s both for the sample with C and for the reference sample. All the profiles after RTA, as shown in Fig. 3, were also satisfactorily simulated by solving the Fick's law equation, by using as input the profile

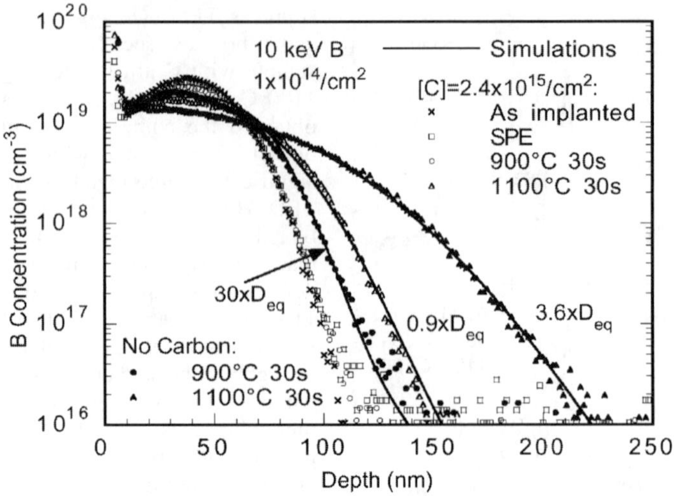

Figure 3 *SIMS profiles of B implanted at 10 keV, $1x10^{14}/cm^2$ in preamorphized Si before (x) and after (□) SPE, after SPE plus 900°C RTA for 30s in sample with (○) and without (●) C, and after SPE plus 1100°C RTA for 30s in sample with (Δ) and without (▲) C. Simulated profiles are shown as continuous lines together with the average diffusivity used in the calculations.*

measured by SIMS after SPE, and assuming a diffusion coefficient constant with time and depth which was used as a free parameter. Without C, the dissolution of {311} defects produces a significant TED. For example, as shown in Fig. 3, the reference sample was satisfactorily fitted after 30s at 900°C by assuming an average diffusivity higher by a factor of 30±10 with respect to equilibrium [12].

As Fig. 3 demonstrates, the presence of C has a dramatic influence on B diffusion. The profile after 900°C with the C-rich layer, compatibly with equilibrium diffusion [12], is undistinguishable from the sample after SPE. Therefore, a substitutional C-rich Si layer with a C dose of 2.4 $x10^{15}/cm^2$ completely suppresses the TED produced by the complete dissolution of {311} defects from the EOR damage. This, added to the fact that the diffusion during the SPE is extremely small, results in a profile broadening during the whole process (i.e. between the as implanted sample and after SPE plus RTA) causing a shift of less than 3 nm at the concentration level of $1x10^{17}/cm^3$.

The profiles in Fig. 3 after 30 s at 1100°C show what happens after complete dissolution of the EOR defects. A huge diffusion is observed in the sample without C, supported by the complete dissolution of the EOR damage. The diffusion enhancement in the 30 s, 1100 °C RTA was estimated by comparison with the equilibrium diffusivity measured in a sample obtained by MBE growth of B deltas in Silicon and annealed simultaneously with the regrown samples. This gives a diffusion enhancement of 3.6±0.5 at 1100 °C. Also in this case the introduction of C suppresses dramatically the diffusion, as shown by the sample with C where the diffusion enhancement is 0.9±0.2, i.e. the average diffusivity is compatible with that of equilibrium. Thus, the C-rich layer completely prevents the interstitials wind generated by the complete dissolution of EOR damage by reaching the shallow B.

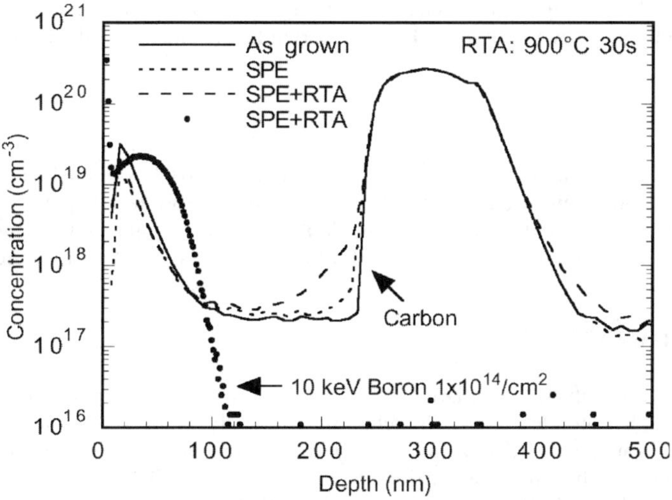

Figure 4 *Carbon profiles before (solid) and after (short dashed) SPE, and after SPE plus RTA at 900°C (dashed), compared to the B profile after SPE plus RTA at 900°C (●).*

In order to realize a shallow junction, the B must be electrically active. We performed four-point probe analyses in order to measure the sheet resistance and therefore evaluate the dopant activation. If fully active, the B in the sample with C after SPE or after SPE plus RTA at 900°C, would produce a sheet resistance of about 830 Ω/square. After SPE we obtained a value of 1240 Ω/square, indicating that the regrowth process is not enough to fully activate B. Complete activation was instead reached after SPE plus RTA at 900°C, where we obtained a sheet resistance of 850±50 Ω/square.

The question now is whether the C-rich layer will overlap the B profile, because of C diffusion which may occur during the RTA necessary to fully activate B, i.e. the 30 s 900 °C RTA. Such overlapping might be detrimental for the junction electrical properties. In Fig. 4, the C profiles before and after SPE, and after SPE plus the RTA at 900°C are reported and compared to the B profile after SPE plus RTA at 900 °C. The SPE process leaves the C profile almost unchanged, while a diffusion tail is observed in the Si cap after 30s at 900 °C. Such a tail starts at a concentration level of about $2x10^{18}/cm^3$, and decays reaching the SIMS background of $2x10^{17}/cm^3$ at a depth of about 120 nm. Therefore, the C which diffuses out of the C-rich layer will certainly overlap the B profile well below the concentration level of $2x10^{17}/cm^3$, i.e. more than one order of magnitude lower than the C concentration used by P. A. Stolk. et al [6]. It must be pointed out that the significant C surface peak with a pronounced tail does not correspond to a real C contamination in the bulk. It is in fact a well known SIMS artifact produced during analysis by the relocation by the sputtering beam of the C contamination which is always present at the surface of Si exposed to air. In fact, similar features are usually observed in a virgin silicon sample. As a further check we performed analyses with a much less energetic beam, allowing a much better depth resolution (but with a worse detection limit), which showed a significantly steeper tail, with the C signal confined within 10 nm from the surface. Similar considerations can be done for the deeper interface of the C-rich layer, which is artificially broadened by the sputtering beam. As a matter of fact, the exponential decay of such interface has the same slope of the feature which is present at the surface. Therefore, we can conclude that after 30s at 900°C, where the B is fully active without any TED, the C-rich layer remains well spatially separated from the dopant.

CONCLUSIONS

In conclusion, we demonstrated that by introducing a C-rich silicon layer between B implanted in preamorphized Si and the EOR damage it is possible to completely recover the crystallinity of the material and fully activate the dopant with TED completely suppressed. This method might be considered for the realization of ultrashallow junctions for far future generation devices.

ACKNOWLEDGMENTS

The authors wish to thank A. V. Drigo, M. Berti, A. Terrasi, S. Pulvirenti, G. Bisognin, for helpful discussions, S. Pannitteri and C. Spinella for TEM analyses, M. G. Grimaldi for expertise in four-point probe analyses, R. Storti and A. Marino for technical assistance. This work has been partially supported by Consiglio Nazionale delle Ricerche (CNR) through the project "5% Microelettronica" and by MURST through the project COFIN 2000.

REFERENCES

1. International Technology Roadmap for Semiconductors (Semiconductor Industry Association, Austin, TX, 2000).
2. V. Privitera, E. Napolitani, F. Priolo, S. Moffatt, A. La Magna, G. Mannino, A. Carnera, A. Picariello, Mater. Sci. Semicond. Process. 2, 35 (1999).
3. E. Napolitani, A. Carnera, E. Schroer, V. Privitera, F. Priolo, S. Moffatt, Appl. Phys. Lett. 75, 1869 (1999).
4. K. S. Jones, L. H. Zhang, V. Krishnamoorthy, M. Law, D. S. Simons, P. Chi, L. Rubin, R. G. Elliman, Appl. Phys. Lett. 68, 2672 (1996).
5. L. S. Robertson, M. E. Law, K. S. Jones, L. M. Rubin, J. Jackson, P. Chi, D. S. Simons, Appl. Phys. Lett. 75, 3844 (1999).
6. P. A. Stolk, D. J. Eaglesham, H. - J. Gossmann, J. M. Poate, Appl. Phys. Lett. 66, 1370 (1995).
7. I. Ban, M. Öztürk, K. Christensen, D. M. Maher, Appl. Phys. Lett. 68, 499 (1996).
8. F. Priolo, G. Mannino, M. Miccichè, V. Privitera, E. Napolitani, A. Carnera, Appl. Phys. Lett. 72, 3011 (1998).
9. J. W. Strane, H. J. Stein, S. R. Lee, S. T. Picraux, J. K. Watanabe, J. W. Mayer, J. Appl. Phys. 76, 3656 (1994).
10. C. Bonafos, D. Mathiot, A. Claverie, Appl. Phys. Lett. 83, 3008 (1998).
11. S. Mirabella, A. Coati, D. De Salvador, E. Napolitani, A. Mattoni, G. Bisognin, M. Berti, A. Carnera, A. V. Drigo, S. Scalese, S. Pulvirenti, A. Terrasi, F. Priolo, Phys. Rev. B 65, 045209 (2002).
12. P.M. Fahey, P.B. Griffin, J.D. Plummer, Rev. Mod. Phys. 61, 289(1989); R. B. Fair, in *Impurity Doping Processes in Si*, edited by F. F. Y. Wang (North-Holland, Amsterdam, 1981), p. 315; Y. M. Haddara, B. T. Folmer, M. E. Law, T. Buyuklimani, Appl. Phys. Lett. 77, 1976 (2000).

Mat. Res. Soc. Symp. Proc. Vol. 717 © 2002 Materials Research Society C5.5

Self-interstitials and substitutional C in silicon: Interstitial- trapping and C- clustering

S. Mirabella, S. Scalese, A. Terrasi, and F. Priolo
INFM and Dept. of Physics and Astronomy, University of Catania, Corso Italia 57, 95129 Catania, Italy
A. Coati, D. De Salvador, E. Napolitani, and M. Berti
INFM and Dept. of Physics, University of Padova, Via Marzolo 8, 35131 Padova, Italy

ABSTRACT

The interactions between self-interstitials (*I*'s) produced by 20 keV silicon implantation, and substitutional carbon in silicon have been studied using a $Si_{1-y}C_y$ layer interposed between a near surface *I* source and a deeper B spike used as a marker for the *I* concentration. The $Si_{1-y}C_y$ layer behaves as a filtering membrane for the interstitials flowing towards the bulk. This trapping ability is related to the total C amount in the $Si_{1-y}C_y$ membrane. Substitutional carbon atoms interacting with self-interstitials are shown to trap *I*'s, to be removed from their substitutional sites, and to precipitate into the C-rich region. After precipitation, C atoms are not able to further trap injected self-interstitials. The atomistic mechanism leading to Si-interstitial trapping has been investigated by developing a simulation code describing the migration of injected interstitials. By a comparison with the experimental data it was possible to derive quantitative indications on the trapping mechanism. It is shown that one Si-interstitial is able to deactivate about two C traps by means of interstitial trapping and C-clustering reactions.

INTRODUCTION

The point defects in crystalline silicon, interstitials (*I*'s) and vacancies, and their interactions with the most commonly used dopants have been attracted great interest in the semiconductor scientific community [1-4]. In fact, changes in the point defects equilibrium populations cause deviations from the equilibrium diffusion coefficients of the dopants and induce anomalous diffusion, such as the well-known transient enhanced diffusion (TED), that have several implications on the advanced microelectronic technology [5-7].

In particular, it has been demonstrated that B diffuses exclusively via an *I* mediated process [2,8], and thus B can be used as a sensitive tool to measure the *I* concentration in the different experimental conditions [2,9,10]. On the other hand, C diffusivity has been shown to be *I* mediated [11] but, differently from B, C appears to be a trap for Si self- interstitials and thus to be able to reduce *I* concentration [5,12-14]. Recently, the ability of a remote $Si_{1-y}C_y$ layer to suppress the TED of B implanted in preamorphized silicon was demonstrated [7]. Moreover, a competition between C diffusion and C clustering under interstitial injection by thermal oxidation has been shown [15]. So, the *I* trapping by C involves very complex phenomena.

In this paper we study the effect of a $Si_{1-y}C_y$ layer placed between a deep B spike and a surface implanted region (Si cap layer) on the TED of the deep B spike. Moreover, we characterized the evolution of both the carbon concentration profile and its lattice location under the interstitial wind. In our conditions of high *I* flux, C clustering will be shown to be predominant with respect to C out-diffusion. Moreover, clustered C will be shown to be no more active as an interstitial trap. Finally, the number of *I*'s interacting with C_S, estimated by the model described in Ref. 16, is compared with the amount of the C out-gone from the substitutional sites. Such a comparison, leads to the conclusion that one self interstitial produces the deactivation of 2 C atoms indicating that the I-C clusters are particularly rich in C.

EXPERIMENTAL DETAILS

We grew by molecular beam epitaxy (MBE) a Si structure containing a deep B spike (depth=660 nm, peak concentration~6×10^{18} B/cm^3, FWHM~8 nm) and a $Si_{1-y}C_y$ layer capped by a 200 nm Si cap layer. The C fraction was in the range 0.015 - 0.03 at. %, and the $Si_{1-y}C_y$ layer thickness (*d*) was varied between ~10 and 250 nm (total C dose ranging from $\sim 7 \times 10^{12}$ to 4×10^{14} C/cm^2). A similar structure without carbon was also grown, as a reference sample. The role of the B spike is that of acting as a marker for the interstitial flux.

The chemical concentration depth profiles of C and B were then obtained by secondary ion mass spectrometry (SIMS), using a CAMECA IMS-4*f* instrument, with a 3 keV O_2^+ analyzing beam. Using 14.5 keV Cs^+ primary ions, the background C contamination in the MBE silicon layers was measured to be 2×10^{17} C/cm^3. High resolution X-ray diffraction (HRXRD) measurements were collected with a Philips MRD diffractometer (Bartels Ge (220) 4-crystals monochromator) to have a quantitative measure of the total substitutional carbon dose [17].

Ion implantation of 20 keV Si was performed at doses of 2×10^{13}, 5×10^{13} and 1×10^{14} ions/cm^2 (the implant damage region being contained into the Si cap layer). All the samples were subjected to rapid thermal annealing (RTA) at a temperature of 800 °C for 5, 5 and 10 min for the three implant doses, respectively (to fully extinguish the TED of boron for each dose).

DISCUSSION

In Figure 1 we show the SIMS profile of the deep B spike after the highest dose implantation and annealing. The continuous line (——) is the profile of the B spike in the as-grown reference sample (which does not contain C on purpose). The open circles (○) represent the spike annealed at 800 °C for 5 min without any implant: the thermal broadening of the profile is very small (mean diffusion length is < 2 nm) [10].

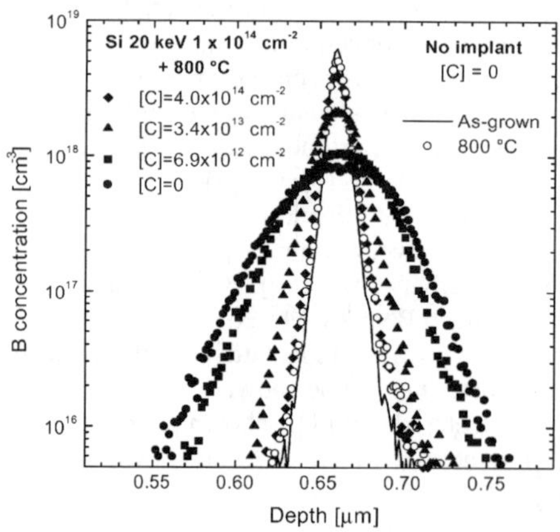

Figure 1. SIMS profiles of B spike. The line (——) and the open circles (○) are the reference sample as-grown and annealed at 800 °C, respectively. The solid symbols refer to 20 keV, 1×10^{14} cm^{-2} Si implanted and 800 °C annealed samples with different C doses: 4×10^{14} C/cm^2 (◆), 3.4×10^{13} C/cm^2 (▲), 6.9×10^{12} C/cm^2 (■) and reference- no C (●).

The profiles of the boron spike, for the implanted samples, are plotted with filled symbols: the largest broadening is obtained in the reference sample (no C, ●), as expected if we assume that C acts as a trap for Si-interstitials. The effect of the $Si_{1-y}C_y$ membrane on the boron enhanced diffusion depends on the total C dose: for the lowest C dose, 6.9×10^{12} at/cm^2 (■), we obtain a slight reduction of B TED with respect to the reference sample (no C), while for the highest dose, 4.0×10^{14} at/cm^2 (◆), an almost complete suppression of the enhanced diffusion is registered. So, the migration of interstitials towards the B spike is affected by the $Si_{1-y}C_y$ layer.

To have quantitative measurements of the *I* concentration we used the Boron peak broadening. The difference $\sigma^2 - \sigma_0^2$ of the variances of the as grown (σ_0^2) and post- TED (σ^2) B peaks (calculated by SIMS concentration distributions) is the variance of the Boron distribution after deconvolution with the as grown profile. This quantity is related to the I concentration through the relation:

$$W = \frac{\sigma^2 - \sigma_0^2}{2} = D_B^{eq} \int \frac{c_I}{c_I^{eq}} dt \qquad [1]$$

where W is the integrated diffusivity, D_B^{eq} the B equilibrium diffusivity, and c_I and c_I^{eq} the *I* concentration at time t and the equilibrium one, respectively [2].

The results of such an exercise are shown in Figure 2, where W is plotted as a function of the total carbon dose for the different implant doses, together with the case of no implant (○); the dotted lines (······) are only guides for the eyes, while the error bars, where not drawn, are to be considered lower than the symbols- size. The effect of the lowest implant dose (2×10^{13} ions/cm^2, ▼) on the B spike broadening is undistinguishable from that measured after the pure thermal annealing (○). This is consistent with *I* diffusion in a trap containing medium[10]. If the implant dose is increased to 5×10^{13} ions/cm^2 (✻) or to 1×10^{14} ions/cm^2 (●), we measure an increasing broadening with respect to the equilibrium condition. This effect becomes smaller the higher is the total carbon dose, and it vanishes for the highest C dose (4×10^{14} cm^{-2}): for this case none of the investigated implant doses produces spike broadening distinguishable from the equilibrium one, this means that the $Si_{1-y}C_y$ membrane, because of its large carbon dose, remains totally opaque to the interstitials flow.

So, the $Si_{1-y}C_y$ layer behaves as an efficient filter for the interstitial flow, reducing the TED of boron. Moreover, this TED reduction depends on the total C dose in the $Si_{1-y}C_y$ membrane.

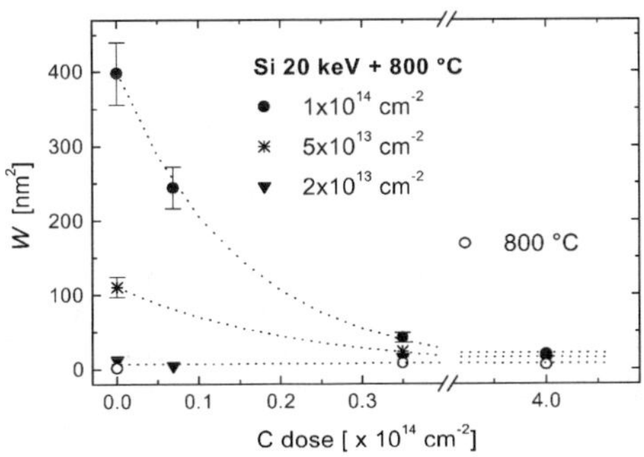

Figure 2. Broadening of the B spike versus total C dose for different implant doses: 1×10^{14} cm^{-2} (●), 5×10^{13} cm^{-2} (✻), 2×10^{13} cm^{-2} (▼). The open circles (○) represent the pure thermal broadening. The dotted lines (······) are guides for eyes.

In order to understand the behavior of C atoms under the interstitials flow, we studied in detail the sample with the medium carbon dose (3.4×10^{13} C/cm^2). By HRXRD strain analyses (not shown), we saw that, for all the samples, all the C incorporated during the MBE growth is substitutional, even after the thermal treatment at 800 °C in equilibrium conditions. But, after the implantation of 1×10^{14} Si/cm^2 and annealing, no strain due to C is revealed, which means that either C is completely out- diffused or C is no more on substitutional sites, as we expect if occurs the "kick-out" trapping reaction $I + C_s \rightarrow C_i$ (an *I* pushes out a substitutional carbon, C_s, creating a highly mobile interstitial carbon, C_i)[12]. For this implant dose we saw (Figure 1) that the migration of interstitials towards the bulk reaches the B spike by passing through the $Si_{1-y}C_y$ membrane. The HRXRD analysis clearly indicates that the interstitials flow, as a result of the trapping reaction, has strongly modified the original arrangement of the carbon atoms: C loses its original substitutional position and does not contribute anymore to the measured strain.

The SIMS analyses on the medium carbon dose sample after implantation and/or annealing are reported in Figure 3, where the as-grown C profile is the continuous line (——), while the circles refer to an RTA (800 °C, 10 min) performed on the as-grown sample (○) or on the 1×10^{14} Si/cm^2 implanted sample (●). As it clearly appears, both the thermal process and the implantation + annealing do not induce strong carbon diffusion: both C profiles after the annealing processes nearly overlap the as-grown sample. On the other hand, for these samples HRXRD analysis show that while the thermal process does not move the carbon from its substitutional position, the implantation + annealing process leads to a total loss of C induced strain. Therefore, *I*'s interact with the substitutional carbon atoms in such a manner that they remove C from the substitutional sites but, once removed, the C atoms are not able to significantly diffuse out of the $Si_{1-y}C_y$ layer. To test the thermal stability of these carbon atoms, we further processed the sample, previously implanted with 1×10^{14} Si/cm^2 and annealed, in a furnace at 850 °C for 4 hours in N_2 atmosphere: the SIMS profile is shown in Figure 3 by closed squares (■). Most of the C atoms remain into the $Si_{1-y}C_y$ layer, indicating a great stability of this precipitated/clustered carbon, in contrast with the equilibrium diffusion that would cause the disappearance of the original C box below the detection limit of about 10^{18} at/cm^3 (Figure 3, open squares,).

To investigate the residual trapping capability, if any, we performed a further Si implantation followed by annealing at 800 °C on the sample previously implanted and annealed. SIMS results (not shown) demonstrate that the $Si_{1-y}C_y$ layer seems to be almost ineffective for the trapping of the second interstitials flow, resulting in a unreduced TED of the deep B spike.

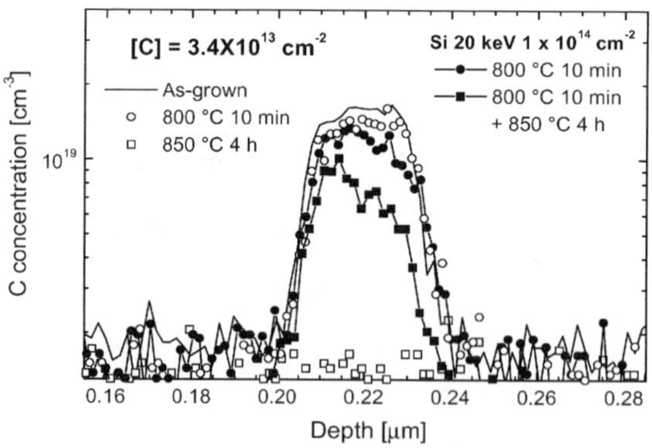

Figure 3. SIMS profiles of C in the 3.4×10^{13} C/cm^2 sample: the line (——) refers to the as-grown sample, while the closed symbols are the C profile after 20 keV, 1×10^{14} cm^{-2} Si implantation and annealing at 800 °C for 10 min (●) and 850 °C 4 hours (■). The relative open symbols refer to pure thermally annealed samples.

Summarizing, substitutional carbon atoms interacting with Si self-interstitials are able to trap them, are removed from their substitutional sites, are forbidden to diffuse and, in this condition, they are unable to trap further implanted self-interstitials.

Finally, in Ref. 16, it is described a model for the diffusion of *I*'s, injected by implantation, in a trap containing medium, taking into account the surface recombination. This calculation allows to estimate the amount of *I's* that actually interacts with C_S in our $Si_{1-y}C_y$ layers and then the number of *I's* trapped by each C atoms. In Fig.4 the number of interstitials interacting with C, as predicted by the model, is compared with the amount of C_S reduction as measured by XRD. Fit of the data allows to show that each *I* deactivates about 2 substitutional C indicating that the C-*I* clusters formed under *I* injection by implant are particularly reach in C. An independent estimate of such ratio, obtained by the analysis of the B-delta doping broadening, is shown in Ref. 16.

CONCLUSIONS

The mechanism of self-interstitials trapping by substitutional carbon in silicon was studied. In particular, the ability of a $Si_{1-y}C_y$ layer, interposed between the near surface self-interstitials source and a deep B spike, to filter the interstitials flow towards the bulk and, consequently, to strongly reduce the boron enhanced diffusion, has been shown. This trapping ability is related to the total C dose in the $Si_{1-y}C_y$ membrane. Moreover, once having trapped the self-interstitials, the substitutional carbon atoms lose their substitutional sites but cannot diffuse away. In fact, C atoms, that have interacted with self-interstitials, precipitate into the C-rich region showing a high thermal stability. C atoms result to be an efficient sink for Si-interstitials only if they are on substitutional sites, moreover, after the trapping reaction this condition is lost, and the C atoms are no more able to interact with new self-interstitials. In order to characterize the interactions between the C-traps and the Si-interstitials, a model describing the implanted interstitials migration was developed, taking into the presence of intrinsic traps in the material and the presence of C traps. By a comparison with the experimental data, the amount of *I*'s that actually interact with C atoms was estimated, and an atomistic interpretation of the C-*I* reaction was given. In fact, by fitting experimental data with the model, it was found that one Si-interstitial is able to deactivate two C traps. The interaction between Si-interstitials and C is supposed to proceed sequentially by means of I-trapping and C-clustering reactions.

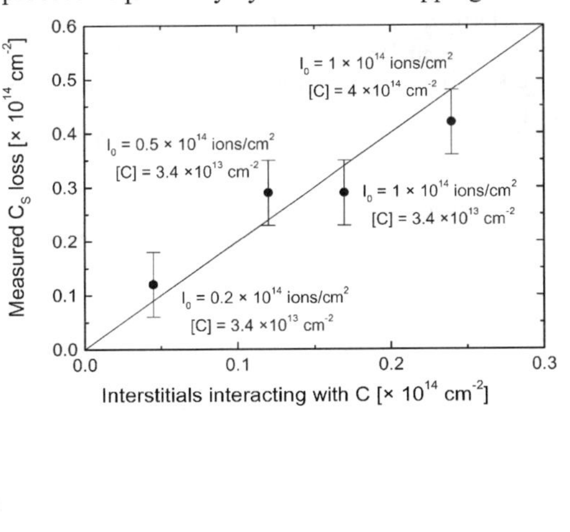

Figure 4. HRXRD measured loss of substitutional carbon versus the amount of predicted *I*'s interacting with C. The continuous line (——) is a straight line with slope 2. A linear fit (not shown) of the data points by means of a straight line through the origin of the axes yields a slope of 1.8 ± 0.1.

ACKNOWLEDGEMENTS

This work has been partially supported by MURST through the project COFIN 2000.

REFERENCES

1. P. M. Fahey, P. B. Griffin, and J. D. Plummer, Rev. Mod. Phys. **61**, 289 (1989).
2. N. E. B. Cowern, K. T. F. Janssen, G. F. A. van de Walle, and D. J. Gravesteijn, Phys. Rev. Lett. **65**, 2434 (1990).
3. H. Bracht, E. E. Haller, and R. Clark-Phelps, Phys. Rev. Lett. **81**, 393 (1998).
4. H.-J. Gossmann, T. E. Haynes, P. A. Stolk, D. C. Jacobson, G. H. Gilmer, J, M. Poate, H. S. Luftam, T. K. Mogi, and M. O. Thompson, Appl. Phys. Lett. **71**, 3862 (1997).
5. P. A. Stolk, J. H.-J. Gossmann, D. J. Eaglesham, D. C. Jacobson, C. S. Rafferty, G. H. Gilmer, M. Jaraiz, J. M. Poate, H. S. Luftman and T. E. Haynes, J. Appl. Phys. **81**, 6031 (1997) and references therein.
6. E. Napolitani, A. Carnera, E. Schroer, V. Privitera, F. Priolo, S. Moffatt, Appl. Phys. Lett. **75**, 1869 (1999).
7. E. Napolitani, A. Coati, D. De Salvador, A. Carnera, S. Mirabella, S. Scalese and F. Priolo, Appl. Phys. Lett. **79**, 4145 (2001).
8. B. Sadigh, T. J. Lenosky, S. K. Theiss, M.-J. Caturla, T. Diaz de la Rubia, and M. A. Foad, Phys. Rev. Lett. **83**, 4341 (1999).
9. P. A. Stolk, D. J. Eaglesham, H.-J. Gossmann, and J. M. Poate, Appl. Phys. Lett. **66**, 1370 (1995).
10. S. Mirabella, A. Coati, D. De Salvador, E. Napolitani, A. Mattoni, G. Bisognin, M. Berti, A. Carnera, A. V. Drigo, S. Scalese, S. Pulvirenti, A. Terrasi, and F. Priolo, Phys. Rev. B **65**, 045209 (2002).
11. J. P. Kalejs, L. A. Ladd, and U. Gösele, Appl. Phys. Lett. **45**, 268 (1984).
12. R. Scholz, U. Gösele, J.-Y. Huh, and T. Y. Tan, Appl. Phys. Lett. **72**, 200 (1998).
13. H. Rücker, B. Heinemann, W. Röpke, R. Kurps, D. Krüger, G. Lippert, and H. J. Osten, Appl. Phys. Lett. **73**, 1682 (1998).
14. H. Rücker, B. Heinemann, D. Bolze, R. Kurps, D. Krüger, G. Lippert, and H. J. Osten, Appl. Phys. Lett. **74**, 3377 (1999).
15. D. De Salvador, E. Napolitani, A. Coati, M. Berti, A. V. Drigo, M. S. Carroll, J. C. Sturm, J. Stangl, G. Bauer, and L. Lazzarini, in *Si Front-End Processing-Physics and Technology of Dopant-Defect Interactions III*, edited by K. Jones *et al.*, Mater. Res. Soc. Symp. Proc. No. 669 (Material Research Society, Warrendale, 2001).
16. D. De Salvador, A. Mattoni, E. Napolitani, S. Mirabella, and F. Priolo, presented at 2002 MRS Spring Meeting, San Francisco, CA, 2002 (unpublished).
17. D. De Salvador, M. Petrovich, M.Berti, F. Romanato, E. Napolitani, A.V. Drigo, J. Stangl, S. Zerlauth, M. Muhlberger, F. Schaffler, and G. Bauer, Phys. Rev. B **61**, 13005 (2000).

Mat. Res. Soc. Symp. Proc. Vol. 717 © 2002 Materials Research Society C5.6

Modeling of self-interstitial diffusion in implanted molecular beam epitaxy silicon

D. De Salvador[1], A. Mattoni[1], E. Napolitani[1], A. V. Drigo[1], S. Mirabella[2], F. Priolo[2]
[1]Dept. of Physics, University of Padova and INFM, Via Marzolo 8, 35131 Padova, ITALY
[2] Dept. of Physics and Astronomy, University of Catania and INFM, Corso Italia 57, 95129 Catania, ITALY

ABSTRACT

In this work a rate equations model describing the interstitials (*I*) diffusion in a trap containing medium is presented. The model takes into account the interstitial injection by implantation and annealing and the surface evaporation. We found an analytical approximated solution of the model which allows clarifying the interplay between the parameters involved and a simple comparison with experimental data obtained by the analysis of boron delta doping arrays broadening. The calculations allow to demonstrate that the *I* injected into the bulk and toward the surface at the end of the *I* clusters dissolution does not depend on the detailed time evolution of the *I* clusters, but only on the total amount of *I* produced by the implantation. The fitting of the experimental data allows to easily quantifying important physical parameters such as the *I* evaporation rate at the surface and the density of intrinsic interstitial traps. Applications of the model are shown in the case of MBE materials intentionally doped with substitutional C. The model successfully predicts the TED reduction by MBE intrinsic *I*-traps and allows to estimate the average composition of Interstitial-Carbon clusters.

INTRODUCTION

The quantification of the number of silicon self interstitials (*I*) evolving inside a sample by implantation and subsequent annealing is a debated matter. This is a crucial problem since the interstitials regulate important physical processes such as the dopant transient diffusion and the formation of *I*-dopant and *I*-impurity clusters [1]. A proper description of the phenomena involves, in principle, the knowledge of many non-trivial physical processes: i) the dissolution dynamics of the *I* clusters produced by the implantation ii) the evaporation of interstitials through the surface iii) the interaction of the interstitials with dopants, with interstitial traps present in intrinsic MBE [2,3] silicon or with traps intentionally incorporated during growth (such as Carbon).

In this work we will show a simple rate equation calculation which describes the evolution of a finite amount of interstitial generated by implantation and annealing just bellow the surface. The model, by introducing proper parameters, allows to calculate the amount of interstitials evaporating at the surface and the amount injected into the bulk. Moreover, the interaction of the interstitial with intrinsic traps characteristic of the MBE silicon will be taken into account.

The model is tested on an MBE sample implanted with Si. Information about the interstitials population injected in the sample after annealing at 800 C is obtained by the SIMS analysis of the broadening of a B-delta doping series present in the sample. Such experiment allows to calibrate the interstitials trap density and the surface evaporation rate. The knowledge of this parameters allows to apply the model to the evaluation of the average composition of Carbon-Interstitial clusters formed by interaction of the *I* flux with substitutional C.

MODEL

It is well established that the *I* diffusion in MBE material is mediated by the interaction with intrinsic traps which are present in the material and reduce the effective diffusivity of the interstitials [2]. This phenomenology is satisfactorily described by the rate equations described in reference [3] when a constant injection of interstitials is provided by surface oxidation.

In order to extend such description to the case of interstitial supersaturation obtained by implantation, it is necessary to introduce in the *I* diffusion equation proper terms which can take into account both the *I* injection by the damaged part of the sample and the recombination of such interstitials to the surface.

The proposed set of equations is the following:

$$\frac{\partial c_I(x,t)}{\partial t} = \nabla(D_I \nabla c_I) + \phi\,\delta(x - r_p) - \frac{D_I}{\lambda}(c_I - c_I^{eq})\delta(x) - (4\pi a_t D_I c_I c_t - G) \qquad (1.a)$$

$$\frac{\partial c_t(x,t)}{\partial t} = -(4\pi a_t D_I c_I c_t - G)/n_t \qquad (1.b)$$

The first equation describes the time evolution of the *I* concentration c_I following the Fick diffusion according to the diffusion coefficient D_I. The term $\phi\,\delta(x - r_p)$ in equation 1a describes the interstitial injection from the implanted zone. The interstitial injection is generated by a flux ϕ which acts only at the depth r_p (ion implantation range). The term $-D_I(c_I - c_I^{eq})/\lambda\ \delta(x)$ allows to take into account the recombination of the interstitials to the surface. As proposed in reference [4], the amount of interstitials lost through the surface by unit time is proportional to the excess at the surface of interstitial concentration c_I with respect to the equilibrium concentration c_I^{eq} and to the surface evaporation rate D_I/λ, where λ is defined as a surface recombination length. When λ goes to infinity the surface acts as a perfect mirror for the interstitials, while the surface is a perfect sink when λ is equal to zero. Moreover the possibility of the interstitials to disappear by the interaction with traps is described by the term $-(4\pi a_t D_I c_I c_t - G)$, where c_t is the trap density, a_t is the capture radius, and G is the rate of *I*-trap backward reaction. The time evolution of the traps is described by equation 1b , where n_t is the number of interstitial captured by each trap.

These equations contain a lot of parameters which must be set in order to obtain a quantitative results. To this aim in the following, we briefly describe an approximated analytical approach to the equations in order to have an easy handling of the parameters.

The basic idea of equations 1 is that the Si interstitials both diffuse and react with traps inside the Si matrix. The interaction term ($4\pi a_t D_I c_I c_t$) causes a reduction of both the interstitials and traps population. As a result, the interstitials are obliged to fill the traps before penetrating inside the sample. The basic approximations and steps to reach the analytical solution are the following: i) the active trap density at time t is described by an edge function $N_T\theta[x-R(t)]$, being θ the Heaviside function (see figure 1). N_T is the trap density before the I injection process, x is the depth below the surface and $R(t)$ is the depth at which the active trap edge is located at the time t. This abrupt trap distribution corresponds to the assumption that the mean free path of the interstitials inside the trap zone is negligible. Comparison of numerical and analytical calculations will show the reliability of such approximation. ii) The I concentration profile evolves through quasi stationary states. In other words, $\partial c_I/\partial t$ can be considered equal to zero in the equation 1, thanks to the very high value of the interstitial diffusivity which allows the interstitial population to rapidly reach the stationary state at every change of the boundary conditions. Under these hypotheses, equation 1 allows to calculate the I concentration c_I at time t in the range from the surface to the depth $R(t)$. $c_I(x,t)$ turns out to have a triangular shape as sketched in figure 1. Furthermore, on the basis of c_I, it is possible to calculate the two flux components ϕ_{surf} (towards the surface) and ϕ_{in} (towards the bulk) in which the interstitial flux generated by I-cluster dissolution is split. iii) The evolution with time of the interstitial concentration is therefore regulated by the evolution of the boundary condition $R(t)$. This can be obtained by equating the quantity of deactivated traps, $N_T\ dR$, and the number of interstitials needed to deactivate them, $(\phi_{in}/n_t)\ dt$. The solution of such simple differential equation allows to obtain $R(t)$ and therefore the full dynamic of the system as a function of the injected interstitials.

The integration of $c_I(x,t)$ over the annealing time T provides the following equation for the integrated interstitial supersaturation in the depth range $r_p < x < R$:

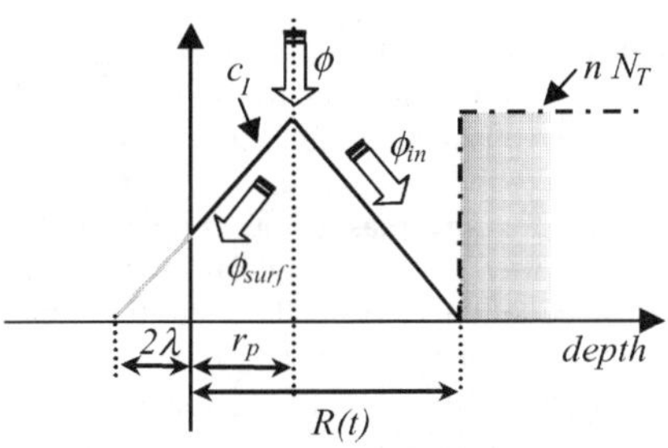

Figure 1 Schematic picture of the c_I concentration (continuous line) and of the traps concentration (dot-dashed line) as a function of depth following the analytical approximated approach.

$$S = B(x-R)^2 \quad (2)$$

where:

$$B = \frac{n_t N_t}{2} \Big/ D_I\, c_I^{eq} \quad (3.a)$$

$$R = \sqrt{\frac{2 I_0 (r_p + 2\lambda)}{n_t N_t}} \quad (3.b)$$

S is the supersaturation integrated over the whole time of injection T, while:

$$I_0 = \int_0^T \phi(t)dt$$

is the number of injected interstitials in the time T. Equation 3b is obtained under the approximation $r_p \ll R$.

Equation 2 demonstrates that the experimental observation of the integrated supersaturation as a function of depth can provide no more than two independent parameter (B and R).

Such parameters are a function of the following quantities: i) the injected interstitials I_0 up to the time T of duration of the annealing (the detailed time evolution of the system which has brought to the final situation at time T is not relevant); ii) the product $n_t N_t$, which corresponds the total number of sites where the interstitials can be trapped; iii) the geometrical factor $r_p + 2\lambda$, which is the effective distance of the I source from a perfect surface sink (see figure 1); iv) the interstitial self-diffusion coefficient $D_I\, c_I^{eq}$ [5,6].

By setting the number of injected interstitials and the self-diffusion coefficient, it is possible to obtain information about the density of trap and $r_p + 2\lambda$.

COMPARISON WITH EXPERIMENTAL DATA

In figure 2a, the results of an experiment on a boron-delta doping array grown by MBE is reported. The MBE layer is 1100 nm thick and it was grown keeping the substrate at 500 °C while the Si deposition rate was 1 Å/s. The sample was then implanted with 5 10^{13} a/cm^2 of silicon at the energy of 20 keV and annealed for 10 minutes at 800 C up to a complete running out of the TED process [7,8]. From the SIMS analysis the B-spikes broadening it is possible to obtain the integrated B diffusivity over time (W) which is proportional to the integrated supersaturation ($W = S\, D_B^{eq}$ [7,8], D_B^{eq} is the equilibrium B diffusivity [9,10]). In figure 2b the experimental values of W are reported as a function of the depth, together with an analytical fit to the data by means of equation 2. A comparison with a numerical calculation of the full set of equations 1 is also reported showing the good degree of approximation of the analytical solution. As previously pointed out, the numerical solution implies the assumption of many more physical quantities than the analytical one; the results of both methods strictly agree for any reasonable choice of the physical quantities providing the same values of B and R in equation 2. Both the analytical and numerical fit give $n_t N_t$ = (5.5 ± 2.0) × 10^{16} cm^{-3} and λ = (10 ± 10) nm. The error bars are obtained by taking into account the range of the coefficient $D_I\, c_I^{eq} / D_B^{eq}$ given in the literature [5,6,9,10] and assigning a 30% error to the ion projected range r_p (~ 30 nm) as calculated by the TRIM simulation [11]. Since the TED process is concluded at the end of the annealing, the injected interstitials, I_0, are assumed to be equal to the implantation fluence, as given by the plus-one model [1]. The obtained results indicate that the number of traps in our material is relatively low if compared to other proposed experiments [3,11] testifying the relatively good quality of the growth also at this low temperature.

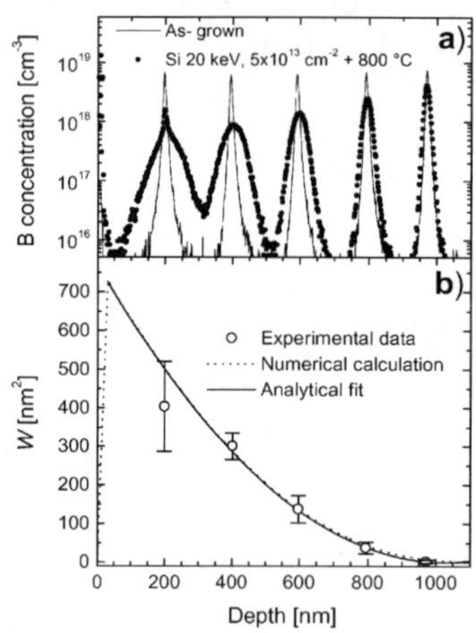

Figure 2 B spikes. (a) SIMS profile of the B spikes as-grown (—) and after 20 keV, 5 × 10^{13} cm^{-2} Si implantation and subsequent annealing at 800 °C for 5 min (♦). (b) Broadening of the B spikes after implantation and annealing (O). The continuous line (—) refers to an analytical fit of the data, while the dotted line (·····) is a numerical calculation.

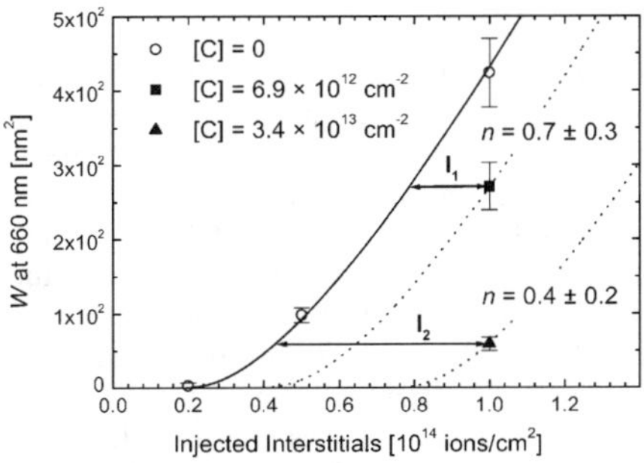

Figure 3 Broadening of a 660 nm deep B spike versus injected interstitials dose. The symbols refer to experimental data for a pure Si sample without C (O) and for samples containing C with a dose of les 6.9×10^{12} C/cm^2 (■) and a dose of 3.4×10^{13} C/cm^2 (▲). The continuous line (—) represents the model results for the sample without C. The dotted lines (·····) are the model fit in the case of C-rich samples. I_1 and I_2 quantify the excess in interstitials injection needed to deactivate the excess of silicon trapping sites due to Carbon.

The λ value is very close to zero indicating that the surface is a very efficient interstitial sink at this temperature. This fact qualitatively confirms the results obtained by Lim and coworkers in reference [4], attesting the fact that a great part of the interstitials is annealed at the surface during TED process. As a matter of fact, our experiment gives a λ upper limit of about 20 nm, which is lower than the value proposed by Lim.

In figure 3 a further application of the model is shown. We show, as a function of the I fluence, the *W* broadening of a boron spike experimentally placed at a depth of 660nm. Also in this case the annealing times of each sample are up to a complete dissolution of the source. As can be seen, the TED of the boron spike (open circles) can be observed only if the implantation dose exceeds $2{\cdot}10^{13}$ I/cm^2. In other words there is a sort of TED suppression below this limit. The effect is perfectly described by the model (continuous line) and is due to the presence of traps in the MBE silicon. In fact, according to the model (by using the parameters obtained from the experiment of figure 2) the trap deactivation depth *R* is of about 450 nm after the injection of $2{\cdot}10^{13}$ I/cm^2 and therefore the spikes can not undergo anomalous diffusion.

The close symbols in figure 3 are relative to samples in which a small amount of C is introduced between the source and the boron spike at the depth of 200nm. The details about this experiment can be found in reference [12]. The presence of C introduces a further delay in the TED onset. Moreover, such delay is higher the higher is the C dose. As demonstrated in references [8,12] the C atoms acts, in this case, approximately as a not diffusing interstitials trap. Therefore the substitutional C can be considered in the model in the same way as the intrinsic traps. The delay introduced by C is due to the fact that a greater amount of interstitial trapping sites is present in the C-containing sample with respect to the pure Si sample. Such excess of trapping sites is given by $n{\cdot}N_C$ where N_C is the C concentration and n is the number of interstitials gettered by each C atoms. Since the N_C is known by SIMS measurements [12], an estimate of n can be performed. Dotted line in figure 3 represent the results of the model obtained by fitting n. The results reported in the figure attest that about 0.5 interstitials are trapped by each carbon i.e. the carbon-*I* clusters formed by *I*-C interaction appear to be rich in C.

CONCLUSIONS

A model is proposed in order to evaluate the number of interstitials injected into the bulk and towards the surface by implantation and annealing in trap-containing silicon. An analytical approximated approach to the problem allows to point out the number of independent parameter which can be gained from a typical experiment on the interstitial diffusion by means of delta doping broadening analysis. The validity of the model is tested on experimental data which allows, in turn to estimate two important physical data such as the number of intrinsic traps in the our MBE material and the recombination length at the surface. The very low upper limit of the recombination length (20 nm) suggests that the role of the surface is fundamental in understanding the interstitials evolution under implantation. The model, after this calibration, allows to correctly predict the effect of TED suppression due to the MBE intrinsic traps. Moreover, we used the present calculation to investigate the interaction of the interstitial injected by implantation with substitutional C intentionally incorporated into the sample. It turns out that the C-*I* clusters formed during the C-I interaction are more rich in Carbon than in interstitials. This indication can allows to deeper understand the precise role of Carbon as a self-interstitial trap.

ACKNOWLEDGEMENTS

This work has been partially supported by MURST through the project COFIN 2000.

REFERENCES

1. P. A. Stolk, J. H.-J. Gossmann, D. J. Eaglesham, D. C. Jacobson, C. S. Rafferty, G. H. Gilmer, M. Jaraiz, J. M. Poate, H. S. Luftman and T. E. Haynes, J. Appl. Phys. **81**, 6031 (1997) and reference therein.
2. N. E. B. Cowern, K. T. F. Janssen, G. F. A. van de Walle, and D. J. Gravesteijn, Phys. Rev. Lett. **65**, 2434 (1990).
3. N. E. B. Cowern, Appl. Phys. Lett. **64**, 2646 (1994).
4. D. R. Lim, C. S. Rafferty, and F. P. Clemens, Appl. Phys. Lett. **67**, 2303 (1995).
5. H. Bracht, N. A. Stolwijk, and H. Mehrer, Phys. Rev. **B 52**, 16 542 (1995).
6. A. Ural, P. B. Griffin, and J. D. Plummer, Phys. Rev. Lett. **83**, 3454 (1999).
7. N. E. B. Cowern, G. Mannino, P. A. Stolk, F. Roozeboom, H. G. A. Huizing, J. G. M. van Berkum, F. Cristiano, A. Claverie, M. Jaraiz, Phys. Rev. Lett. **82**, 4460 (1999).
8. S. Mirabella, A. Coati, D. De Salvador, E. Napoletani, A. Mattoni, G. Bisognin, M. Berti, A. Carnera and A.V. Drigo, S. Scalese, S. Pulvirenti, A. Terrasi, and F. Priolo, Phys. Rev. **B 65**, 045209 (2002).
9. P. M. Fahey, P. B. Griffin, and J. D. Plummer, Rev. Mod. Phys. **61**, 289 (1989).
10. Y. Haddara, B.T. Folmer, M.E. Law, T. Buyuklimanli, Appl. Phys. Lett. **77**, 1976 (2000).
11. J. P. Biersack and L. G. Haggmark, Nucl. Instrum. Methods **174**, 257 (1980).
12. S. Mirabella, S. Scalese, A. Terrasi, F. Priolo, A. Coati, D. De Salvador, E. Napolitani presented at the 2002 MRS Spring Meeting, San Francisco, CA, 2002 (unpublished).

Mat. Res. Soc. Symp. Proc. Vol. 717 © 2002 Materials Research Society C5.7

Thermal Evolution of Extrinsic Defects in Ion Implanted Silicon: Current Understanding and Modelling

Fuccio Cristiano, Benjamin Colombeau*, Bernadette de Mauduit, Caroline Bonafos,
Gerard Benassayag and Alain Claverie
Ion Implantation Group CEMES-LAAS/CNRS Toulouse
- CEMES/CNRS, 29, rue J.Marvig, 31055 Toulouse, France
- LAAS/CNRS, 7 Ave. Colonel Roche, 31077 Toulouse, France
*Current address: School of Electronics, Computing and Mathematics, Univ. of Surrey, UK

ABSTRACT

We present an extensive study of the thermal evolution of the extended defects found in ion implanted Si as a function of annealing conditions. We will first review their structure and energetics and show that the defect kinetics can be described by an Ostwald ripening process whereby the defects exchange Si atoms and evolve in size and type to minimise their formation energy. Finally, we will present a physically based model to predict the evolution of extrinsic defects during annealing through the calculation of defect densities, size distributions, number of clustered interstitials and free-interstitial supersaturation. We will show some successful applications of our model to a variety of experimental conditions and give an example of its predictive capabilities at ultra low implantation energies.

INTRODUCTION

The continuous evolution of ICs towards smaller sizes requires the reduction of both the lateral and the vertical device dimensions. In particular, the junction depths of source and drain extension must decrease well below 50 nm [1]. While new strategies of shallow junction formation are being investigated (plasma doping; laser anneals, heavy ions…) [2,3], it is expected that the conventional ion "implantation + RTA" procedure will still be used (and pushed to its limits) for the next future. This will be achieved by the use of ULE implants (< 1 keV), preamorphisation with Ge^+ or Si^+ ions and "spike" anneals with ultra fast ramp up times[4]. In these conditions, the major obstacle to the fabrication of such shallow junctions remains the well known phenomenon of Transient Enhanced Diffusion (TED) of dopants, in conjunction with the formation of extended defects of various types (including dopant-interstitial clusters) following an ion implantation step [5]. At the same time, physical models are needed to improve the reliability of predictive process simulation tools.

In this paper, we will focus on the various aspects related to the formation of extended defects in ion implanted silicon. The aim is, on one hand, to review our knowledge of the structure and thermal behavior of the different types of extended defects which can be formed by ion implantion and thermal annealing in silicon and, on the other hand, to understand the interactions between these defects and the dopant atoms that are at the origin of the TED phenomenon.

In a first part, we will present an extensive study of the thermal evolution of the extended defects and we will discuss their thermal behaviour on the basis of TEM analysis. We will show that the defect kinetics can be described by an Ostwald ripening process whereby the defects exchange Si atoms and evolve in size and type to minimise their formation energy. In the second

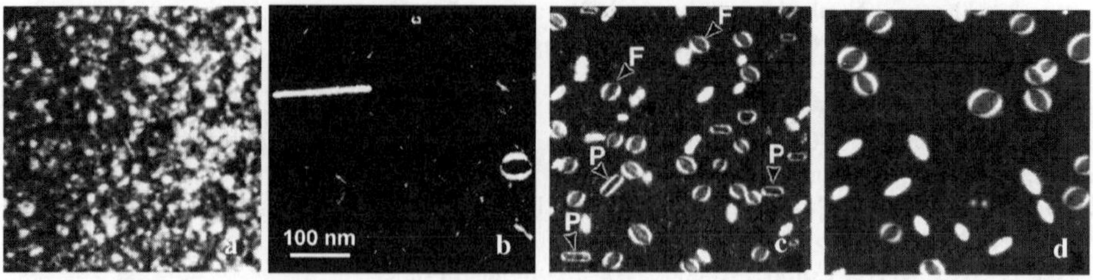

Figure 1. Plan view (400) WBDF TEM micrographs from samples implanted with 150 keV Ge^+ ions to a dose of $2x10^{15}$ ions/cm^2 after RTA annealing in a N_2 ambient. *(a)* T = 800°C, t = 10 sec. *(b)* T = 800°C, t = 100 sec. *(c)* T = 900°C, t = 10 sec. *(d)* T = 900°C, t=400 sec.

part, we will present a physics based model to predict the evolution of these extrinsic defects during annealing. Our model makes use of the combined physical concepts of non-conservative Ostwald ripening and defect formation energy and calculates the time evolution of defect densities, size distributions, number of clustered interstitials and free-interstitial supersaturation. We will present some applications of our model to a variety of experimental conditions and we will give an example of its predictive capabilities.

EXTENDED DEFECTS IN ION IMPLANTED SILICON

Type of defects

Depending on experimental conditions, up to 4 types of extended defects can be detected by TEM [6]. They range, depending on the number of Si atoms they contain, from clusters of about 2 nm in size, to {113} defects and to dislocation loops of two types, perfect dislocation loops (DPLs) and faulted dislocation loops (FDLs). The possibility for the extended defects to evolve up to their most stable forms requires that the initial concentration of Si interstitial atoms ($S_{int}s$) is sufficiently large and that the surface of the wafer is far enough from the defect region. Such conditions are fulfilled after most amorphising implants by Si and Ge. A typical example of such an evolution is shown in figure 1. Depending on annealing conditions, extended defects of various types are formed. For low annealing time/temperature combinations clusters and {113}s are observed (figure 1a), for intermediate combinations {113}s and dislocation loops (figure 1b) or dislocations loops of the two types coexist (figure 1c), while for the highest thermal budgets only faulted DLs are seen (figure 1d). It has to be noted that all these defects are of extrinsic character i.e., they consist of extra Si atoms precipitated in the form of clusters, 1 or 2D objects of increasing symmetry and size. In the following we will recall what we objectively know about the atomic structure of these different defects.

Defect structure

The {113} defects. They are introduced by various processes such as oxidation, electron irradiation, and ion implantation followed by thermal annealing and depending on the process, they can assume two morphologically different structures. After ion implantation and annealing, long narrow defects are usually observed: they are named "rod-like defects". These defects are too narrow to allow accurate analysis by conventional or high resolution electron microscopy. In

constrast, during electron irradiation, wider defects of apparently the same overall structure can be observed with sizes that have permitted extensive studies of their atomic structure. It is important to note that our apparent knowledge of the elongated {113} defects heavily relies on the hypothesis that they have the same atomic structure than the planar {113} defects.
It is now believed [7,8,9] that these defects are agglomerates of <110> rows of Si self-interstitials as suggested by early researchers [10,11]. The relative displacement vector of the planar defects, ***R***, was estimated from HRTEM images [12,13] to be *a*/25<116>, almost perpendicular to the defect plane. The accurate analysis of diffraction contrast described in [10] has yielded the same value.

Eaglesham et al. [14] have observed typical rod-like defects with a {113} habit plane in the case of Si wafers implanted at room temperature with 5×10^{13} cm^{-2}, 40 keV Si^+, after a 15 s post-implant annealing at 815°C. In order to count the total number of Si atoms agglomerated within these defects, they had to measure three parameters: the density and the length of the defects (from plan-view Weak-Beam Dark-Field, WBDF, images), and their width, from cross-sectional HRTEM. Based on their observations, it has been concluded that the mean width of their {113}s did not depend too much on annealing conditions and was constant at about 4 nm. The areal density of interstitials in {113} defects (5.1 nm^{-2} – 5.5 nm^{-2}) has been taken from Takeda's structure [9,15]. Since the work of Eaglesham, the same procedure has been generally used in many different studies and this whatever the experimental conditions, i.e. whatever the true structure of the rod-like defects. More recently, Venezia et al. [16] have finally shown that significantly larger defects could be observed after only slightly different annealing conditions.

<u>The {111} faulted circular Frank loops</u>. These loops (FDLs) are the most common and well known type of End-Of-Range (EOR) defects [6,17]. They were observed and studied many years ago using weak-beam dark-field methods [18,19] in annealed ion implanted silicon wafers. An alternative method for the unambiguous determination of their structure has been more recently described [20]. These loops are interstitial in nature i.e., they are extrinsic defects. They consist of two circular extra {111} net planes of Si atoms, with a stacking fault displacement vector ***R*** = *a*/3<111> and a planar density of interstitials $d = 15.66\ nm^{-2}$. The fault is bounded by a Frank partial dislocation whose ***b*** = *a*/3<111>. FDLs are therefore pure edge dislocations that can only grow by climb. There are four variants of such loops. When plane-viewed in a (001) Si wafer, with ***B*** = [001], they project as ellipses whose long axes (i.e. diameters) are parallel to the two perpendicular [110] and $[1\bar{1}0]$ directions of the foil. Counting the total number of interstitial Si atoms agglomerated within these loops only requires the measurements of these long axes and of the areal loop density.

<u>The {111} perfect elongated loops</u>. Perfect *a*/2<110> loops (PDLs) have early and often been mentioned in the literature in Si wafers implanted with various ions. They have been reported to be either elongated or more or less circular or hexagonal objects [21-26], and even once referred as "Rectangular Elongated Defects" (REDs) [27]. We have observed and rigorously identified such elongated perfect defects with hexagon-shaped ends in Ge amorphized (001) Si wafers (Ge^+: 150 keV, 2×10^{15} ions cm^{-2}) after thermal annealings [20,28,29]. They have nearly {111} habit planes and are elongated along <110> directions perpendicular to their ***b***. Each (111) plane containing three <110> directions, there are twelve variants of such defects. The planar density within such defects is thought to be the same as that of faulted dislocation loops, i.e. $d = 15.66\ nm^{-2}$.

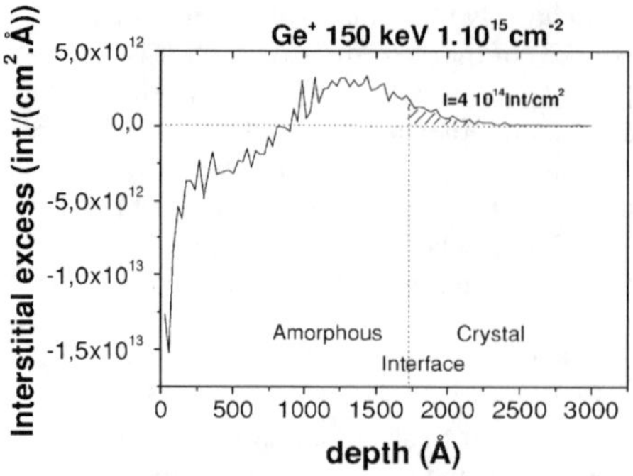

Figure 2. Depth profile of the interstitial excess after a Ge^+ implant (150 keV, $1x10^{15}$ ions/cm^2) using Monte Carlo simulation.

Position and origin of defects

After non amorphising implants, for example after boron implants at room temperature, extended defects are always seen as a band of defects centred on the projected range of the ions. The width of this band increases as the dose increases. It is thought that immediatly after the implantation, probably during the ramping up of the anneal, all vacancies and interstitials initially created by the slowing down of the ions recombine. The total number of vacancies being equal to the number of recoil interstitials from the Si matrix, excess interstitials are created by the activation of the dopant, i.e. the ejection of "regular" Si atoms due to the dopant incorporation on substitutional sites. This model is called the "+1" model [30] and predicts that the defects form at the projected range of the ions and contain a number of Si interstitials equal to the dose. This model works fine at least for light ions at low implantation energies because the spatial separation between the vacancies and the recoils atoms generated by the implantation is small. However, deviations from this model have been evidenced for heavy ions at higher energies [31]. In this last case, the spatial separation between vacancies and interstitials is too large to allow total recombination of the different point defects.

Following high dose P or As implantation, amorphisation occurs and after annealing, defects are still found but this time just below the original amorphous-crystal (*a/c*) interface. As shown in figure 2 after an amorphising Ge implant, the region below the *a/c* interface is heavily supersaturated with Si recoil interstitials. During annealing, both defect growth and solid phase epitaxial regrowth occur. The former being the fastest phenomenon and the self-diffusivity of Si being smaller in amorphous Si than in the crystal, the defects stay located below the original interface [32]. The total number of Si atoms bound to the defects is approximately equal to the number of excess interstitial atoms left behind the *a/c* interface [28]. This model has received the name of "excess interstitials" model [33].

It has to be noted that the "+1" and the "excess interstitials" models are mathematically equivalent, i.e. the total number of Si atoms bound to the defects is equal to the integral of the difference between interstitials and vacancies created by the implantation and the dopant activation. The only difference is that, in the case of amorphising implants, the lower bound of this integral is not zero but the depth of the *a/c* interface. This is based on the assumption that no point defects are left within the amorphous phase.

Defect evolution and transformation: from clusters to dislocation loops

Immediatly following implantation and/or during the ramping-up of the anneal, most of the "excess interstitials" are stored as di-interstitials [34]. After annealing at moderate temperature (600-800°C) the smallest defects that the TEM can resolve are already 2 nm-long {113} defects containing at least 40 atoms [35]. Understanding the mechanism by which these di-interstitials transform into {113} defects has been an important issue in the recent past. From TED experiments, it has been suspected that "precursors" of the {113}s might exist with binding energies intermediate between those of the di-interstitials and of extended defects [36].

Once these {113}s are formed, further annealing makes them grow in size and reduce their density. While they can dissolve if a highly recombining surface is close enough to them [5,14], they can transform into dislocation loops of two types (PDLs and FDLs) if they grow up to a certain size [37]. This transformation has been proposed to be due to some unfaulting reactions and will be discussed later on.

Further annealing of these defects shows that PDLs and FDLs are both involved in a competitive ripening process at the end of which, either FDLs or PDLs will "survive" depending on experimental conditions. This competition has been extensively studied by Cristiano et al [38,39] and, in most practical cases, only large FDLs are found after long annealings. Once only FDLs exist, they are again involved in a ripening process in which the large FDLs are sinks for the small ones and the mean size of the population increases.

Defect growth mechanism: the Ostwald ripening

Historically, the understanding of the mechanism responsible for the thermal evolution of these defects has been gained through TEM studies of the kinetics of the Faulted DLs [40]. Upon annealing at 1000°C, these DLs grow in size and reduce their density, while the number of Si atoms bound to them stays constant (conservative ripening). This phenomenon has been interpreted as being due to the interchange of Si atoms between defects of different sizes following an Ostwald ripening process, similarly to what is observed for the growth of precipitates in a solid matrix [33,40]. The concept is schematically shown in figure 3. This theory relies on the Gibbs-Thomson equation which states that every defect is surrounded by a supersaturation of free Si interstitials, S, directly equal to $exp(E_f/kT)$, where E_f is the formation energy of the defect, i.e., the energy cost to add one extra atom to it. Since the formation energy of a defect typically decreases as its size increases, there exist supersaturation gradients between defects of different sizes, and hence a flux of interstitials from small defects to large ones that makes the whole population evolve (see figure 3). Moreover, the continuous interchange of Si atoms, during this diffusion assisted defect growth, maintains a "mean supersaturation" in the region where the defects are located whose amplitude depends on the size distribution of the defects.

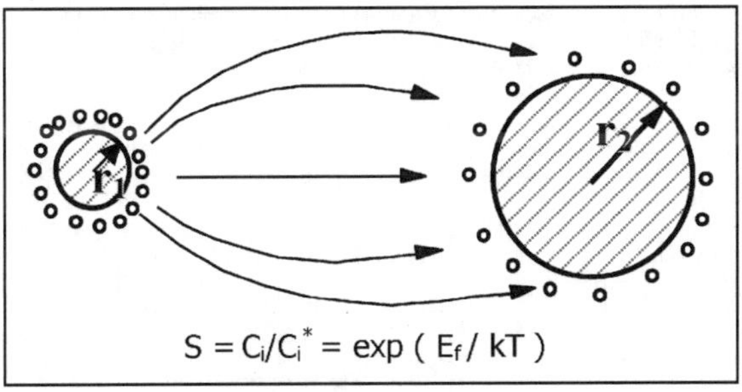

Figure 3. Schematic description of the $Si_{int}s$ interchange between defects of different sizes.

This mechanism has been found to drive also the growth of other defect types, including the clusters of small sizes [41], the {113}s [5,14] and the competitive growth of perfect and faulted dislocation loops [38]. In all cases, the driving force for the evolution of a population of defects or the transformation of a defect from one type to another is the reduction of the formation energy of the defects, i.e.,

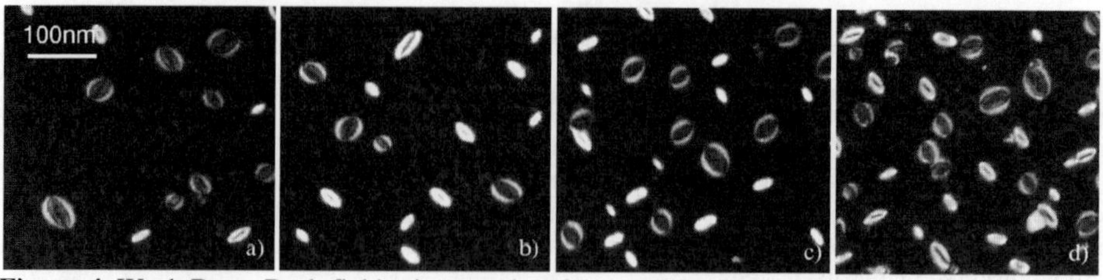

Figure 4. Weak Beam Dark field micrographs of the samples implanted with Ge 150 keV to a dose of *(a)* $1x10^{15}$ ions/cm^2, *(b)* $2x10^{15}$ ions/cm^2, *(c)* $4x10^{15}$ ions/cm^2 and *(d)* $8x10^{15}$ ions/cm^2, after RTA annealing at 1000°C for 100 sec.

the increase of the binding energy (E_b) of the Si atoms attached to them as they grow in size ($E_f + E_b + E_m = E_{sd}$, where E_m and E_{sd} are the Si migration and self-diffusion energies, respectively). In the mean time, the concentration of Si interstitial atoms in dynamical equilibrium with the defects decreases back to the equilibrium value.

A fundamental consequence of this description is that the concept of "defect nucleation" is not valid in ion implanted silicon. From di-interstitials (stable at room temperature) to dislocation loops obtained after very energetic annealings, all these defects only evolve through diffusion assisted growth and eventually undergo "phase" transformation for certain sizes.

The dose effect. Further insight on the characteristics of the Ostwald ripening mechanism comes from the quantitative study of the effect of the implantation dose on the thermal evolution of extended defects. In particular, we have studied a series of 4 samples preamorphised with increasing doses of 150 keV Ge^+ ions (1 to $8x10^{15}$ cm^{-2}) and subsequently implanted with 3 keV B^+ to a dose of $1x10^{14}$ cm^{-2}. The implantation of increasing doses of Ge results in the formation of amorphous layers of increasing thicknesses from 173 to 200 nm. During annealing of these preamorphised structures (at 1000°C), solid phase regrowth of the amorphous layers occurs while a band of EOR defects is formed below the *a/c* interfaces. These defects, which after a few seconds at 1000°C mostly consist of faulted dislocation loops [20], grow in size and reduce their density during annealing following an Ostwald ripening mechanism [40], as seen in the previous

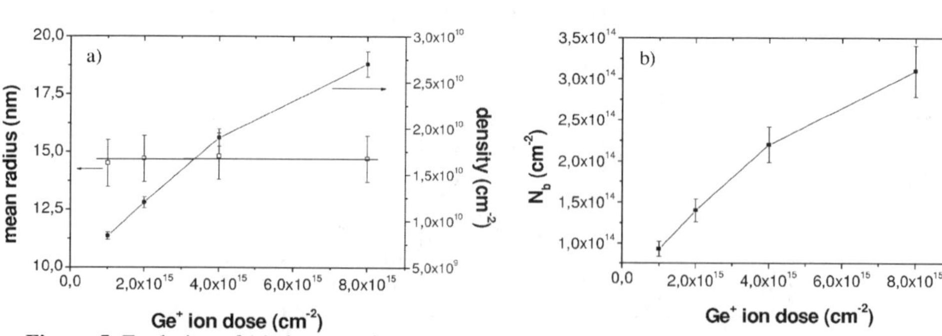

Figure 5. Evolution of (a) the mean loop radii, loop densities and (b) the total number (N_b) of interstitials bound to the loops, after RTA annealing at 1000°C for 100 sec as a function of the Ge ion dose (Implantation energy: 150 keV).

section. Figure 4 shows plan view TEM micrographs of these samples after an annealing time of 100 sec at 1000°C. The quantitative analyses of these images are plotted in figure 5. It is striking to note that increasing the Ge dose only results in the increase of the loop density (figure 5a, right axis) while the defect mean radius remains unchanged (figure 5a, left axis) as well as the shape of the entire size distribution [42]. Consequently, the total number of $Si_{int}s$ bound to the loops increases with the Ge dose (figure 5b). Similar results have been obtained after a shorter anneal (1000°C, 30 sec).

The invariance of the mean size of "precipitates" with respect to the initial supersaturation introduced into the matrix is an expected characteristic of the Ostwald ripening of *diluted* systems. Indeed, in such a case, the growth rate of a precipitate only depends on its size and on the supersaturation mean field in the region and not on the relative proximity of other precipitates. Several theoretical studies have shown this effect [32,43], but until now, no experimental results have clearly confirmed these predictions. This quite surprising result requires that the diffusive fields surrounding the precipitates do not overlap and thus, that their growth is driven only by the interchange of atoms between the precipitates and the supersaturation mean field. This behaviour can be observed only if the distance between the particles is large enough, i.e. their density is small enough and if the diffusivity of the precipitate atoms in the matrix is large enough. These two conditions are exceptionally fulfilled in the case of EOR defects. On the contrary, in many systems (Si in SiO_2 for example), it is observed that increasing the ion dose does results in a size increase of the precipitates [44]. This is always observed when the densities are very large and/or the diffusivity of the impurity is small. In such cases, as the density of the particles increases, their diffusive fields overlap more and more and thus, their growth rate increases.

MODELLING OF DEFECT GROWTH AND TED

Defect energetics

From an energetic point of view, the thermal evolution of all types of extended defects reflects the defect hierarchy in terms of formation energy, as shown in figure 6. The formation energy of the very small clusters (less than 12 atoms) has been experimentally deduced by Cowern et al [41]. In that work the authors used the concept of Ostwald ripening to simulate the diffusion of boron delta-doped layers after low dose Si implants and during moderate temperature annealing. The most important finding of their work is that oscillations of the formation energies occur for small, discrete, "magic" sizes. Clusters of 4, 8 and, to a lesser extent, 12 atoms are superstable. For larger cluster sizes, the formation energy of the clusters "gently" tends towards 0.8 eV, the value expected for {113} defects of small sizes. This observation shows that clusters of more than 20 atoms are similar at least in terms of formation energy to the {113}s. Recent theoretical calculations support this idea [45].

Once the {113}s are formed, they grow in size and reduce their density, again as a result of ripening. In order to access the formation energies of the {113}s, we have calculated the total energy of these defects on the basis of their crystallographical characteristics i.e., by taking into account the two edge dislocations plus the two mixed dislocations plus the stacking fault energy which altogether define them [46]. They have been assumed to be planar, rectangular and of constant width (4 nm). They contain 20 atoms per nm. This energy curve gently tends towards its asymptotical limit at 0.65 eV. The total energies of the PDLs and FDLs can be calculated as a function of their sizes and depend on the defects geometry, i.e. their Burgers vector and the

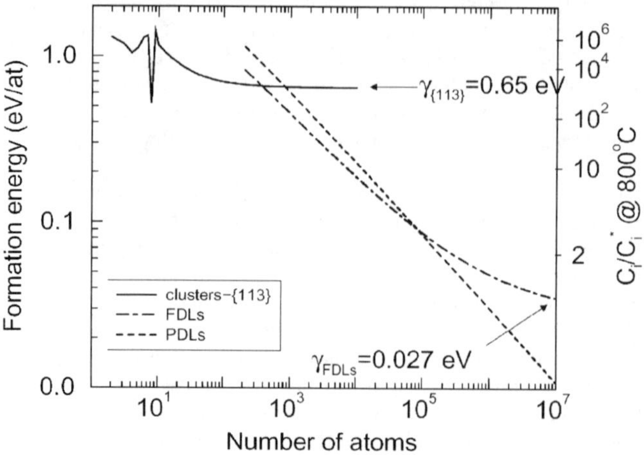

Figure 6. Formation energy (left axis) of the different types of extrinsic defects as a function of their size and corresponding values of the Si(int)s supersaturation (right axis).

presence or absence of a stacking fault. Details of the calculations are given elsewhere [38]. We have plotted these curves for defects from 5 nm in diameter to their asymptotic limit for infinite size. These curves have been used to explain the apparent contradiction between results obtained by different groups after slightly different experimental conditions. Indeed, these plots show that for small sizes, FDLs are more stable than PDLs while there is a stability inversion for radii of about 40nm [38]. Moreover, the formation energies of both DLs cross the {113}'s one for sizes of about 500 atoms i.e. for a 20 nm-long defect, suggesting that a {113}-to-dislocation rearrangement might occur above this size (as experimentally found by Li and Jones [37]).

All the intriguing characteristics of the concomitant thermal behaviour of all these defects can be understood from this figure. For example, a PDL of 1000 atoms can be a sink for a {113} but is a source for a FDL of the same size.

Defects, surface and TED

The Ostwald ripening of such defects maintains in the region a supersaturation of $Si_{int}s$ which decreases as the formation energy of the extended defects decreases (cf. figure 3). For this reason, we have plotted on the right vertical axis of figure 6 the supersaturation of $Si_{int}s$ in dynamical equilibrium with these defects for a temperature of 800°C. Keeping in mind that the diffusivity of a dopant such as boron is directly proportional to this supersaturation, this curve also shows a hierarchy of levels of non-equilibrium diffusion, ranging from supersaturations C_i/C_i^* of 10^6 in the presence of small clusters through 10^3 in the presence of {113}s, to C_i/C_i^* in the range 100 down to 1 as the loops evolve and eventually evaporate.

While this curve determines the amount of supersaturation in the defect region, its spatial extension towards the surface depends on the characteristics of that surface. A highly recombining surface will compete with the defects during Ostwald ripening if situated close enough to the defects. Recent experiments by Omri et al. [29] and Venezia et al. [16] have independently shown that both {113}s and dislocation loops dissolve faster when a recombining surface is put closer to the defects. This explains well a clear difference between high energy and low energy implants. After high energy implants, the Ostwald ripening of the defects is conservative and thus the defects grow until reaching their most stable form, the dislocation

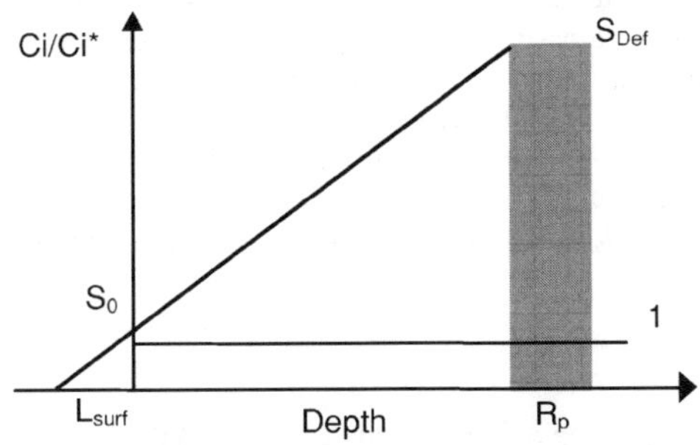

Figure 7. Depth variation of the amplitude of the Si(int) supersaturation in the surface region.

loops. In contrast, after low energy implants, the situation can be depicted as shown in figure 7. From the defect region, the supersaturation S_{EOR} linearly decreases to reach 0 at the fictitious distance L (recombination length) from the surface [41] and thus the supersaturation at the surface (S_0) also depends on the defect types and sizes.

This necessary pinning of the supersaturation at the surface has two major effects. On one hand, it creates a flux of $Si_{int}s$ leaving the defect region. The ripening is no longer conservative and, for example, dissolution of the {113}s can occur before their transformation into loops. More generally, the extent to which the defects evolve before finally evaporating to the surface, depends on the implant dose (total number of available $Si_{int}s$) and distance between the "defect" layer and the surface. On the other hand, since the amplitude of the supersaturation is depth dependent, boron diffusivity will also depend on the depth position of the profile with respect to this gradient.

The dose effect. A further evidence of the relation between the Ostwald ripening of defects and boron TED is given by the study of the effect of the preamorphisation dose on the enhanced diffusivity of boron. Figure 8 is a set of SIMS measurements showing both the effect of annealing time and Ge preamorphisation dose on boron diffusivity. It is striking to note that the preamorphisation dose has no effect on boron diffusivity. However, the comparison of the measured profiles with state-of-the-art simulations shows that boron encompasses enhanced diffusivity during annealing. This average enhancement factor is 3 for a 30 s anneal and reduces to 2 for a 100 s anneal. This clearly shows that boron TED exists in preamorphised layers.

It is very often thought that it is the "dissolution" of extrinsic defects that creates a flux of $Si_{int}s$ towards the surface and is responsible for TED [14]. Here, the defects do not "evaporate" since the total number of $Si_{int}s$ bound to the EOR defects remains constant during annealing. But boron still encompasses diffusivity enhancements. This is in accordance with some recently reported experiments [47] and clearly disagrees with earlier experiments [14,48] that have been the reference in the past. Thus, our results show that the dissolution of defects is not a prerequisite condition to observe TED. Indeed, the interpretation of these characteristics is relatively straightforward if one considers the

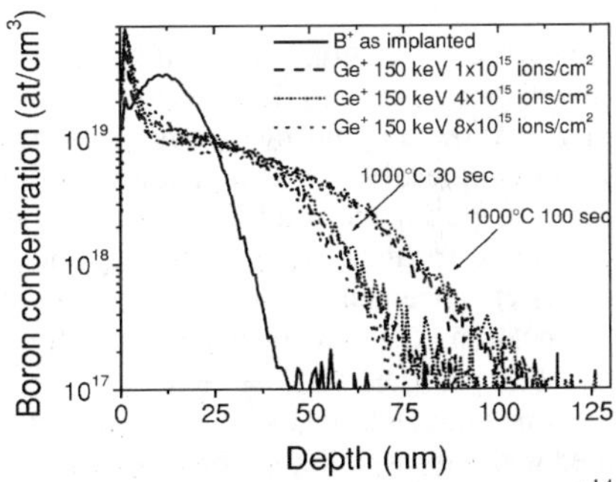

Figure 8. SIMS profiles of B 3keV 10^{14} cm^{-2} implant into preamorphised samples by various Ge ion doses, after annealing at 1000°C 30 and 100s under N_2.

dynamical equilibrium that exists between point and extended defects involved in an Ostwald ripening mechanism: the amplitude of the $Si_{int}s$ supersaturation in the defect region only depends on the size distribution of the loops and on the temperature but not on their density (cf. figure 6). Since the Ge ion dose has no effect on the size distribution of the defects (cf. figures 4 and 5), it has no effect on the amplitude of the $Si_{int}s$ supersaturation mean field. This supersaturation decreases only as the defects grow in size during annealing. This behaviour suggests that the defect region provides a time-decaying source of free $Si_{int}s$. It is the coupling between this source and the boron atoms that is responsible for TED.

Modelling

Our model describes the concomitant time evolution of extrinsic defects and of the supersaturation of $Si_{int}s$ in the defect region, according to the Ostwald ripening theory. Each defect (of size n from 2 up to thousands of atoms) is characterised by its formation energy $E_f(n)$ (the energy required to add one extra atom to the defect) and capture efficiency A_n/R_{eff}. Both $E_f(n)$ and A_n/R_{eff} depend on the geometry of the defect (type and size). The growth of a given defect is calculated by the difference between its capture rate, F_n, and its emission rate, R_n. The capture rate F_n depends on the defect environment and is directly proportional to the mean interstitial supersaturation, S, in the defect region, while the emission rate R_n depends only on the defect formation energy $E_f(n)$ [49]. The evolution of the whole defect population is mathematically expressed as follows:

$$\frac{dN_n}{dt} = F_{n-1}N_{n-1} - F_nN_n + R_{n+1}N_{n+1} - R_nN_n \qquad (1)$$

where N_n is the number of defects of size n, while the mean interstitial supersaturation (free $Si_{int}s$ in dynamical equilibrium with the defects) is given by

$$S = \frac{C_i}{C_i^*} = \frac{\sum_{n=2}^{\infty} \beta_n R_n N_n}{D_i C_i^* \left(\left(\sum_{n=2}^{\infty} \frac{A_n}{R_{eff}} N_n \right) + \frac{1}{L_{surf} + R_p} \right)} \qquad (2)$$

β is the number of Si atoms released by the break-up of a cluster (β=2 for n=2, β=1 otherwise). R_p is the distance between the defect region and the surface, while the surface recombination length, L_{surf} is the only fitting parameter of the model.

Typical results of our simulations are given in figure 9, where we have tested our model against the experimental study by Cowern et al. [41], concerning the TED of CVD-grown boron delta layers after low dose Si^+ implantation. The measured evolution of the boron diffusivity enhancement, i.e. of the $Si_{int}s$ supersaturation, is well reproduced by the simulated curves, including the two "plateaus" associated to small stable clusters (size 4 and 8) and {113} defects observed at 600 °C. At higher temperatures, this evolution is faster and only the {113}s survive long enough to be detected at this time scale. It is noticeable that such fits cannot be obtained without assuming the presence of at least two efficient traps in the 2-10 atoms range. We have also been able to perfectly fit published data concerning the dissolution of {113} defects [50].

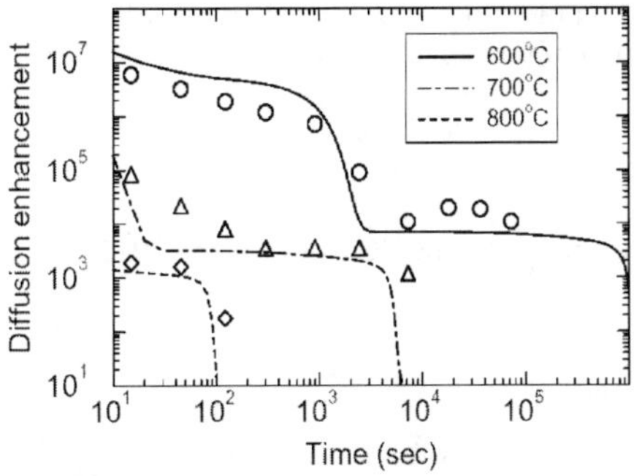

Figure 9. Time variation of the supersaturationof $Si_{int}s$ during annealing: comparison between "marker" layer

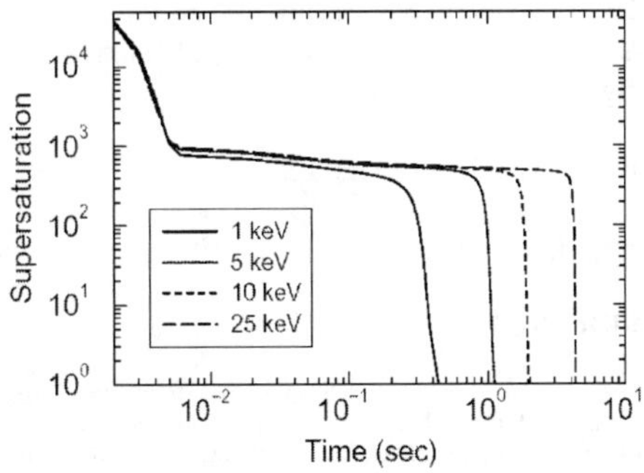

Figure 10. Simulation of the $Si_{int}s$ supersaturation evolution for $1x10^{14}$ cm^{-2} boron implants in the 1-25 keV range at an annealing temperature of 950°C.

Specifics of ultra shallow implants

Having gained some confidence on the ability of our model to describe *both* defect growth and Si supersaturation evolution, we now study through computer simulation the effect of reducing the energy of a 10^{14} ions/cm^2 boron implant in Si from 25 keV to 1 keV. For this, we assumed the "+1" model to hold and we "annealed" these implants at 950°C. Figure 10 shows the evolution of the $Si_{int}s$ supersaturation in the defect region after these implants. At very short times (<0.01 s), the supersaturation shows a "pulse" when only clusters of small sizes exist and rapidly evolve towards {113}s or dissolve. For longer times, a large plateau is observed and corresponds to the dynamical equilibrium between the supersaturation mean field and the {113} defects. This behaviour is common to all energies. Reducing the implant energy from 25 to 5 keV only affects the time after which dissolution starts. It is worth to be noted that for the 1 keV implant, the "pulse" component (driven by small clusters) weighs near about 40% of the overall diffusivity enhancement that boron will experience. This result demonstrates how the vicinity of the surface to the boron profile plays a crucial role in determining the TED decay time, especially at ultra low implantation energies.

CONCLUSIONS

In this paper we have recalled the structure of the extrinsic defects often found in silicon after low energy implantation and annealing. They grow in sizes by interchanging Si atoms following an Ostwald ripening mechanism which can be conservative or not depending on the eventual presence of a strong sink such as a surface in their vicinity. The driving force for such an evolution is provided by the reduction of the formation energy of the defects as they grow in size or change their structure. This hierarchy in terms of defect energy induces an even stronger hierarchy of levels of Si supersaturations at the origin of the non-equilibrium diffusion of dopants observed in such implanted materials. Finally, we have presented a physically based modeling of the growth of extended defects. This model describes the non-conservative Ostwald ripening of defects in presence of a recombining surface. We have tested the model against some published experimental studies of TED and defect evolution. We have successfully applied our model to a variety of experimental conditions including ultra low energy implants. The strength of the model relies on the accurate physical description of the defects formation and evolution during annealing.

ACKNOWLEDGEMENTS

This work is part of the FRENDTECH Project and is funded by the EC as IST/2000-30129.

REFERENCES

[1] International Technology Roadmap for Semiconductors (SIA, Austin, 2000)
[2] S.Baek, T.Jang and H.Hwang, Appl. Phys. Lett. 80, 2272 (2002).
[3] G.Kerrien, J.Boulmer, D.Debarre, D.Bouchier, A.Grouillet and D.Lenoble, Appl. Surf. Sci. 186, 45 (2002).
[4] S.Thomson, P.Packan and M.Bohr, Intel Technol. J., Q3 (1998).
[5] P.A.Stolk, H.-J.Gossmann, D.J.Eaglesham, D.C.Jacobson, C.S.Rafferty, G.H.Gilmer, M.Jaraiz, J.M.Poate, H.S.Luftman, T.E.Haynes, J. Appl. Phys. 81, 6031 (1997).
[6] A.Claverie, B.Colombeau, G.Ben Assayag, C.Bonafos, F.Cristiano, M.Omri and B.de Mauduit, MSSP, 610, (2000).
[7] J.L. Hutchison, A.L. Aseev and L.I. Fedina, Inst. Phys. Conf. Ser. 134, 41 (1993).
[8] D.J.Eaglesham, P.A.Stolk, J-Y.Cheng, H.-J.Gossmann, T.E.Haynes and J.M. Poate, Inst. Phys. Conf. Ser. 146, 451 (1995).
[9] S. Takeda, Proc. Microsc. Semicond. Mater. Conf. (Oxford, 7-10 April, 1997) Inst. Phys. Conf. Ser. 157, 25 (1997).
[10] C.A. Ferreira Lima and A. Howie Phil. Mag. 34, 1057 (1976).
[11] I.G. Salisbury and M.H. Loretto Phil. Mag. A39, 317 (1979).
[12] S. Takeda Jpn. J. Appl. Phys. 30, L639 (1991).
[13] S. Takeda, M. Kohyama and K. IbePhil. Mag. A70, 287 (1994).
[14] D.J. Eaglesham, P.A. Stolk, H.J. Gossmann and J.M. Poate, Appl. Phys. Lett., 65, 2305 (1994).
[15] S. Takeda, S. Muto and M. Hirata Mat. Res. Soc. Symp. Proc. 262, 209 (1992).
[16] V.C. Venezia, R. Kalyanaraman, H.-J.L. Gossmann, and C.S Rafferty and P. Werner, Appl. Phys. Lett. 79, 1429 (2001).
[17] A. Claverie, L.F. Giles, M. Omri, B. de Mauduit, G. Ben Assayag and D. Mathiot, Nucl. Instr. Meth. B, 147, 1 (1999).
[18] R. Bicknell J. of Microsc. 98, 165 (1973).
[19] M.L. Jenkins, D.J.H. Cockayne and M.J. Whelan, J. of Microsc. 98, 155 (1973).
[20] B. de Mauduit, L. Laânab, C. Bergaud, M.M. Faye, A. Martinez, and A. Claverie, Nucl. Inst. and Meth.. B, 84, 190 (1994).
[21] L.J. Chen and I.W. Wu, J. Appl. Phys. 52, 3310 (1981).
[22] I.G. Salisbury Acta Metall. 30, 627 (1982).
[23] D.K. Sadana and J. Washburn Phil. Mag. B46, 611 (1982).
[24] I.W. Wu and L.J. Chen J. Appl. Phys. **58**, 3032 (1985).
[25] K.S. Jones, S. Prussin and E.R. Weber Appl. Phys. A45, 1 (1988).
[26] G.Z. Pan, K.N. Tu and A. Prussin J. Appl. Phys. 81, 1 (1997).
[27] R. Raman, M.E. Law, V. Krishnamoorthy and K.S. Jones Appl. Phys. Lett. **74**, 700 (1999)
[28] L. Laânab, C. Bergaud, MM. Faye, J. Fauré, A. Martinez and A. Claverie, Mat. Res. Soc. Symp. Proc., 279, 381 (1993).
[29] M. Omri, L. F. Giles, B. de Mauduit and A. Claverie, Mat. Res. Soc. Proc., (1999).
[30] M.D. Giles, J. Electrochem. Soc., 138, 1160 (1991).
[31] B.Herner, H.-J.Gossman, L.Pelaz, G.H.Gilmer, M.Jaraiz, D.C.Jacobson and D.J.Eaglesham, J. Appl. Phys. 83, 6182 (1998).
[32] E. Lampin, V. Senez and A. Claverie, J. Appl. Phys., 85, 8137 (1999).
[33] A. Claverie, C. Bonafos, D. Alquier and A. Martinez, Solid State Phenomena, V47-48 , 195 (1996).
[34] J.W. Corbett, J.P. Karins and T.Y. Tan, Nucl. Intr. and Methods in Phys. Res., B182-183, 457 (1981).
[35] F. Cristiano, B. Colombeau, N.E.B Cowern and A. Claverie, private communication.
[36] H.G.A. Huizing, C.G.G Visser, N.E.B Cowern, P.A. Stolk and R.C.M de Kruif, Appl. Phys. Lett, 69, 1211 (1996).
[37] J. Li and K.S. Jones, Appl. Phys. Lett. 73 (25), 3748 (1998).
[38] F.Cristiano, J.Grisolia, B.Colombau, M.Omri, B.deMauduit, A.Claverie, F.Giles, N.Cowern, J. Appl. Phys. 87, 8420 (2000).
[41] N.E.B. Cowern, D. Alquier, M. Omri, A. Claverie and A. Nejim, Nucl. Instr. Meth. in Phys. Res. B148, 257 (1999).
[42] B.Colombeau, F.Cristiano, J.-C.Marrot, G.Ben Assayag and A.Claverie, Mat. Res. Soc. Symp. Proc. 2001, in press.
[43] H.Schroeder, P.Fichner, H.Trinkaus, *Fundamental aspects of inert gases u solids*, Plenum Press New York, 279, 289 (1991).
[44] C.Bonafos, B.Colombeau, A.Altibelli, M.Carrada, G.Ben Assayag, B.Garrido, and A.Claverie, NIM B, 178, 17 (2001).
[45] M.M.De Souza, C.K.Ngw, M.Shishkin and E.M.Narayanan, Phys. Rev. Lett., 83, 1799 (1999).
[46] B.Colombeau, PhD Dissertation, Toulouse, France, 2001.
[47] D.Alquier, N.Cowern, P.Pichler, C.Armand, A.Martinez, D.Mathiot, M.Omri, A.Claverie, MRSSP, 532, 67 (1998).
[48] H.S.Chao, S.W.Crowder, P.B.Griffin, J.D.Plummer, J. Appl. Phys., 79, 2352 (1996).
[49] A.Claverie, B.Colombeau, F.Cristiano, A.Altibelli and C.Bonafos, MRSSP 2001, in press..
[50] B.Colombeau, F.Cristiano, A.Altibelli, C.Bonafos, G.Ben Assayag, A.Claverie, Appl. Phys. Lett., 78, 940 (2001).

Mat. Res. Soc. Symp. Proc. Vol. 717 © 2002 Materials Research Society C5.8

Annealing behavior of locally confined dislocation loops under inert and oxidizing ambient

C. Tsamis, D. Skarlatos, I. Raptis, D. Tsoukalas, P. Calvo[1], B. Colombeau[1], F. Cristiano[2] and A. Claverie[1]
IMEL/NCSR "Demokritos", Terma Patriarchou Grigoriou, P.O.BOX 60228, 15310 Aghia Paraskevi, Athens, Greece
[1] CEMES/CNRS, 29 rue J. Marvig, BP 4347 31055 Toulouse cedex 4, France
[2] LAAS/CNRS, 7 av. Col. Roche, 31077 Toulouse, France

ABSTRACT

In this work, we report data for the growth kinetics of locally confined dislocation loops organized within lines of controlled dimension and periodicity. The dislocation loop lines were formed in the crystalline substrate after local Si implantation and annealing in predefined areas. The distance between the lines ranges between 0.2 μm to 5 μm. It is shown that the kinetics of the DLs depends on the distance between them. When the distance is less than 1 μm, the DLs behave in a similar fashion like those grown in a continuous layer, under inert and oxidizing conditions. However, when the distance between the lines is increased (eg. 5 μm), the behavior of the loops is changed. Fast dissolution of the dislocations loops is observed during annealing in inert ambient, due to enhanced interstitial losses, while under oxidizing conditions the loops grow faster.

INTRODUCTION

Dislocation loops are two dimensional extended defects, formed in the silicon crystal after annealing of an amorphizing implant. The interaction of point defects with the dislocation loops is of general interest for front-end Si processing technology, since their presence in specific areas of modern semiconductor devices strongly influences the device performance. For that reason systematic study has been done over the past years, in order to determine the physical mechanisms that underlie the defect formation [1,2] and their influence on phenomena such as the Transient Enhanced diffusion (TED) of dopants [3].

On the other hand, DLs have been used as point defect detectors to study point defect injection processes and to extract point defect parameters in silicon [4-8]. In that case their use has been proven a very efficient technique since a continuous dislocation loop layer can absorb up to 85 % of the silicon interstitials that are generated during surface oxidation [9]. Another interesting situation is the study of the growth kinetics of dislocation loops when the defects are formed locally in silicon areas. In that case the growth kinetics of the defect might be influenced from the limited dimensions of the area where the loops are grown, both under oxidizing and under inert conditions. The behavior of the defects under such conditions would be much more representative of the situation encountered during S/D formation in a real device. Moreover, locally confined dislocation loops, in the form of lines or boxes, could be used to monitor two-dimensional distribution of silicon interstitials during local oxidation of silicon.

In this work, we report data for the growth kinetics of dislocation loops organized within the lines of controlled dimension and periodicity. The annealing behavior of the defects is studied under inert and oxidizing conditions. The total number of silicon atoms bounded into the dislocations loops is calculated for each case and the net number of silicon atoms that are trapped in the loop lines is estimated.

EXPERIMENTAL

The fabrication process of the test samples in shown schematically in figure 1. The starting material used for the current experiments was CZ <100>, p-type silicon wafers. After the formation of a thin thermal oxide (~20nm), a 250 nm thick polysilicon layer was deposited and patterned by e-beam lithography. Lines were defined on the polysilicon, the width of each line being 1μm, while the distance between the lines varied from 0.2 to 5 μm. Etching of the polysilicon layer was performed in an RIE plasma reactor, using an anisotropic process that is selective in respect to the oxide under the polysilicon layer. Subsequently the samples were implanted with Si with a dose of $2\ 10^{15}cm^{-2}$ and energy 50 KeV (Fig1a). After annealing at 900°C for 10min dislocation loops are formed only in the areas where the silicon implantation took place, while no dislocation loops are formed in the areas that were protected by the polysilicon mask during implantation. Polysilicon was selectively removed from the top of the samples by using an Ethylenediamene-Pyrocatechol (EDP) solution. This solution is highly selective in respect to silicon dioxide (the etch rate is less than 0.5nm/min). Part of the samples were annealed at 900°C for 2h in N_2, while the rest of the samples were oxidized in dry O_2 for the same conditions. Prior to the oxidation, the thermal oxide that existed on the surface of the samples was removed (Fig 1b).

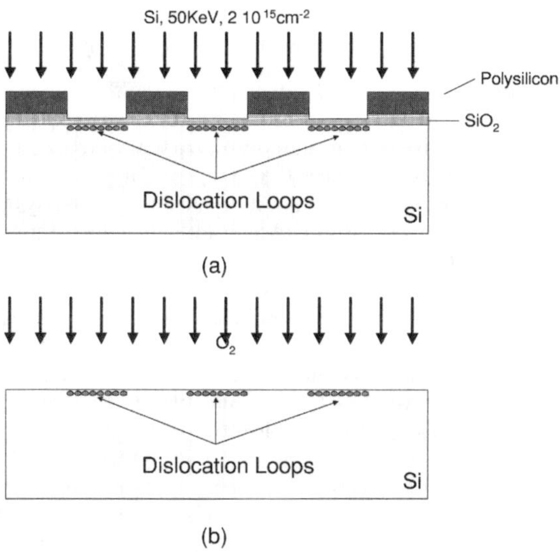

Figure 1. Schematic drawing describing the fabrication process of the test structures

The imaging of the dislocation loops was performed with Plan-view Transmission Electron Microscopy. Since the conventional two-beam or weak-beam imaging technique is unable to reveal simultaneously all the dislocation loops having different habit planes and Burgers vectors [10], the symmetrical bright-field technique was used for the imaging of the dislocation loops. All the TEM pictures were taken at 100K magnification. Each image was then further magnified in dimensions 20cm x 24cm to achieve a very clear view of the EOR defects. Their size (radius and surface) was measured with the aid of a pattern recognition programme.

RESULTS AND DISCUSSION

Figure 2 (a-b) show two representative plan-view TEM images of dislocation loops lines separated by 0.2 μm and 1 μm respectively. As we can see that dislocation loops are formed only within the lines that were implanted with Si. From the analysis of the defect morphology, we can identify two different types of loops that are formed under the conditions of our experiment: a) the Frank faulted loops, with elliptical projections on <100> plane (in two vertical to each other directions), smooth edges at the two sides of their long axis and **a**/3<111> Burgers vector (**a** is the Si lattice constant) and b) the perfect prismatic loops, with more irregular shapes and **a**/2<110> Burgers vector. The crystallographic morphology of the EOR defects is the same as in Ref. 7 and 10. Both types of defects can act as interstitial sinks and will be taken into account in the analysis that follows.

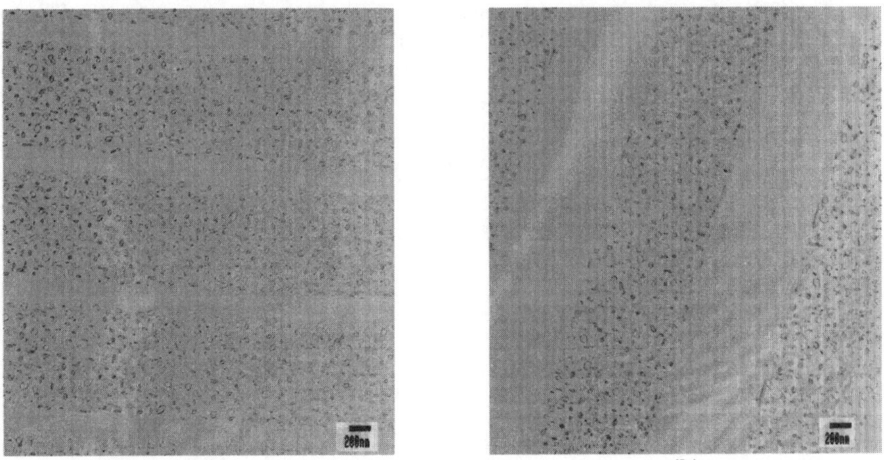

(a) **(b)**

Figure 2. TEM plan view images of Dislocation loops grown in 1um lines that are separted by (a) 0.2 μm and (b) 1 μm.

The number of interstitials bounded in the loops was calculated separately for the circular Frank faulted loops and the perfect prismatic loops. For the Frank loops, the density D, the mean radius $\bar{r}$ and its standard deviation σ_f was measured and the total number of atoms per cm^2 incorporated in the loops was estimated from the equation :

$$N_F = D \times \pi \times \left(\bar{r}^2 + \sigma_f^2 \right) \times d_{111} \qquad (1)$$

where d_{111} (=1.57 10^{15} cm^{-2}) is the atomic density on the <111> plane. The mean radius and the standard deviation were extracted by measuring the longest dimension of the loops.
For the prismatic dislocation loops the density D and the mean area $<S>$ (projected on the <111> plane) were measured and the total number of bounded atoms per cm^2 is given by the equation :

$$N_P = D \times <S> \times d_{111} \qquad (2)$$

For each sample, at least 200 dislocation loops were measured. The accuracy of the mean loop radius measurements is 2nm. The error of the above calculated quantities varies from 10 to 20% depending on the loops mean size.

Figure 3 (a) shows the number of bounded atoms per surface area that are incorporated in the dislocation loops lines as a function of the distance between them for inert (Fig. 3a) and oxidizing conditions (Fig. 3b). In the same figure we have also indicated the number of bounded atoms that was obtained for a continuous layer under the same conditions [7]. There are two observations we can make from these figures. At first we notice that when the distance between the lines is less that 1 μm the number of bounded atoms under inert conditions is almost equal to the number obtained for a continuous layer. However, when the DLs are separated by 5 μm, the number of the total bounded atoms decreases significantly compared to the continuous layer case, indicative of increased interstitial fluxes in that case.

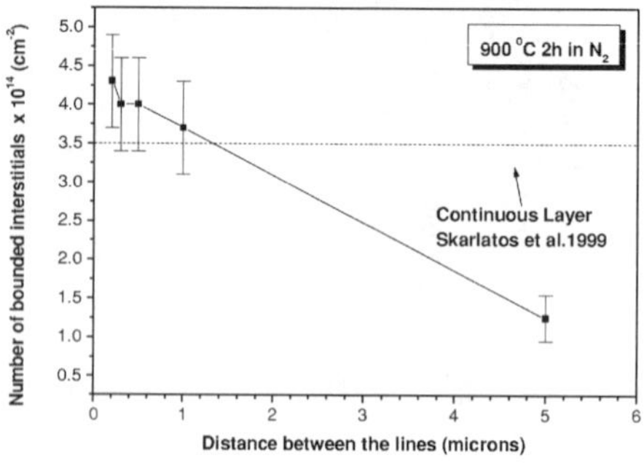

(a)

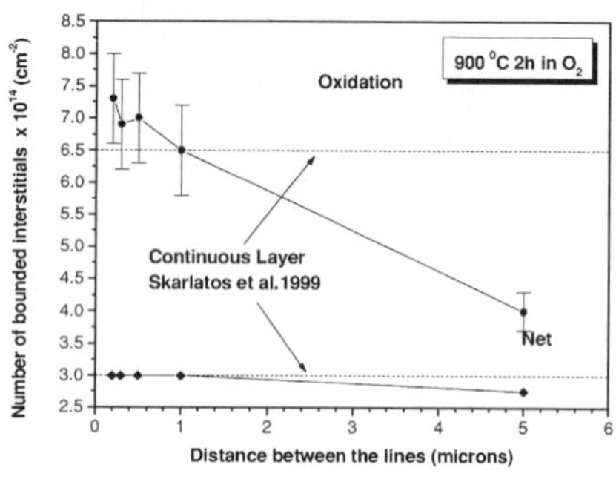

(b)

Figure 3. (a) Number of bounded atoms in the dislocation loops as a function of the distance between the lines for inert and oxidizing conditions. The dash line shows the number of bounded atoms that was obtained under the same conditions for a continuous layer by Skarlatos et al [7]. (b) Number of net trapped atoms as a function of the distance between the lines.

From the same figure (Fig. 3b) we can also notice that the number of bounded atoms under oxidizing conditions is higher compared to that of inert conditions, for all distances, indicating that silicon interstitials that are generated during the oxidation process are captured by the dislocation loops, as it is expected.

From the difference between the number of interstitials in the two cases we can estimate the net amount of interstitials trapped during oxidation, as shown in Fig3b. We can see that the net number of trapped atoms when the distance between the lines ranges from 0.2μm to 1μm, is of the same order as the net number trapped for a continuous layer. However, the situation is completely different when the lines are separated by 5μm. Figure 4a and 4b are TEM images of the dislocation loops after inert and oxidizing condition. We observe that under oxidizing conditions the perfect dislocation loops within the lines have disappeared (Fig. 4b) although their presence is evident under inert conditions, as can be seen from fig. 4a. Such a behavior is not observed for a continuous layer of dislocation loops treated under the same oxidizing conditions, where the prefect prismatic loops are clearly seen after oxidation at the same conditions, as can be seen for fig 4c [7]. Furthermore, we can see that the Frank faulted loops grow faster within the lines (Fig. 4b), compared to the continuous layer case (Fig.4c) and finally intercept the surface, having a semicircular shape.

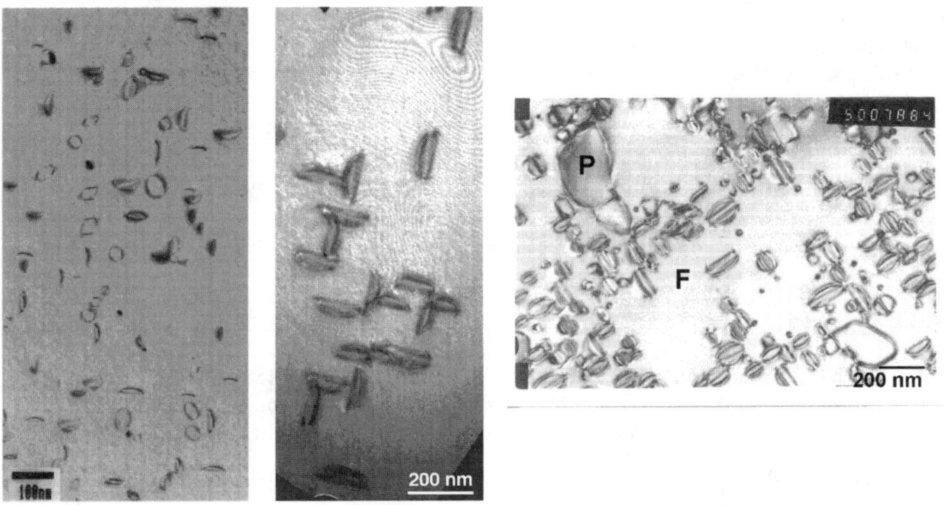

(a) (b) (c)

Figure 4. TEM view image of DLs grown in lines that are separated by 5 μm after (a) inert annealing and (b) dry oxidation at 900°C for 2h and (c) within a continuous layer after dry oxidation at 900°C for 2h. We see that perfect dislocation loops disappear under oxidizing conditions, while Frank loops grow faster, compared to a continuous layer of loops.

Such phenomena are expected at much higher interstitial injection and have been observed experimentally at higher temperatures (fe. 950°C for 1h in the case of a continuous layer). This means that the value that is estimated for the lines that are separated by 5 μm, is a lower limit of the number of interstitials that are trapped in the lines. If we assume that the total number of

silicon interstitials that are stored in the perfect loops, just before they are lost by gliding to the surface, are equal to the value that we obtain for the lines that are 1μm apart, then the total number of trapped atoms is expected to be much higher than the case of a continuous layer, indicating that the dislocation loops are efficient sinks for silicon interstitials. The systematic study of the transition region between 1 and 5 μm as well as experiments in appropriately selected experimental conditions, where both types of dislocation loops will be are present, are required, in order to have a more accurate estimation of the number of trapped interstitials during oxidation and arrive at more definitive conclusions.
It is important to note that during the experiments it became clear that one of the parameters that have to be evaluated also is the width of the line. As the width of the line decreases, the absolute value of atoms stored within the lines decreases and the interstitials fluxes to the loop layer have more pronounced effects.

CONCLUSIONS

In summary, we have presented experiments on the annealing behavior of locally confined dislocation loops under inert and oxidizing conditions. Our results show that the behavior of dislocation loops in that case differs from the case of the continuous layer of defects. Enhanced interstitial losses can be observed during inert annealing, when the loops are formed locally, while under oxidizing conditions the defects act as interstitial sinks, absorbing an increased number of atoms. More systematic work of the dependence of the defects on the dimensions of the area that are grown will highlight further the physical phenomena that take place.

ACKNOWLEDGMENTS

The work was supported by EU through the "FRENDTECH" project (IST 2000-30129)

REFERENCES

1. J. Liu, M. E. Law and K. S. Jones, Solid-State Electronics, **38** (7), 1305 (1995)
2. S. Chaudhry, J. Liu, K. S. Jones and M. E. Law, Solid-State Electronics, **38** (7), 1313 (1995)
3. C. Bonafos, M. Omri, B. de Mauduit, G. Ben Assayag, A. Claverie, D. Alquier and D. Mathiot, J. Appl., Phys. **82** (6), 2855 (1997)
4. H. S. Chao, P. G. Griffin and J. D. Plummer, Appl. Phys. Lett., **68** (25), 3570 (1996)
5. D. Skarlatos, L. F. Giles, C. Tsamis, A. Claverie and D. Tsoukalas, MRS Spring Meeting (1999)
6. H. L. Meng, S. Prussin, M. E. Law and K. S. Jones, J. Appl. Phys., **73,** 955 (1993)
7. D. Skarlatos, M. Omri, A. Claverie and D. Tsoukalas, J. Electrochem. Soc. **146** (6), 2276 (1999)
8. D. Skarlatos, D. Tsoukalas, L. F. Giles and A. Claverie, J. Appl. Phys., **87** (3), 1103 (2000)
9. D Tsoukalas, D. Skarlatos and J. Stoemenos, J. Appl. Phys., **87** (12), 8380 (2000)
10. G. Z. Pan, K. N. Tu, and S. Prussin, J. Appl. Phys., **81**, 78 (1997)

Mat. Res. Soc. Symp. Proc. Vol. 717 © 2002 Materials Research Society

Modeling Dislocation Loop Nucleation and Evolution in Germanium, Arsenic and Boron Implanted Silicon

Ibrahim Avci and Mark E. Law
Swamp Center, Department of Electrical Engineering,
NEB Room # 535, University of Florida, Gainesville FL 32611

ABSTRACT

A loop nucleation and evolution model in Si^+ implanted Silicon was previously introduced [1]. In this study, the model is extended to predict end of range (EOR) and projected range defect nucleation and evolution created by different ion implant species such as Germanium, Arsenic and Boron. The model assumes that all the nucleated loops come from {311} unfaulting and the loop density and average loop radius follow a log normal distribution. The model is verified with the experimental data obtained from literature for Germanium [2], Arsenic [3] and Boron [4] implanted Silicon for different implant doses and energies. Modeling results are in agreement with the experimental results.

INTRODUCTION

Ion implantation is one of the primary ways of introducing impurity atoms into the silicon substrate. Jones *et al.*[5] showed that implant species play an important role on the defect formation and defect evolution. Light ions, such as boron, cannot produce enough damage to cause amorphization. Point defects created during implant usually form {311} defects and dislocation loops around the projected range after annealing which are classified as type I defects. Meanwhile, arsenic and germanium are heavy ions and they can produce amorphous layers if the implant dose exceeds the critical dose. EOR defects are the product of these heavy ions. Since most of the defects are extrinsic in nature, the amount of the excess interstitials generated by the implanted species will affect the defect densities.

In contrast to silicon self implants, some of these implanted species may interact with the excess interstitials and vacancies generated during the implantation process. They may pair with interstitials and vacancies [6] and diffuse away from the damaged region. This reduces the super saturation of excess interstitials and vacancies in the damaged region. Thus, they indirectly affect the defect densities.

To improve the device performance, it is very important to understand and model the unique material aspects of damage/defect accumulation during implantation and the subsequent annealing cycle. A model would also help device/process engineers to change and adjust their device structures and process conditions without having to build costly test lots. In this article we verify previously introduced loop nucleation and evolution model [1] for different implant species. Although the model was calibrated using Si^+ implanted Silicon defect data, it was successfully used to predict loop nucleation and evolution in Germanium, Arsenic and Boron implanted Silicon. Interactions between Silicon interstitials implanted species (ie. Arsenic [3] and Boron [7]) were also incorporated into the model. Simulation results were validated against the experimental data [2][3][4]. The results are in good agreement with the data.

MODEL

Ion implantation generates large amount of excess interstitials and vacancies. Upon annealing, defects such as {311}'s and dislocation loops form. Excess interstitials provide the bases for the nucleation of these defects. To simulate defect nucleation and evolution, UT-Marlowe [8] program is used to determine the excess interstitials and vacancies created in silicon substrate because of ion implantation. These profiles are used to nucleate interstitial clusters, vacancy clusters and {311} defects [9]. While some of these {311} defects unfaults to dislocation loops, some of them will dissolve. {311}'s become the source of dislocation loops [10]. In the developed loop nucleation model, unfaulting process is called the nucleation stage of the dislocation loops and the evolution of the loops is called the Ostwald ripening stage.

The model is described by equations 1, 2 and 3 [1]. Equation 1 describes the change in total number interstitials trapped by loops (N_{all}) with time. N_{rate}^{Nall} is the unfaulting rate. K_{IL} is the constant of a reaction between the interstitials and the dislocation loop. C_I is the concentration of interstitials. C_{Ib} is the effective equilibrium concentrations of interstitials at the loop layer boundaries.

$$\frac{dN_{all}}{dt} = N_{rate}^{Nall} + K_{IL}\int_{0^+}^{\infty}(C_I - C_{Ib})f_D(R)dR\Bigg|_{\text{at loop layer boundaries}} \quad (1)$$

Assuming that all loops are circular, the density of dislocation loops (D_{all}) can be expressed by:

$$\frac{dD_{all}(R)}{dt} = N_{rate}^{Dall} - \frac{2D_{all}(R)}{R^2}K_R \quad (2)$$

The constant, K_R, is the coarsening rate of dislocation loops and R is the loop radius. The first term in equation 2 is the nucleation term and the second term is the coarsening term. It is also possible to extract a relation between σ (standard deviation of data) and average loop radius R from TEM data as an analytic function of R (cm) as follows:

$$\frac{\sigma}{R} = 0.33 + 5\times10^4 R \quad (3)$$

The nucleation terms shown in equations 1 and 2 are given by

$$N_{rate}^{Nall} = K_{311}C_{311} \quad (4)$$

$$N_{rate}^{Dall} = K_{311}D_{311} \quad (5)$$

C_{311} and D_{311} are the number of interstitials in {311}'s and density of {311}'s, respectively. K_{311} is the unfaulting rate. K_{311} is the same in both equations because when a {311} defect unfaults to dislocation loop, the number of interstitials bounded by that {311} is transferred to the unfaulted loop.

DEFECTS IN GERMANIUM IMPLANTED SILICON

The effects of Ge^+ implant energy and dose on the evolution of defects created via the annealing of the implant damage were studied by Gutierrez *et al.* [2]. Gutierrez performed PTEM analysis on 30 keV, 1×10^{15} cm^{-2} Ge^+ implanted samples. For the short annealing times (< 120 sec), a Rapid Thermal Anneal (RTA) was employed. A conventional furnace was used for longer annealing times. All anneals were carried out in a nitrogen ambient at 750 °C or at 825°C.

In order to carry out the simulations, damage profiles are obtained for each implant condition using UT-Marlowe [8] and truncated to the amorphous depth (i.e. $C_I^*=1.0$ in amorphous region). It is assumed that implanted Ge atoms were in the substitutional position as an initial condition and there was no interaction between Ge and Si atoms. Simulations are

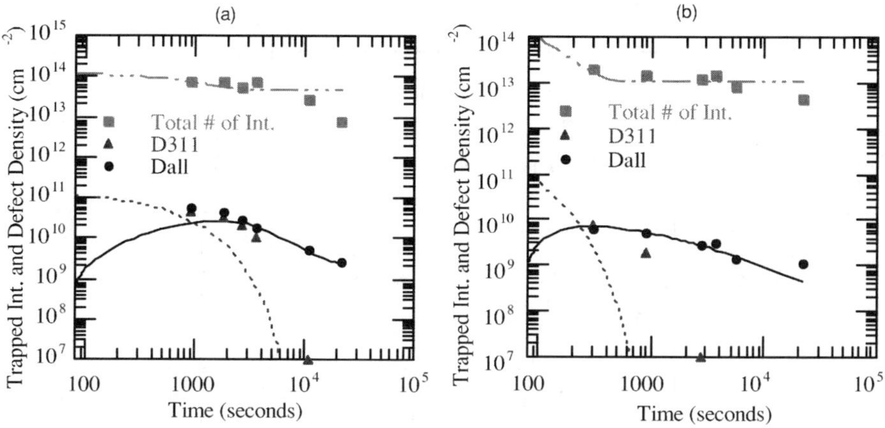

Figure 1. Experimental (symbols) and simulation (lines) results for the defect evolution for a 30 keV, 1×10^{15} cm^{-2} Ge^+ implant on silicon, annealed at (a) 750 °C and (b) 825 °C.

performed at 750 °C and 825 °C for 360 minutes.

Figure 1 shows both the experimental and simulation results for the defect evolution of a 30 keV, 1×10^{15} cm^{-2} Ge^+ implant. Symbols represent the experimental data and lines represent the simulation results. Gutierrez observed that both {311} defects and dislocation loops were present for this case. It is clear from the figure that {311} defects rapidly dissolve with increasing time, whereas the loop population stabilizes and becomes the dominant defect type. Although simulation results closely predict the total number of interstitials trapped in defects for short annealing times, there is some discrepancy between the data and the simulation results at longer annealing times. There is a good correlation between the loop density data and simulation results at longer annealing times. The difference seen between them in short annealing times could be attributed to counting errors, since it is very difficult to discern between {311}s and loops on short anneals. A qualitative fit is all that can be done. There is also some discrepancy between the {311} data and simulation results. Although no interaction between germanium and silicon is assumed in the simulations, it is possible that there is such an interaction and contributes to the differences between data and simulations.

DEFECTS IN ARSENIC IMPLANTED SILICON

Brindos *et al.* [3] investigated enhanced diffusion of high concentration, low energy arsenic implants (3 keV at doses of 5×10^{14} cm^{-2}, 1×10^{15} cm^{-2} and 5×10^{15} cm^{-2}) when annealed at 800 °C for 1 hour. Brindos observed that EOR defects did not form for these implant conditions. Also, in previous experiments by Jones et al. [11] it was suggested that interstitials present at EOR were responsible for a lack of extended defects. To test this theory, Brindos's As-I (Arsenic Interstitial) pair formation model was combined with the developed loop model and applied to low energy, high dose simulations. Equation 6 represents Brindos' As-I interaction model as a basic pair reaction where As is an arsenic atom and $Si_{(I)}$ is a silicon interstitial atom.

$$As + Si_{(I)} \Leftrightarrow AsI \tag{6}$$

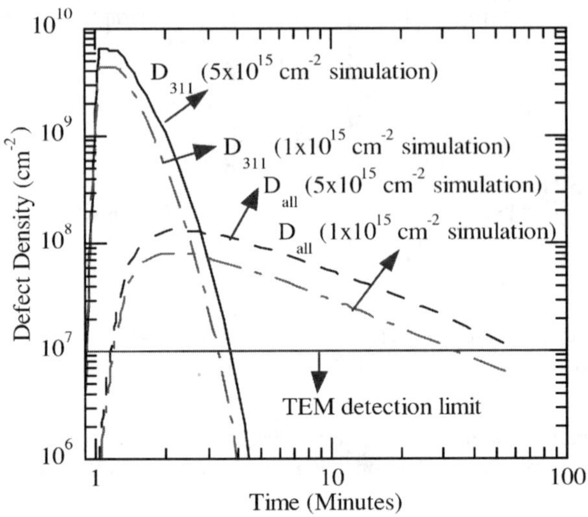

Figure 2. Simulation results for the defect evolution for 3 keV, 5×10^{14}cm^{-2}, 1×10^{15} cm^{-2} and 5×10^{15} cm^{-2} As implants on silicon, annealed at 800 °C for 60 minutes. Experiments showed no defects at the end of the annealing cycle. Therefore, no data are shown.

Arsenic implants were simulated using UT-Marlowe to obtain damage profiles for 3 keV implants at doses of 5×10^{14} cm^{-2}, 1×10^{15} cm^{-2} and 5×10^{15} cm^{-2}. Since the implants were amorphizing implants, the initial damage profiles were truncated to the amorphous depth. Diffusion of As-I is set to the values obtained by Brindos [3]. Figure 2 shows the change in defect density with respect to time at 800 °C for three implant conditions. It is seen that {311} defects nucleate very fast at the short annealing times and their density drops below TEM detection limit of 1.0×10^{7} cm^{-2} very quickly. There is no loop nucleation for the lowest dose (5.0×10^{14}). For higher doses, the density of loops also falls below the TEM detection limit at the end of the annealing cycle. Since experiments showed no defects at the end of the annealing cycle, the simulation results are in direct correlation with the Brindos' experimental results.

DEFECTS IN BORON IMPLANTED SILICON

Liu *et al.* [4] systematically studied the defect formation threshold in the low energy implant regime in order to understand how the sub-amorphization defects influence dopant diffusion. Liu implanted n-type wafers with boron ions. The implant matrix is shown in table 1. 750 °C, 5 min. anneals are used to study the formation threshold of {311} defects. 900°C, 15 min. anneals are used to study sub-amorphization dislocation loops. PTEM analyses were

	5 keV	10 keV	20 keV	30 keV	40 keV
$5\times10^{13}cm^{-2}$	None	None	None	None	None
$1\times10^{14}cm^{-2}$	None	None	{311}'s	{311}'s	{311}'s
$2\times10^{14}cm^{-2}$	None	{311}'s	{311}'s	{311}'s	{311}'s
$5\times10^{14}cm^{-2}$	{311}'s & Loops	{311}'s & Loops	{311}'s & Loops	{311}'s & Loops	{311}'s & Loops
$1\times10^{15}cm^{-2}$	{311}'s & Loops	{311}'s & Loops	{311}'s & Loops	{311}'s & Loops	{311}'s & Loops

Table 1. Simulation results for the types of extended defects formed in B^+ implanted silicon after an anneal at 750°C and 900 °C.

performed to determine defect densities. Liu chose a defect density of 1.2×10^7 cm^{-2} to distinguish between samples with and without extended defects. Table 1 lists the formation threshold for both {311} defects and dislocation loops for the whole implant matrix after anneals at 750 °C for 5 minutes and at 900°C for 15 minutes. The results showed that the critical dose for forming {311} defects decreased with increasing implant energy. The defect density also increased with increasing energy. It was concluded that the interstitial supersaturation necessary to nucleate {311} defects was far less than that for dislocation loops, as well. Liu observed that 40 keV samples produced dislocation loops at a dose of 2×10^{14} cm^{-2}. 20 keV and 5 keV implants did not show loops until the dose was increased to 5×10^{14} cm^{-2}. Results showed that the critical dose for type I loop formation appeared to decrease with increasing implant energy.

Formation of defects in boron implanted silicon is simulated in FLOOPS. While implant damage is present, diffusion of boron is not trivial. It is shown that implanted boron ions can form clusters in which boron atoms are bound to silicon interstitials into thermally stable and immobile complexes [7]. They also pair with silicon interstitials and diffuse away. Consumption of excess interstitials by immobile clusters and mobile pairs reduces the concentration of free interstitials available to form {311} defects and dislocation loops.

To study defect formation in boron implanted silicon, Lilak's [7] model is coupled with the developed loop model. As an initial condition, all of the implanted boron is assumed to be in the substitutional position. Simulations are carried out for each implant condition at 750 °C for 5 minutes and at 900°C for 15 minutes. In order to determine whether the simulation results predict enough {311} defects, the average {311} length is studied for each implant condition. If the average length of {311}'s is greater than 330 Å, then it is assumed that {311} defects exist. Although the chosen average length is large, it is within the error limits of Liu's TEM pictures where the scale is set to 2000 Å and visible defects are around 300 Å. A similar approach is followed to determine whether loops exist for an implant condition or not. First, the density of loops whose radii greater than 100 Å is calculated from the distribution functions. The minimum radius value is chosen from TEM pictures as well. Then, in order to determine if loops are present, a defect density of 1×10^9 cm^{-2} is chosen. If the density of dislocation loops is greater than 1×10^9 cm^{-2}, it is assumed that the loops exist. The chosen defect density is much greater than the defect density selected by Liu, but visible defect densities reported are usually above 1×10^9 cm^{-2}. Table 1 shows that the simulation results for {311} formation are in good agreement with Liu's observations in almost all cases. The discrepancy exist between the data and simulation results for 5×10^{14} cm^{-2}, 30keV and 40keV implant conditions (shaded areas). It is clearly seen that simulation and experimental results for loops are in good agreement for all other implant conditions. Simulation results also agree with the experiments in that the critical dose for forming {311} defects decreases with increasing implant energy. They also follow the observation that the interstitial supersaturation required to nucleate {311} defects is far less than that for dislocation loops, and loops are more likely to nucleate with increasing implant energy.

CONCLUSIONS

Previously developed loop nucleation and evolution model [1] was used to predict end of range (EOR) and projected range defect nucleation and evolution created by different ion implant species such as Germanium, Arsenic and Boron. Experimental data are taken from the literature [2][3][4]. A good agreement between the experimental and simulation results is obtained. The discrepancies between the data and simulations are given whenever they exist.

REFERENCES

1. I. Avci, Mark E. Law, Erik Kuryliw and Kevin S. Jones, IEDM 2001 pp. 835-838.
2. A. F. Gutierrez, K. S. Jones an D. F. Downey, MRS 2001, Vol 669, pp. J5.11.1-5, 2001
3. R. Brindos, K. S. Jones and M. E. Law, MRS 2001, Vol 669, pp. J5.2.1-6, 2001
4. J. Liu, V. Krishnamoorthy, B. Beaudet, K. S. Jones, M. E. Law, J. Shi, and J. Bennett, IEEE Ion Implantation Technology Proceedings, Vol 96, pp. 626-629, 1997.
5. K. S. Jones, S. Prussin, and E. R. Weber, Applied Physics, Vol. A 45, pp. 1-34, 1988
6. P. M. Fahey, P. B. Griffin, and J. D. Plummer, Reviews of Modern Physics, Vol. 61, No 2, April 1989
7. A. D. Lilak, M. E. Law, K. S. Jones and M. D. Giles, IEDM 1997, pp. 493-496, 1997
8. B. Obradovic, G. Wang, Y. Chen, D. Li, C. Snell, A. F. Tasch, UT-MARLOWE 5.0 with tomcat, 1999
9. M. E. Law and K. S. Jones, IEDM 2000, pp. 511-514, 2000
10. J. Li, and K. S. Jones, Applied Physics Letters, Vol. 73, No 25, pp. 3748-3750, 1998.
11. K. S. Jones, D. Downey, H. Miller, J. Chow, J. Chen, M. Puga-Lambers, K. Moller, M. Wright, E. Heitman, J. Glassberg, M. E. Law, L. Robertson, and R. Brindos, Proceedings of the International Conference on Ion Implantation Technology, Vol. 2, p. 841, 1998

Mat. Res. Soc. Symp. Proc. Vol. 717 © 2002 Materials Research Society

EXPERIMENTAL STUDY ON THE MECHANISM OF CARBON DIFFUSION IN SILICON

N.E.B. COWERN*[,1], B. COLOMBEAU*, F. ROOZEBOOM[†], M. HOPSTAKEN[†], H. SNIJDERS[†], P. MEUNIER-BEILLARD[‡] and W. LERCH[§]

*School of Electronics, Computing and Mathematics, University of Surrey, Guildford, Surrey, GU2 7XH, United Kingdom
[†] Philips Research Laboratories, Prof. Holstlaan 4, 5656 AA Eindhoven, The Netherlands
[‡] Philips Research, IMEC, Kapeldreef 75, B-3001 Leuven, Belgium
[§] Mattson Thermal Products GmbH, Daimlerstrasse 10, D-89160 Dornstadt, Germany

ABSTRACT

CVD-grown lightly C-doped superlattices with peak C concentrations of $2.10^{18}/cm^2$ and $2.10^{19}/cm^2$ were annealed in NH_3, N_2/H_2, N_2, and O_2 ambient gases to investigate the influence of a range of point-defect conditions on C diffusion at the nanometer scale. C profiles were measured by secondary-ion mass spectroscopy. The profiles exhibit exponential-like diffusion consistent with a 'long hop' diffusion process with a characteristic migration length λ (=19 ± 3 nm at 850 °C). Within experimental errors the value of λ is the same for all the ambient gases used, whereas the migration frequency g increases by two orders of magnitude as the ambient gas is changed from NH_3 ambient (interstitial undersaturation) to O_2 ambient (interstitial supersaturation), and decreases as a function of C concentration in the as-grown superlattice. The results confirm that C diffuses predominantly by a kick out mechanism under near-equilibrium diffusion conditions. Initial results support the chemical-pump model for suppression of diffusion in C-doped silicon.

INTRODUCTION

Carbon plays a potentially important role as a suppressor of interstitial-mediated dopant diffusion and clustering in advanced silicon-based devices. There are two ways in which C is believed to influence dopant diffusion. First, since C is a fast interstitial-mediated diffuser it acts as a 'chemical pump' expelling a flux of interstitial atoms from the C-doped region, thus creating a localised undersaturation of interstitials [1]. Second, the clustering of C atoms at high C-concentrations can consume excess interstitials [2]. Both effects inhibit the motion of dopants that rely on interstitial-mediated diffusion mechanisms. In order to get a proper understanding of the relative importance of these two phenomena in different device applications, quantitative data on the fundamental mechanism and parameters of C-diffusion are required.

This paper presents a fundamental study of the C-diffusion mechanism using an approach first developed in Ref. [3]. Essentially all substitutional impurities in silicon diffuse by converting for brief intervals into a fast-diffusing point-defect complex (of vacancy or interstitial type), a process which has been described as 'intermittent' or

[1] E-mail: Nick.Cowern@surrey.ac.uk

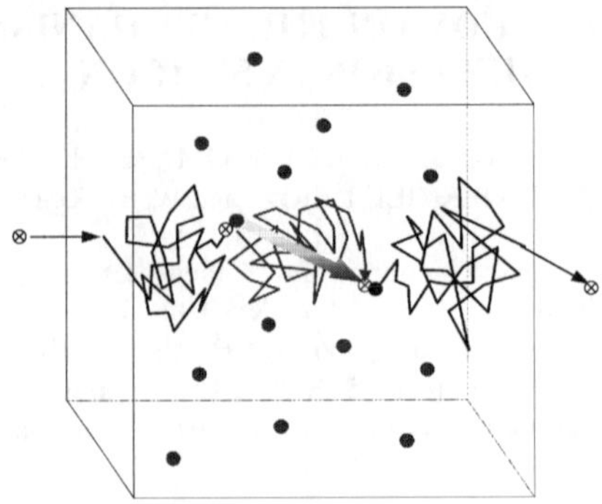

***Figure 1**: Illustration of a single dopant migration step. A native point defect (dashed line) reacts with a substitutional dopant atom to form a short-lived fast-migrating species. After traveling some distance the migrating dopant species reverts to a substitutional site, releasing the native point defect to continue its diffusion.*

'long-hop' diffusion. For example, Sb diffuses by converting briefly into a Sb-V pair following a capture reaction with a vacancy, whilst B diffuses by converting briefly to an interstitial B species following capture of a self interstitial. The process is illustrated in Fig. 1, for the case where the migrating species is formed by an interaction with a diffusing native point defect (interstitial or vacancy) and subsequently dissociates to release the same type of diffusing point defect at another location.

Generally the times and distances over which the migrating species moves are assumed to be small, and diffusion is treated as a more or less continuous process. However, it has been shown in the case of B that the migrating species travels several nanometers before reverting to a substitutional position [3]. For short diffusion times this leads to typically exponential-like diffusion profiles, even at low dopant concentrations where point defects are undisturbed by chemical-pump effects.
In all models of diffusion, the diffusivity, D, is essentially the product of a jump frequency and the square of a jump length. For a simple diffusion mechanism in which an atom jumps between adjacent lattice sites the relevant jump length would be the interatomic distance and the diffusion would appear to all intents and purposes as continuous. However, in the actual situation in silicon, long-hop diffusion applies. In this case the relevant jump length is the mean projected path length, λ, travelled by the migrating species between its initial displacement from a fixed subsitutional position and its subsequent return to a substitutional site. The relevant frequency is the frequency, g, at which the substitutional atom is converted to the migrating form. Thus the migration length is given by the relation $\lambda^2 = D / g$.
C is generally believed to diffuse by a kick out diffusion mechanism in which the substitutional C atom is converted to a mobile C-interstitial by capturing a silicon self-interstitial atom. This is the same mechanism that has been shown to operate in the case of B diffusion. Assuming there is no energy barrier to the kick out reaction, as we know to be true in the case of B diffusion [3], the frequency g should depend

only on the arrival flux of self interstitials, and thus be the same for C as it is for B. Since the diffusivity of C is known to be about two orders of magnitude higher than that of B, we can infer that the value of λ for C should be about one order of magnitude higher than that B. This implies values of λ in the range 5-50 nm at typical processing temperatures – a migration length scale that could easily be observed under appropriate diffusion conditions.

EXPERIMENT

In this study we investigate the diffusion of lightly C-doped marker layers in silicon during annealing in ambient gases ranging from NH_3 (which injects vacancies) to O_2 (which injects self interstitials). We use two values for the peak C-concentration in the as-grown marker layers. In the first, the C-concentration is kept as low as possible ($2.10^{18}/cm^2$) to keep disturbance of the point-defect concentrations in the silicon crystal to a minimum, whilst in the second, a higher C-concentration ($2.10^{19}/cm^2$) is used to investigate the influence of chemical-pump effects and C clustering on diffusion. Epitaxial growth is performed in an ASM Epsilon One CVD reactor, which produces material with an extremely low density of traps for point defects. This ensures that, under conditions where point defects are injected from the silicon surface, either by oxidation or nitridation, the depth distribution of the enhancement or retardation of diffusion will be essentially constant as a function of depth in the epitaxially grown silicon.

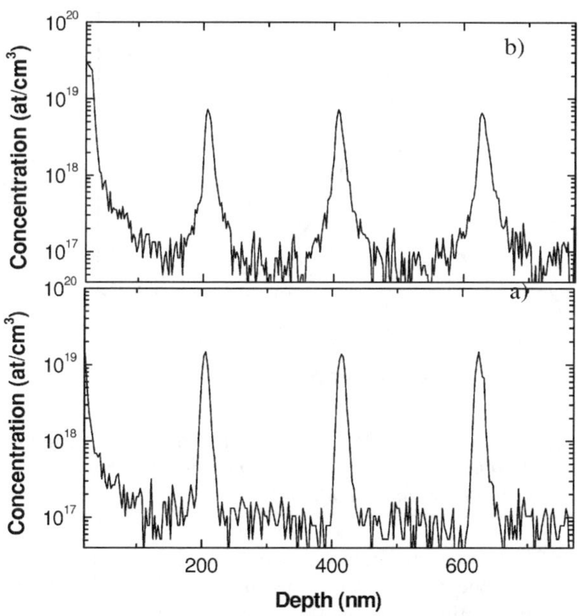

***Figure 2**: SIMS profiles of Carbon (a) as grown and (b) after annealing at 850 °C for 5 s under oxidizing (O_2) ambient, for the highest C-concentration ($2.10^{19}/cm^2$).*

This initial paper reports results on C-diffusion in samples that have been processed by RTP with a set temperature of 850 °C, using annealing times in the range 5 – 240 s under NH_3, N_2/H_2, N_2, and O_2 ambient conditions. The diffusion is analysed by depth profiling with the help of secondary-ion mass spectrometry (SIMS), and detailed information on the migration frequency and path-length of diffusion is extracted by fitting the depth profiles with the 'long-hop' diffusion model described in Ref [3].

SIMS profiles for a C-doped superlattice before and after annealing for 5 s at 850 °C in O_2 are shown in Fig. 2. Observing the qualitative difference between the profiles for the as-grown and annealed samples, it appears that the diffusion in all of the markers is approximately the same, and that there is substantial diffusion in the low-concentration part of the C-profile. To obtain accurate information from the profiles it is necessary to undertake detailed computer fitting. The fitting procedure carries out a non-linear least-squares fit between (i) the as-grown profile which has been convoluted with a trial diffusion function calculated from the 'long-hop' model (with two free parameters, λ and g) and (ii) the annealed profile, scaled as necessary to 'line up' the superlattice peaks and match the C-dose in the as-grown profile (three free parameters – 'shift', 'stretch', and 'dose' scaling). This procedure is applied to all the available data in this study.

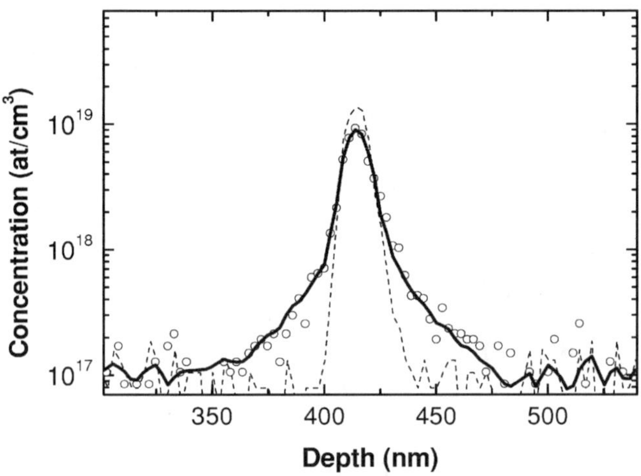

***Figure 3**: Diffusion of a C peak profile in oxidizing ambient at 850 °C. Dashed line represents the as-grown profile and circles represent the profile after 5 s diffusion. Solid line represents "kick-out" fit of the annealed profile.*

An example of a typical profile fit resulting from this procedure can be seen in Fig. 3, where we have focussed on just one of the C-peaks to show the diffusion profile in full detail. There is a clear distinction between the high-concentration part of the profile, where no diffusion has occurred, and the low-concentration part, where a long exponential-like diffusion tail is clearly visible. The match between the data for the annealed profile and the diffusion fit is remarkable. The fit in this case is consistent with a migration length λ=18.5 nm and a migration frequency $g=6.10^{-2}$ s^{-1}. For longer annealing times all of the C in the peak diffuses, leading to a more Gaussian-like diffusion behaviour (not shown in this figure).

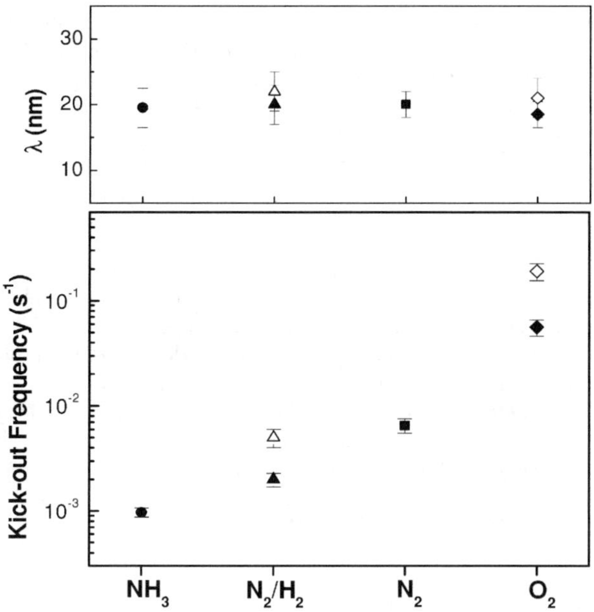

***Figure 4**: Kick-out frequency (g) and migration length (λ) for C diffusion as a function of processing conditions: NH_3 (circles), N_2/H_2 (up triangles), N_2 (squares) and O_2 (diamonds) ambient for both high (fill symbols) and low (open symbols) C-peak concentration.*

The behaviour of the migration length and migration frequency for all the samples in our study is summarised in Fig. 4. The results show a highly consistent pattern, in that the value of λ remains constant within the experimental error, independent of the ambient gas (i.e. independent of the injected point-defect concentrations), and independent of the peak concentration in the as-grown C profile. On the other hand, the migration frequency increases by two orders of magnitude as we change the ambient gas from NH_3 (which injects excess vacancies) through N_2H_2 and N_2 to O_2 (which injects excess self interstitials). This is a clear signature of a kick-out mechanism in which the migrating C-species is set in motion by an interaction with a self-interstitial atom, and ends its migration with a thermally activated kick-in to a substitutional site. This latter event happens without assistance from an external point defect, hence the distance travelled before it occurs is independent of the supersaturation of point-defects in the crystal. Note that our conclusion that the kick-out mechanism is dominant does not conflict with published data on C-outdiffusion which appear to show a very small Frank-Turnbull component of C-diffusion [4].

Such a small contribution would have no significant influence on the migration path length measured in the present study.

It is interesting to note that the migration frequency is substantially reduced in the samples with the higher peak C-concentration. This could be explained in two alternative ways. In principle it could an effect of C-clustering, which reduces the number of C-atoms available for diffusion, or it could be a chemical-pump effect in which C expels self interstitials from the region of the C-peak. The fact that the diffusion in all the C peaks is approximately the same, even under conditions where point defects are being injected, makes it unlikely that C-clustering is taking place to a significant extent, and we suggest that the reduction in migration frequency in our study is a consequence of the chemical-pump effect.

CONCLUSIONS

In this work we have confirmed that C diffuses predominantly by a kick-out mechanism in silicon, and have shown that C clustering appears to have little role in the behaviour of diffusion under the conditions investigated. Diffusion is reduced by increased C concentrations as a result of the chemical-pump effect. The migration length for C in silicon is large (~20 nm at 850 °C), a factor 6-8 larger than that of B at the same temperature. This result suggests that, as in the case of B, the kick-out reaction for C has no significant energy barrier. Our results emphasize that the accurate simulation of C diffusion in Si at the short length scales relevant for modern IC devices requires use of models that explicitly treat the separate species (substitutional and interstitial C) involved in C-diffusion.

ACKNOWLEDGMENT

This work was partly funded by the CEC IST-FRENDTECH project and was initiated as a project of Philips Research. We thank P. Magnee at Philips Research Leuven, M. Caymax at IMEC and R. Woltjer at Philips Research in Eindhoven for support and discussions on carbon diffusion experiments.

REFERENCES

1. R. Scholtz, U. Goesele, J.-Y. Huh, and T.Y. Tan, *Appl. Phys. Lett.* **72**, 200 (1998)
2. P.A. Stolk, *Materials Science and Engineering* B **36**, 275 (1996)
3. N.E.B. Cowern, K.T.F. Jansen, G.F.A. van de Walle and D.J. Gravesteijn, *Phys. Rev. Lett.* **65**, 2434 (1990).
4. R.F. Scholtz, P. Werner, U. Goesele and T.Y. Tan, *Appl. Phys. Lett.* **74**, 392 (1999)

Characterization Using Surface Analysis Techniques

Non-routine Dopant, Impurity and Stoichiometry Characterization of SiGe, SiON and Ultra-low Energy B-implanted Si Using Secondary Ion Mass Spectrometry

Charles W. Magee, Temel H. Buyuklimanli, John W. Marino, Steven W. Novak and M. Alper Sahiner
Evans East (A member of the Evans Analytical Group)
104 Windsor Center Drive, Suite 101, East Windsor, New Jersey 08520 USA

ABSTRACT

New, non-routine metrology issues are addressed for three kinds of materials and processes that are necessary for the fabrication of ultra-high speed devices. We look at the problems and solutions for measuring both stoichiometry and dopant content of SiGe material when using Cs primary ion bombardment. We examine the challenges of determining the N content of ultra-thin SiON gate dielectrics with emphasis on what will be necessary for the measurement of 1nm thick oxides. And finally we show some promising early results of using a new protocol for measuring ULE B ion implant profiles in the top 3nm with emphasis on obtaining a more realist profile shape in this region for TCAD modeling purposes.

INTRODUCTION

As device dimensions shrink, new materials, new processes and novel device structures are needed to obtain the device speed that will be needed in the near and foreseeable future. These new materials and structures oftentimes require new methods of characterization. This is certainly the case with Si-based high-speed bipolar and field-effect transistors being planned for future CMOS and BICMOS technologies. This manuscript will illustrate three examples of how new metrology methods using secondary ion mass spectrometry (SIMS) have been devised to support the new materials and technologies that are in use now or will be needed in the near future.

STOICHIOMETRY AND IMPURITY ANALYSIS IN SiGe HBTs

SiGe is being used currently in the base region of bipolar transistors to increase the speed of devices, and may be used in the future for fabrication elevated source-drains in FETs [1]. Not only does one have to measure the Ge content of the thin layers that are used in these structures, but one must also perform accurate measurements of the dopants and impurities as well. These measurements are complicated when using SIMS because the matrix is no longer a virtually pure material, i.e. 100% Si, but an alloy of Si and Ge with from 5 to 50% Ge. What is more, the Ge composition of the SiGe layers used in HBTs is usually not constant, but graded in composition through the thickness of the base layer. In some cases, this graded Ge composition requires one to make a point-by-point correction for the sensitivity factors for dopants and impurities that takes into account the Ge concentration at every data point. The necessity for applying such correction factors is based on the results from calculating relative sensitivity factors (RSFs) from a series of reference materials of single layers of SiGe that were ion implanted with B, P, C and O [2].

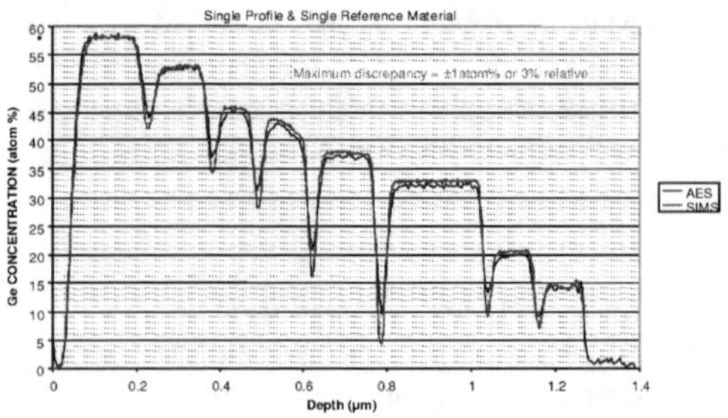

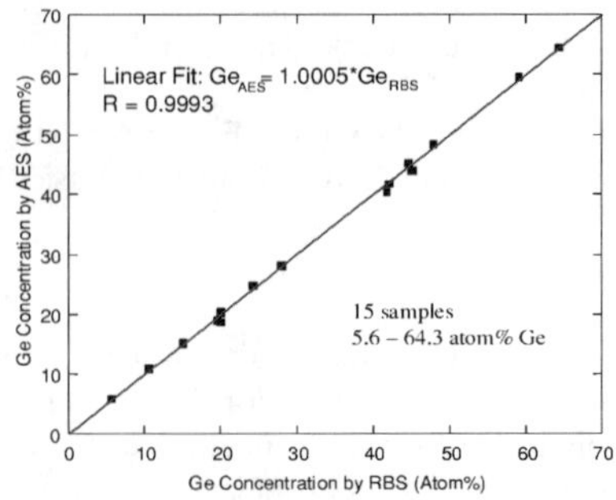

Figure 1a **Figure 1b**

The Ge contents were measured by Rutherford backscattering spectrometry and found to be 6, 10, 15, 20, 24, 28 and 45%. The effect of the Ge concentration on the RSFs was plotted and a correction procedure found to allow one to plot the concentration profiles of Ge, B, P, C or O with a single RSF (but corrected for the Ge concentration at every point). The study was made for both oxygen primary ion bombardment and with cesium primary ion bombardment. Figure 1a shows the agreement of Ge concentrations measured by SIMS (Cs bombardment) with those measured by Auger electron spectrometry (AES). The agreement is within 3% relative over a Ge concentration range of 15 to 58%. Figure 1b shows that Ge quantification by AES has, in turn, been checked against RBS on single layers of SiGe from 5.6 to 64.3% with a correlation coefficient of 0.9993. Figure 2 shows the effect of making this correction with measuring B in SiGe when using Cs bombardment. Note that when bombarding with oxygen, this correction procedure is not needed. However, <u>*Cs*</u> bombardment must be used if C and O are to be measured in the same analysis as B and Ge.

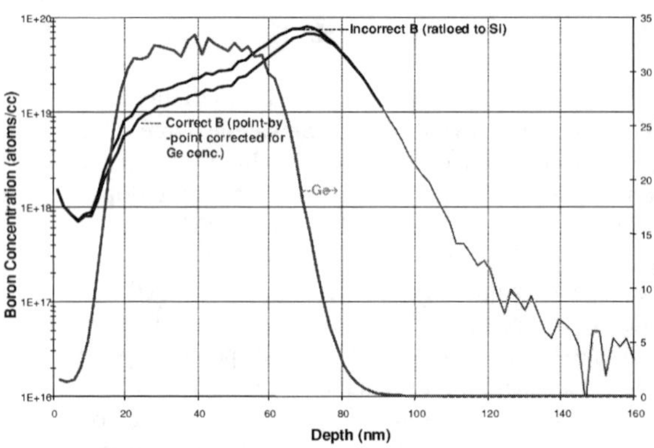

Figure 2

The same correction procedure must be made for Cs bombardment when measuring C in SiGe material. The accurate determination of C concentrations in SiGe is particularly import because C is being added to SiGe to retard B out-diffusion from the base region into the emitter and collector regions of the transistor [3].

MEASURING NITROGEN CONTENT IN 1nm SiON GATE DIELECTRICS

Current MOS technologies, as well as those in the near future, use nitrided SiO_2 as the gate dielectric material. The presence of N at the SiO_2/Si interface retards B penetration from the poly-Si gate material into the channel region. However, the films are currently under 3nm thick and appear to be headed down in thickness towards 1nm, with peak N concentrations at the interface of only a few percent atomic. SIMS has been one of the few analytical techniques with both the depth resolution and the sensitivity to provide N depth profiles in these materials [4-7]. Quantification of the N profiles is calibrated using an N_2O-grown oxide, the N areal density of which has been determined by $^{14}N(d,\alpha)^{12}C$ nuclear reaction analysis. Figure 3 shows the profiles taken on this reference material. The N distribution peaked at the interface is typical of N_2O-grown oxides.

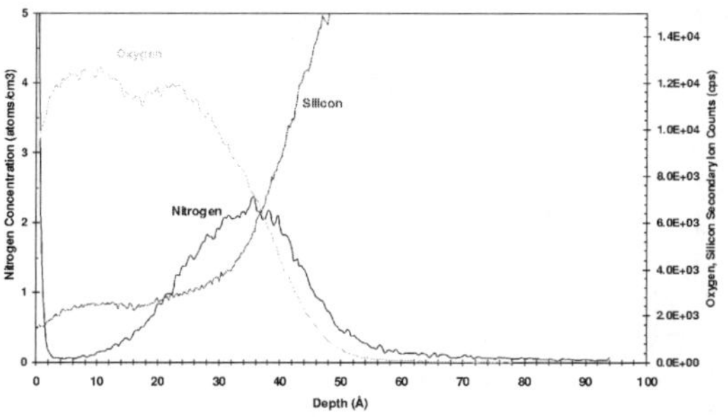

Figure 3

The accuracy with which we can measure N areal densities is shown in Figs 4a and b. Samples in a series of 2.5nm oxides were exposed to a remote N plasma that produced N peaks near the surface in differing amounts. The overlaid SIMS profiles of these samples are shown in Fig. 4a. Figure 4b shows the correlation of the N areal densities determined by SIMS, with the N areal densities determined by X-ray photoelectron spectrometry (XPS). There is good correlation between SIMS and XPS up to approximately 1.5E15 N atoms/cm^2. Above this point, two group of samples appear to give quite different results by the two techniques. It is known that the samples in these two groups have percent levels of H present, and it is thought that this may be affecting either or one of the two techniques.

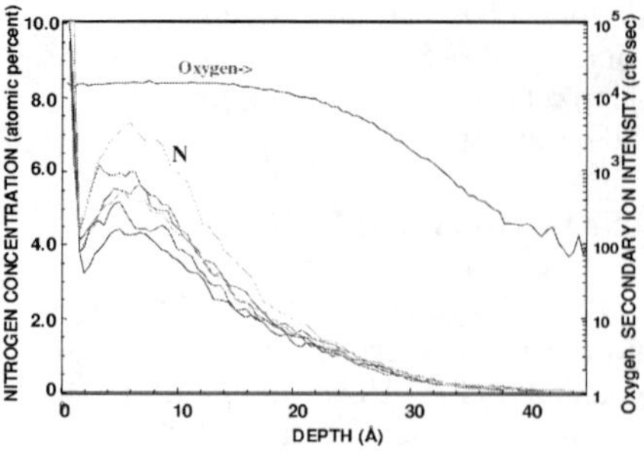

Figure 4a

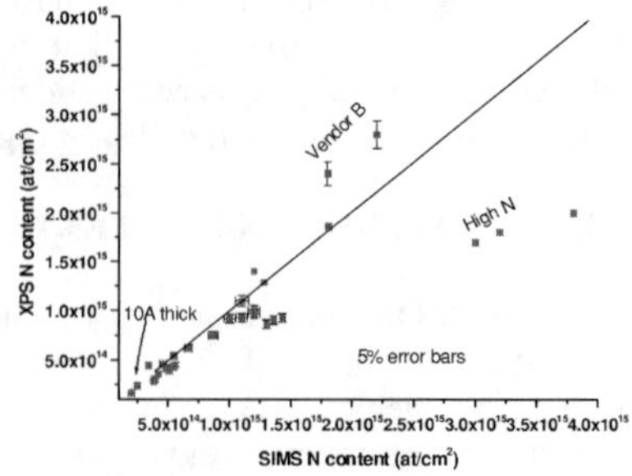

Figure 4b

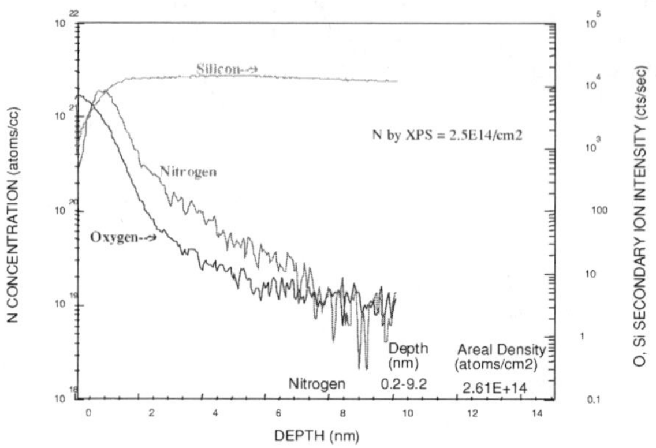

Figure 5

With the quantification issue seemingly well in hand, one must now investigate the problems of extending our analytical methods down to gate oxides that approach 1nm in thickness. The profiles from a SIMS analysis of such a film are shown in Fig. 5. The analysis shows sufficient depth resolution to detect the presence of N only at the SiO_2/Si interface for this 1.1nm thick, N_2O-grown oxide, but better depth resolution would undoubtedly be of use in the analysis of such thin films. One might naturally assume that the depth resolution in Fig. 5 is being limited by the atomic mixing that is introduced into the sample by the primary ion beam. Reducing the Cs primary ion energy from the 500eV used in this analysis to some lower energy might realistically be expected to yield decreased atomic mixing and improved depth resolution. The data in Fig.6 show that this is not the case.

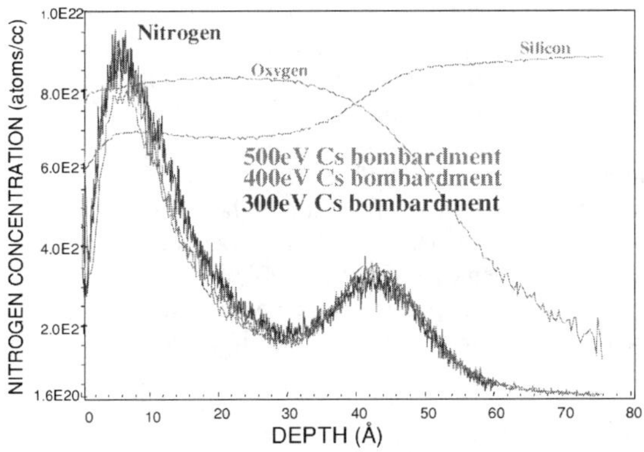

Figure 6

Figure 6 [7] shows overlays of profiles taken on an ammonia-treated SiO_2 film of 4nm thickness, where the energy of the Cs primary ions was 500eV to 400eV and finally to 300eV. There is virtually no difference between the profiles, even though separate experiments in Ref 7 lead one to believe that the depth resolution *should* be better at 300eV than at 500eV. In order to explain this non-improvement in depth resolution with decreasing bombardment energy, we must look at the sputtering process at an atomic scale.

It was pointed out long ago [8] that sputtering does not take place in a layer-by-layer fashion but is influenced by the statistical nature of sputtering. This sequential layer sputtering (SLS) model was later extended and modified [9,10] to agree more closely with observations. However, in it's mature form, the SLS model still predicts that in the initial stages of sputtering, the surface is roughened by the statistical nature of sputtering. In sputter profiling films of more common thickness (<10nm), this sputter-induced roughness is not significant, but there is a contribution for films in the 1-3 monolayer range [10]. Heretofore, one has not been called upon to provide quantitative depth profiles of films this thin, but the advent of 1nm SiON gate dielectrics changes this. We now must profile films where the roughness predicted by the SLS model can no longer be neglected. The onset of surface roughness with sputtering is illustrated in Fig. 7.

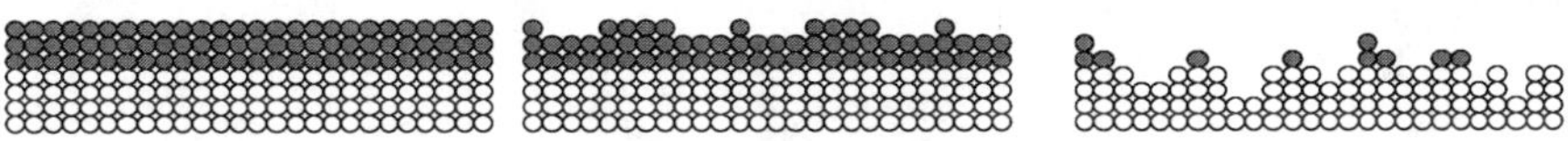

a. b. c.

Figure 7

Panel (a) shows a three-atom-thick film on a substrate prior to sputtering. Panel (b) shows what the sample surface might look like after sputtering only a partial monolayer. The figure assumes that only atoms in the surface layer are sputtered, and that the primary ion beam causes no atomic mixing of the film and substrate atoms. These assumptions are, of course, not true, but their imposition shows that even if these other effects are left out, the surface still roughens on a scale that is significant with respect to the film thicknesses that we are trying to measure. This is shown in panel (c) where the equivalent of all three monolayers have been sputtered off of the sample. The cartoon agrees qualitatively with the predictions of the modified SLS model [10] in that roughness has formed on the surface that is significant with respect to the very thin surface layer that we are trying to analyze. In fact, some film atoms remain on the sample while some substrate atoms have already been sputtered off. If this statistical sputter-induced roughness could be modeled and removed from the data, higher depth resolution profiles might be possible. In order to test the feasibility of this procedure, one had to first show that a roughness parameter could be removed from a profile while obtaining a valid result. To test this, we started with a real profile, (a), convolved a 3Å roughness into it, (b), and then deconvolved the same roughness, (c), to see if the starting profile was obtained (Fig. 8).

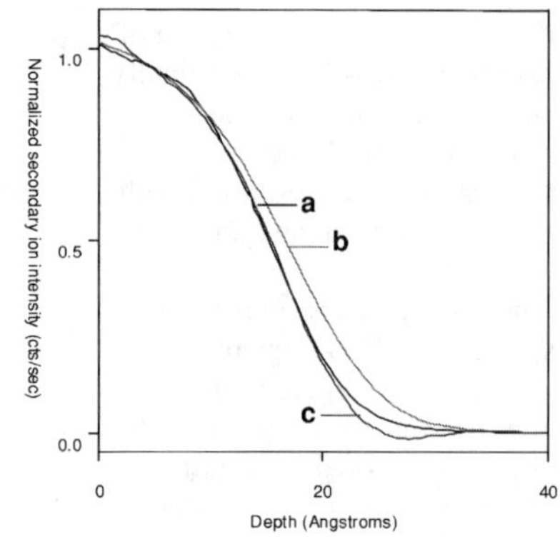

Figure 8

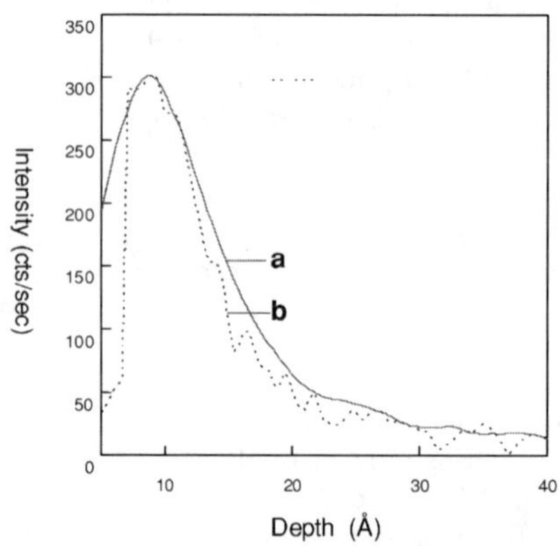

Figure 9

Except for small deviations at the extremities of the profile, good agreement was obtained between the initial profile, (a), and the deconvolved profile, (c). Figure 9 shows the result from deconvolving a 6A roughness from the N profile of the 1.1nm SiON film shown in Fig. 5. The initial N profile, (a) was smoothed before the deconvolution. The N profile after deconvolving the 6Å roughness, (b) is significantly narrower than the starting profile, however, difficulties remain near the surface boundary of the deconvolution. That is why the profiles are not shown to zero depth. In addition, our numerical methods assume a constant roughness for the entire depth of the profile when we know that the roughness at the beginning of the analysis is really zero. However, the SLS model [10] shows that the roughness *does* become constant after sputtering a very few atom layers.

MEASURING THE TOP 3nm OF UTRA-LOW ENERGY B IMPLANTED IN Si

The need for creating shallower junctions as device sizes shrink has led to significant advances in SIMS metrology, especially for profiling ultra-low energy (ULE) boron ion implanted into Si [11,12]. Although the methods used measure correct junction depths and total integrated areal densities [13], the protocols for doing this measurement all result in a B peak at the surface that is not predicted by molecular dynamics calculations (TRIM) or by other analytical techniques [14]. For certain modeling applications, there is a need to know the accurate profile shape in the top 3nm of the B profile where this unpredicted surface peak appears. We believe that this can be obtained by using a different analytical protocol from those in wide use today.

If one acquires SIMS data for ULE B in Si using O_2 primary ion bombardment at an oblique angle of incidence in UHV, the profiles for Si and B will appear as shown in Fig 10a. The oxide on the surface of the Si enhances the sensitivity for both Si and B. Analysts have, until now, believed that quantitative measurements could not be made within this enhancement region. Researchers have instead introduced O_2 gas onto the sample surface during analysis, or orientated the O_2 sputtering beam at normal incidence, which both result in the continuous growth of a thin SiO_2 layer on the surface (full oxidation) as the analysis proceeds. This provides oxide enhancement for the entire profile as shown in Fig. 10b. One can see that the Si matrix profile is indeed virtually flat throughout the entire profile, but the B profile shows a peak at the surface. This surface

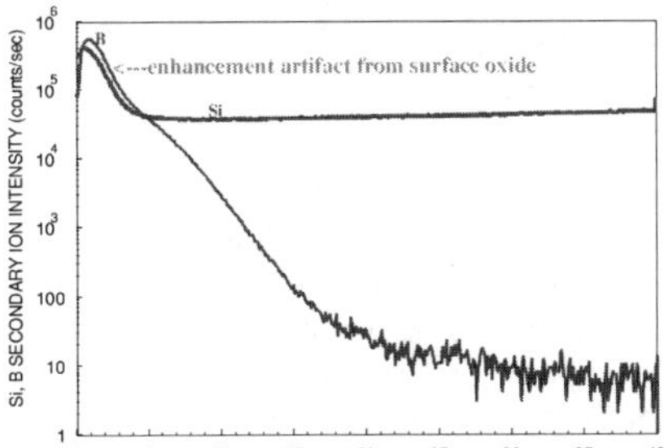

Figure 10a

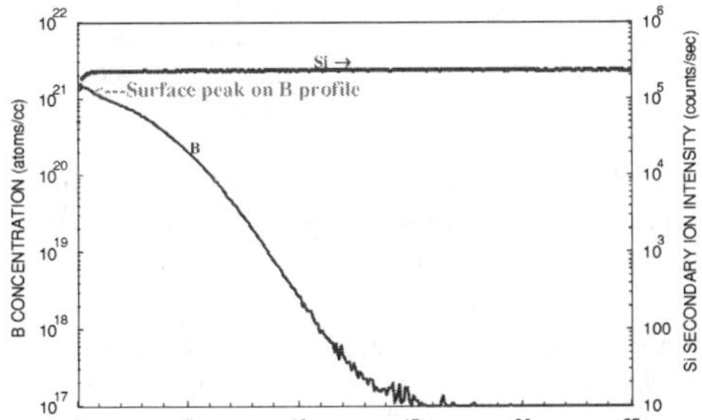

Figure 10b

peak is a ubiquitous feature that is found in all ULE B profiles that are acquired using a "full oxidation" protocol. Because of this, we decided to investigate the possibility of using a non-oxidizing protocol with steps taken to remove the affect of the surface sensitivity enhancement.

In order to use a non-oxidizing protocol for near-surface analyses of ULE B implants in Si, one must do the following: i. correct for ion yield changes across the SiO_2/Si interface, ii. correct for the sputter rate change across the SiO_2/Si interface, iii. find a bombardment energy and angle for minimum roughening while still remaining in a non-oxidizing condition, iv. use interpolation during normalization of B to the Si matrix intensity, and v. correct for detector deadtime loss at high count rates for Si in the surface oxide layer. Figure 11 shows that the first correction is rather simple. The figure shows raw data profiles for both Si and B for a sample with B ion implanted across a structure of SiO_2/Si/SiO_2. The B profile is also plotted in concentration using a single RSF with point-by-point ratioing of the B profile to the Si profile. The quantitative B profile shows no large discontinuity at the two interfaces. This shows that the sensitivity of B relative to Si is the same for both materials. This has not commonly been found to be the case, but we feel the previous results are flawed in that the Si matrix current in the SiO_2 layers was not accurate due to detector saturation problems.

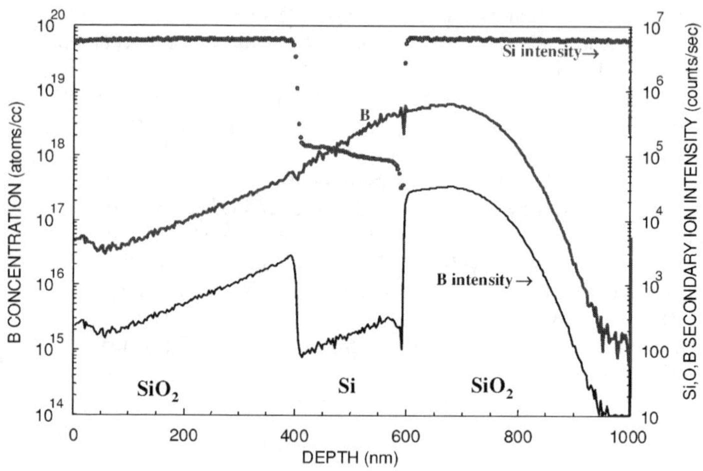

Figure 11

Our dead time-corrected Si profile is accurate in the SiO_2 layers at count rates as high as $5x10^6$ counts/sec (Fig. 11). We also measured the sputtering rates of SiO_2 and Si under low primary ion energy, non-oxidizing conditions and found them to be within 10% of each other. This is in good agreement with data compiled previously but using higher sputtering energies[15]. The wide spacing of the data points in Fig. 11 make interpolation between points necessary so that the value of Si matrix current is used for each B data point is taken at the correct depth. (All elements are acquired sequentially so no two points are acquired at exactly the same depth.) Figure 12a shows the results of processing the raw data B profiles shown in Fig. 10a when the five factors listed above are used to acquire and process the data. Also shown is the B profile taken using the old

O_2 flooding protocol. The B profile taken using the new, non-O_2 flooding protocol shows no peak at the surface, but the profile is comparable to the old, O_2-flooding protocol in the junction region. It should be pointed out, however, that the old protocol data were taken using 700eV O_2 bombardment, while the new protocol data were taken using 500eV O_2 bombardment. The lower primary ion energy for the new, non-oxidizing protocol is necessary to reduce roughening and atomic mixing that are more prominent when full oxidation of the Si is not reached. In fact, the B profile taken using the new, non-O_2 flooding protocol agrees much more closely with a TRIM calculation for 500 eV B in Si than does the profile taken using O_2 flooding. This is shown in Fig. 12b.

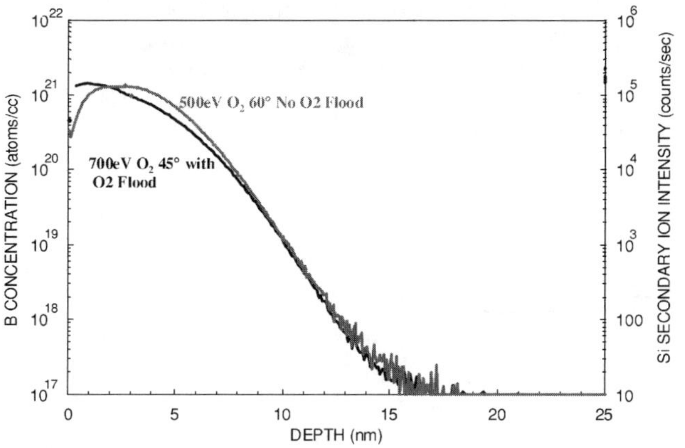

Figure 12a

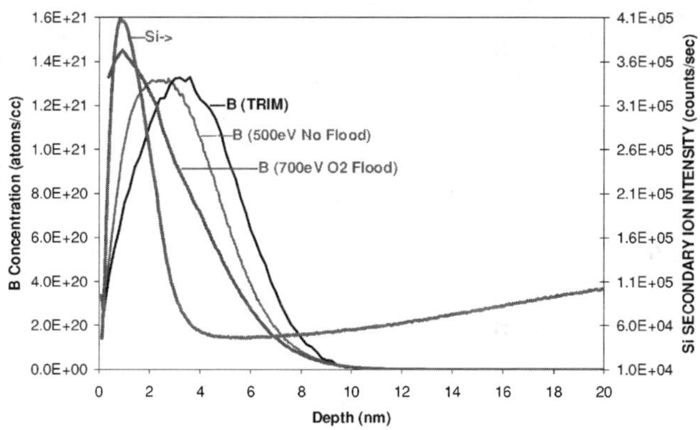

Figure 12b

Clearly the profile taken using the new protocol is much more realistic in its shape than that taken using the old protocol. Also of great importance is the fact that the oxide enhancement of the Si profile gives a clear indication of the location of the SiO_2/Si interface relative to the B profile. This, in turn allows one to gauge whether the implant peak lies within the Si, at the interface, or in the surface oxide. This important information has, heretofore, not been available because protocols that induce full oxidation render the interface invisible.

The utility of this new, non-oxidizing protocol is illustrated in Figs 13a and 13b. Figure 13a shows a comparison of two unannealed B implants, one at 250eV and the other at 500eV. The profiles were acquired using 400eV O_2 bombardments at 45° with O_2 flooding which results in a fully oxidized Si surface throughout the profiles. The profiles are clearly different, but neither has a reasonable shape near the surface, and the Si profiles do not give any indication of the location of the B profile in relation to the interface between the surface oxide that grew during implantation and the Si substrate.

Figure 13b shows the profiles for the same two samples acquired using 1000eV O_2 bombardment at 60° without O_2 flooding. The profiles show the anticipated profile shape in the near surface region, and the Si profiles show the location of the B profiles relative to the SiO_2/Si interface. Unfortunately the profiles in Fig. 13b were not also taken at 400eV bombarding energy. This has resulted in a poorer decay length for the 250eV sample in this figure. Work is underway to make a more direct comparison.

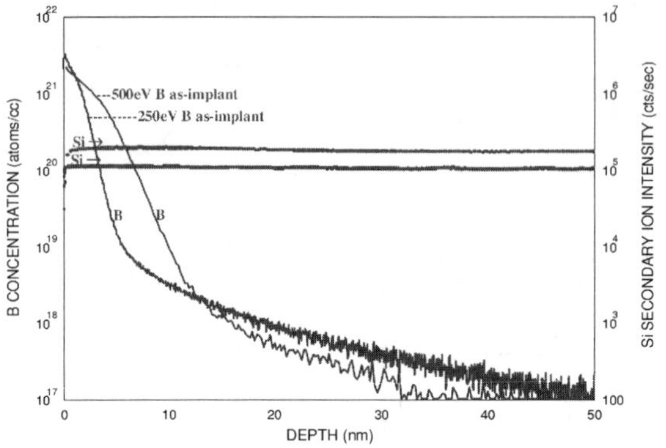

Figure 13a

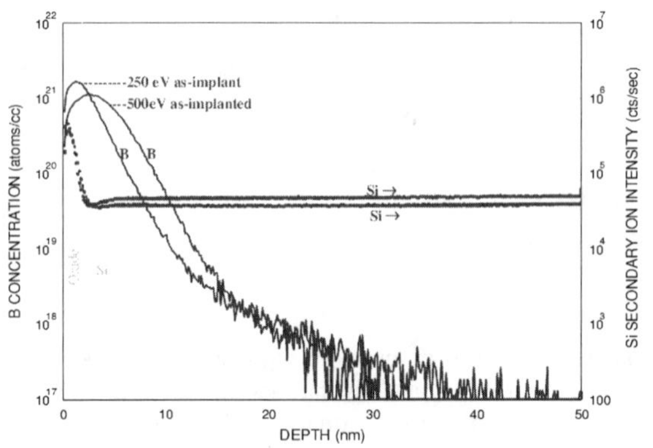

Figure 13b

CONCLUSIONS

New, non-routine metrology issues have been addressed for three kinds of materials and processes that are necessary for the fabrication of ultra-high speed devices: SiGe, ultra-thin SiON gate dielectrics, and ultra-shallow junctions. We looked at the problems and solutions for measuring both stoichiometry and dopant content of SiGe material when using Cs primary ion bombardment. Ge content can be measured with 3% (relative) accuracy. Boron and C profiles must be corrected to take into account the Ge concentration at every B or C data point. We examined the challenges of determining the N content of ultra-thin SiON gate dielectrics and we have shown good agreement between SIMS and XPS. We have also shown that deconvolution of the sputter-induced surface roughness shows promise for improving depth resolution which will help in the analysis of 1nm thick oxides. And finally we have shown some promising early results of using a new protocol for measuring ULE B ion implant profiles in the top 3nm with emphasis on obtaining a more realist profile shape in this region for TCAD modeling purposes. This protocol does *not* result in full oxidation of the Si surface and thus must be carried out using oblique angles of incidence for sputtering.

REFERENCES

1. M.C. Ozturk, N. Pesovic, J. Liu, I. Kang, H. Mo and S. Gannavaram, (Proceedings of this symposium).
2. M. S. Denker, T. Buyuklimanli, J. T. Mayer, Proceedings for the 12th International Conference on Secondary Ion Mass Spectrometry, Brussels, Belgium, 639 (1999).
3. L.D. Lanzerotti, J.C. Sturm, E. Stach, R. Hull, T. Buyuklimanli and C. Magee, Appl. Phys. Lett. **70** (23), 3125 (1997).
4. M.R. Frost and C.W. Magee, Appl. Surf. Sci. **104/105**, 379 (1996).
5. J.R. Shallenberger, D.A. Cole and S.W. Novak, J. Vac. Sci. Technol. A **17** (4), 1086 (1999).
6. S.W. Novak, J.R. Shallenberger, D.A. Cole and J.W. Marino in *Ultra-Thin SiO_2 and High K Materials for ULSI Gate Dielectrics*, edited by M.L Green, T. Hatorri, H.R. Huff, G. Lucovsky and C.A. Richter, (Mater. Res. Soc. Proc. **567**, Pittsburgh, PA, 1999) pp. 579-586.
7. S.W. Novak, E.J. Bekos and J.W. Marino, Appl. Surf. Sci. **175-176**, 678 (2001).
8. A. Benninghoven, Z. Phys. **230**, 403 (1971).
9. S. Hoffmann, Appl. Phys. **9**, 59 (1976).
10. M.P. Seah, J.M. Sanz and S. Hoffmann, Thin Solid Films **81**, 239 (1981).
11. P.K. Chu, V.K.F. Chia, S.P. Smith and C.W. Magee, Mater. Chem. Phys., **52**, 60 (1998).
12. V.K.F. Chia, G.R. Mount, M.J. Edgell and C.W. Magee, J. Vac. Sci. Technol. B **17** (5), 2345 (1999).
13. C.W. Magee, D. Jacobson and H.-J. Gossmann, J. Vac. Sci. Technol. B **18** (1), 489 (2000).
14. J. Fruhauf, R. Lindsay, W. Vandervorst and K.Koch, paper C3.4, these proceedings.

15. *Secondary Ion Mass Spectrometry: A Practical Handbook for Depth Profiling and Bulk Impurity Analysis,* R.G. Wilson, F. A. Stevie and C.W. Magee, (John Wiley and Sons, New York, 1989) p. App. D.1.

Characterization Technologies

Mat. Res. Soc. Symp. Proc. Vol. 717 © 2002 Materials Research Society C7.2

Profile Changes and Self-sputtering during Low Energy Ion Implantation.

W.Vandervorst, T.Janssens, B.Brijs, R.Lindsay, IMEC, Kapeldreef 75, B-3001 Leuven, Belgium
E. J. H. Collart, David A. Kirkwood, Applied Materials UK Ltd, Parametric and Conductive Implant Division,Foundry Lane, Horsham, W-Sussex RH13 5PX, United Kingdom,
G. Mathot, G.Terwagne, Univ. Namur, LARN, rue de Bruxelles 61, B-5000 Namur, Belgium

ABSTRACT

Continued device scaling requires the formation of ever-shallower junctions with low resistance. A desirable option to form these junctions is still the use of conventional ion implantation. However in order to meet the junction depth/sheet resistance goals, a strong reduction in implant energy and increase in implant dose is required. Earlier work for B and BF_2-implants, has suggested that during low energy ion implantation self-sputtering may become an important factor influencing/limiting the retained dose. The basic mechanism for the self-sputtering with increasing dose is the increasing dopant concentration at the surface leading to an increased probability for re-emission by the sputtering process. Simple models describing ion retention in combination with sputtering are based on this concept and indeed predict a self-sputtering process limiting the final retained dose. Unfortunately the theoretical calculations only predict a significant sputtering at doses $>5 10^{16}$ at/cm^2 whereas experimental results already show a limit in retained dose at 5 10^{15}at/cm^2.

In order to confirm the experimental data, low energy B, BF_2, As and Sb implants have been made. Dose retention was monitored using nuclear reaction analysis and RBS whereby details of the dopant profile (redistribution) were studied using high resolution SIMS. For As and Sb no self-sputtering up to a dose of 1 10^{16}at/cm^2 can be found. For B a small dose loss (<10%) is seen significantly below the literature data. For BF_2 a 20 % dose loss is observed. None of the SIMS profiles provide sufficient evidence for enhanced B-surface migration as required to explain the enhanced self-sputtering. On the other hand such a surface migration is reminiscent of the observations in SIMS whereby also an enhanced mobility of B during ion irradiation is required to explain the anomalous B-surface peak in many SIMS profiles. Based on the SIMS profiles a component sputter yield for B can be derived which is significantly higher than the matrix sputter yield suggesting a weak bonding of the segregated species leading to a reduced surface binding energy and thus enhanced sputtering yield.

INTRODUCTION

An important motivation to reduce the dopant implant energy to the even sub-keV regime is to reduce the as-implanted dopant depth as well as the subsequent transient-enhanced diffusion (TED) during annealing. The latter is particularly important for B as its TED is closely linked to the implant damage and the amount (and depth) of the interstitials generated [1]. An important concern for advanced technologies is the reduction in sheet resistance requiring an increase in total activated dopant content. Hence higher as-implant doses may contain the potential to lead to higher active dopant levels.

A concern with decreasing implant energy and increasing dose, is the increasing importance of the sputtering process eventually leading to a limitation in the retained dose. As sputtering constitutes primarily the emission of particles out of the first few monolayers, an increase in the near-surface dopant concentration resulting from the low energy-high dose implant process, inevitably will expose the implanted species to an increased probability of re-emission by sputtering. Hence with increasing dose, increased sputtering of the dopant atoms will occur and eventually one expects, in stationary state, no further increase in retained dose but the emission of one dopant atom per implanted particle i.e. a sputter yield equal to one. Retention curves as a function of implant dose will inevitably level off at a particular saturation level and inherently also provide information on the sputter yield process as well [2].

With increasing dose, the incorporation of a large concentration of "foreign" atoms in the lattice will lead to effects such as range shortening and ion beam induced displacements causing deviations from the "simple" low dose ion implant distribution. In addition since the target atoms are also subject to the sputtering process during the implantation, the standard (range) descriptions of implant profiles will be influenced by the progression of the surface. Hence any description of the implanted dopant profile will have to consider the effects of substrate sputtering as well.

Finally in order to explain enhanced self-sputtering it is necessary to observe a strong increase in near-surface concentration whereby the latter may be caused by any of the above mentioned mechanisms modifying the implant profile.

Experimental observations of dose retention.

Detailed studies on the retention of low energy B, $B_{10}H_{14}$ and BF_2-implants have been reported by Agarwal and Albano [3,4] (fig.1). All reports are based on integral B-doses as measured using nuclear reaction analysis (NRA) and thus expected to be fairly quantitative. Despite the quantification accuracy of NRA, pronounced differences between the results of Agarwal and Albano can be observed (cfr the arrow in fig.1). In addition, we have performed similar experiments again using the $^{11}B\ (p,\alpha)^{8}Be$ reaction at 660 keV [5] to probe the B-content and the $^{19}F(p,\alpha\gamma)^{16}O$ resonance at 340 keV to measure the F-content [6]. B and BF_2-implants were made at two different doses 1 10^{15} at/cm^2 and 5 10^{15} at/cm^2 in an energy range of 0.5-5 keV. The results (fig.2) basically agree with the measurements of Agarwal in the sense that, at reduced energy, effects of self-sputtering become apparent (0.5 versus 1 kev B) and that the

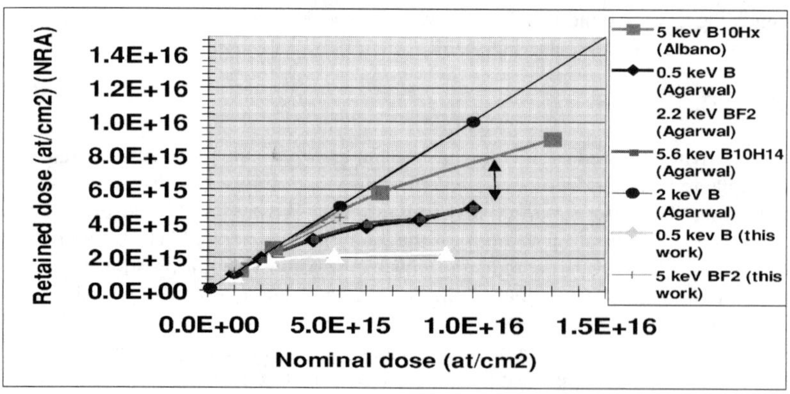

Figure 1 : Retained dose (as measured with nuclear reaction analysis) for low energy B, BF_2 and decaborane implants. After [3,4].

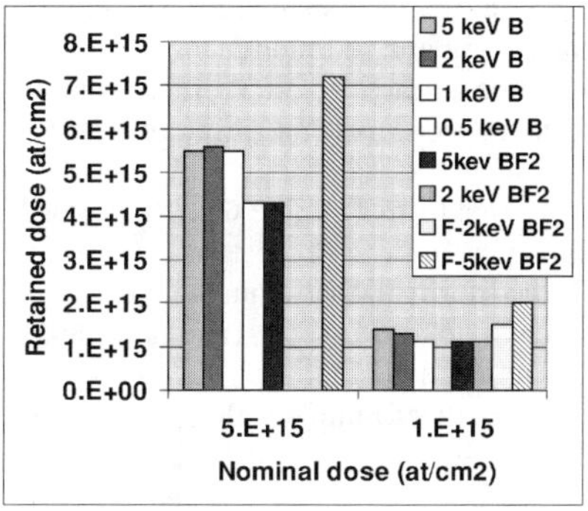

Figure 2: Retained B or F dose (at/cm^2) as measured with NRA for low energy B and BF2-implants.

effects occur sooner (at higher energy) with BF_2 (cfr 5 kev BF_2 versus 1 kev B). The magnitude of the dose loss (<10%) in our experiments is however much smaller than the results reported by Agarwal [3]. The retained F-dose behaves very similar to the corresponding B-dose i.e. a 10% dose loss at 5 kev increasing to 20% at 2 keV. The latter does imply that the mechanism, which is causing the dose loss, is equally operable on the B and the F-atoms.

The B-work has been extended with retention studies of low energy Sb (2 and 5 kev) and As (5,10,30 kev) implants (made at Applied Materials, UK). Retained doses were determined using RBS and details of the profiles using low energy SIMS (Atomika 4500, 500 eV O_2). The results shown in fig.3 indicate that even for a dose up to 1 10^{16}at/cm^2, there is no significant loss of implanted species by self-sputtering. Similar results were obtained for As. It needs to be mentioned that after the high dose Sb-implants a pronounced increase in oxide thickness could be observed (fig.4). The proportionality with implant dose seems to suggest a Sb-enhanced beam-induced oxidation mechanism, although the partial oxygen pressure in the implant chamber is estimated to be less than 1 10^{-7} Torr. The observed energy dependence might reflect the Sb-enhancement on the oxidation, as at lower energies the near-surface Sb-concentration will be higher.

THEORETICAL FRAMEWORK

The theoretical description of ion retention during sputtering has been discussed quite some time ago by Schulz and Wittmaack [2]. The formalism they have developed is quite instructive since it only takes into account the impact of sputtering and the gradual build-up of a stationary implant profile leading to a limit in retained dose. As such the importance of the sputter yield on

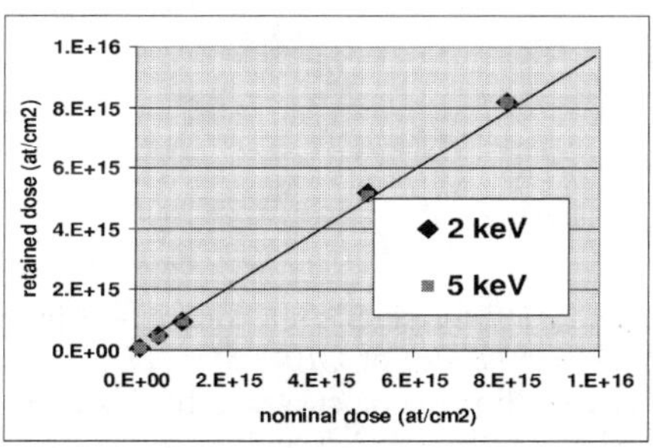

Figure 3 : Retained Sb-dose (at/cm2) as measured with RBS for 2 and 5 keV implants.

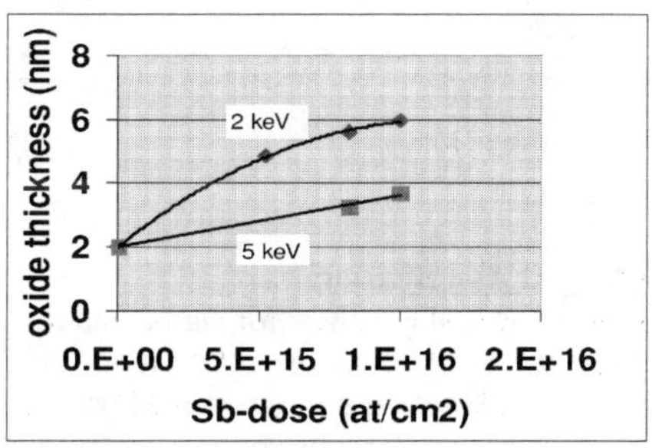

Figure 4 : Oxide thickness (nm) as measured with XPS after the Sb-implant.

the retained dose/profile becomes quite obvious. (More refined treatments/simulation codes which consider the impact of range shortening, collisional mixing and thermodynamics will be not be discussed here [7,8]). The following observations can be extracted from their calculations:

1) in stationary state the surface concentration of the implant species ~ 1/Y (Y=sputter yield of the matrix)
2) with increasing fluence the implant profile gradually evolves from a simple Gaussian profile to an erfc-like function (complementary error function)
3) for the implant energies considered here (< few keV), the (normalized) retained dose ($=NY/n_{Si}$*sigma) increases linearly with implant fluence up to a value of 2. (sigma = range straggling of the implant species, n_{Si}=density of Si, N= retained dose).
4) Maximum retention (saturation) is only reached for a (normalized) implant fluence ($=\phi Y/n_{Si}$*sigma)~2.5-3 with ϕ=implant dose

Based on TRIM [9], sigma varies from 0.7 (0.1 kev) to 10 nm (5 kev) for B. Reasonable estimates for the Si-sputter yield (B→ Si) can also be found from calculations using TRIM as a version of TRIM was used [7] which leads to the correct prediction of the well –known sputter yield values for low energy oxygen bombardment of Si [10]. Based on these values one can also calculate the Si-thickness sputtered during the implant (fig.5). Note that since the sputter yields are smaller than one, accommodation of the implant species (without modification of the target density) will cause a (small) surface swelling instead of a surface recession [7]. The net surface displacement being $(\phi_{retained}-\phi_{nominal}*Y_{Si})/n_{Si}$.

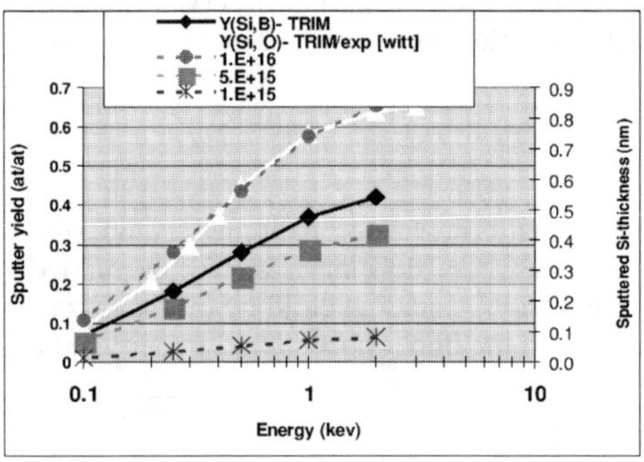

Figure 5 : Calculated Si-sputter yield for B→ Si (TRIM) and corresponding Si-thickness sputtered for three different implant doses. The Si-sputter yield under oxygen bombardment is shown for comparison [10].

DISCUSSION

Sb- and As-retention

Using applicable values for sigma and the sputter yields, the simple retention model predicts a maximum retained dose of ~5 10^{16}at/cm^2 for a saturation fluence of 1 10^{17} at/cm^2. Up to 1 10^{16} at/cm^2 very little self-sputtering should be observed. The Sb-retention curves in fig.5 seem to agree with these predictions confirming the applicability of the model. The detailed Sb-profiles (taken with SIMS at different doses) further indicate the evolution from the simple Gaussian-like profile at low dose towards the erfc-like function at high doses. The (predicted) accumulation in the near –surface region at high doses can quite clearly be observed. Even at high doses the surface concentration remains sufficiently low (<10%) such that sputtering of the Sb does not cause a significant dopant loss.

Completely similar results (not discussed) are obtained for the As-case.

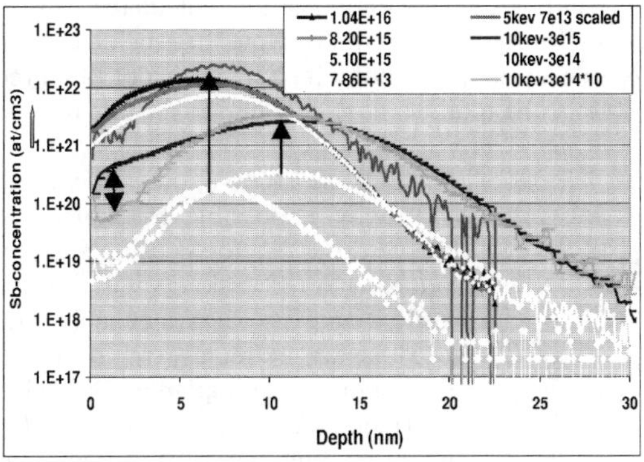

Figure 6 : Sb-profiles (5 and 10 keV) as measured with SIMS (500 eV O_2) for different doses. Scaled low dose profiles are added to highlight the profile shifts and differences in the near-surface region.

B and BF_2-retention

The apparently high dopant loss (as observed by Agarwal) for B and BF_2-implants suggests that significant B-sputtering occurs during high dose implantation. The retention curves of fig.1 (and the extracted B-sputter yields, fig 8) indicate that this occurs at a dose (<1 10^{16} at/cm^2) much smaller than predicted by the retention model (1 10^{17} at/cm^2). Also the maximum retained dose (~ <1-2 10^{16} at/cm^2) is much lower than the theoretical prediction. In order to reconcile these observations with the model predictions, a Si-sputter yield roughly 10x higher (~2-3) than the calculated value would be required. The only parameter which can be changed in the simulations to induce such a high sputter yield, would be the surface binding energy. Since the value used led to good results for sputter yields under oxygen bombardment, and since it should be basically independent from the implant projectile, drastically changing the surface binding energy is not a viable route.

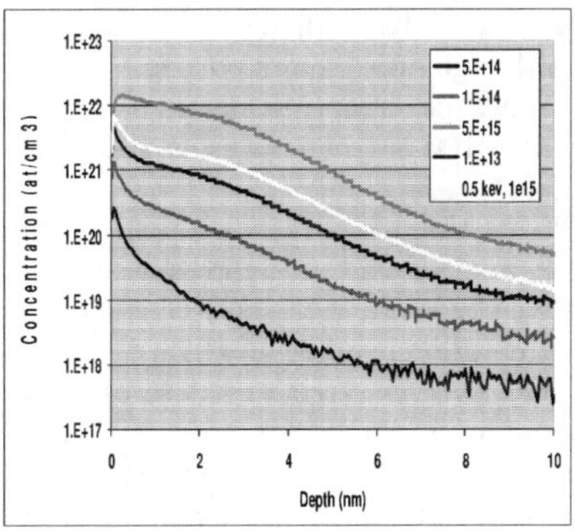

Figure 7 : SIMS profiles of 0.5 keV B-implants. Parameter is the nominal implant dose.

Thus, in order to induce such a high removal rate of B, its surface concentration must be much higher than predicted by the assumed simple Gaussian model, potential mechanisms being surface segregation, ion-beam induced migration, etc…Of course the latter should be evident on the corresponding SIMS profiles as well (fig 7). Using the surface concentrations from the SIMS (ignoring its known overestimate of the near surface concentrations and eventual corrections to remove the surface transient [11]), one can now calculate the intrinsic sputter yield for B as $Y_B(100\%)=Y_B/\text{surf conc}$. These are again (unrealistically) high as compared to the simple TRIM calculations (fig.8).

Alternatively one can try to estimate from the incremental sputtered B-dose, the required surface concentration if all B-atoms contained within the 1 nm would contribute to this dose loss. The latter is an underestimate of the required surface concentration as it assumes a zero binding energy (all atoms sputtered) and overestimates the escape depth of sputtered particles, which normally is ~ 0.2 nm [12]. However even with those assumptions the calculated surface concentrations largely exceeds the observed ones for high doses (fig.8). Note that such a concept also would imply a completely depleted B-profile in the first 1nm which again is not observed. In fact the SIMS profiles roughly show that 30% of the retained dose is contained within the first nm.

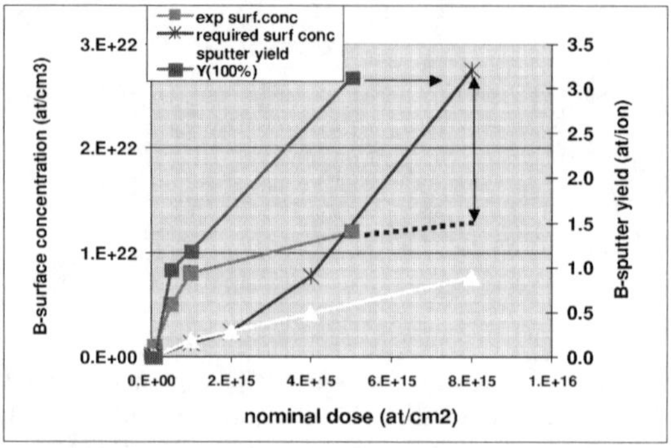

Figure 8 : Intrinsic B-sputter yield (Y100%) calculated from the SIMS B-Surface concentrations and experimental B-sputter yield (based on retention data from Agarwal) and calculated max surface concentration required to match the sputtered dose.

All these observations imply that it is virtually impossible to explain the claimed dose loss for B using standard models of ion retention, sputtering and profile shape. The only mechanism left would be an extremely rapid surface segregation of the implanted B, leading to a (mono?)layer of B with very low surface binding energy. Verification of this model is extremely difficult as such a layer would probably be very hard to be observed with SIMS.

It is true that one can question the values quoted by Agarwal (cfr the inconsistencies of the retention data, fig 1), but even if one were to consider the reduced dose loss of this work (10%), the maximum concentration would only drop by a factor 3. Although this more or less would approach the experimentally observed surface concentrations, the latter calculation is still an underestimate as it is based on the debatable arguments of a 100 % sputtering out of a 1 nm layer.

At last let's turn to the comparison between B and BF_2. Within the present context it is fair to state that we can approximate the case of BF_2 as an B-implant with a higher sputter yield resulting from the coincident impact of 2 extra F-atoms. If we consider the Si-sputter yield due to F or B as equal, the simple retention model predicts a maximum retained dose which should be three times smaller for BF_2 than for B. Experimentally, one observes a retained dose of 2 10^{15} at/cm^2 for 2.2 keV BF_2 versus 4 10^{15} at/cm^2 for 0.5 kev B and 4 10^{15} at/cm^2 for 5 kev BF_2 versus 5 10^{15} at/cm^2 for 1 keV B. In a similar way the increased sputter yield should lead to a smaller saturation dose which again appears apparent in the data (4 10^{15} at/cm^2 for 2.2 keV BF_2 versus >1 10^{16} at/cm^2 for 0.5 keV B).

Finally one can compare the estimated retained dose of B and F from the same BF_2-implant. Theoretically, the retained dose for F should be higher than for B, as B is exposed to three sputter events per implanted B-atom whereas F sees three sputter events per 2 implanted F-atoms. Experimentally a F/B-ratio of 1.5-1.8 is found where we would expect "in saturation" a value of

2 (2/3 versus 1/3). (Note that this reasoning only applies for doses where the B and the F both have reached their maximum retention and thus the higher F-dose no longer matters.).

The above discussion again points to the same observation, that for B the retention model can predict the trends based on variations in sputter yields but that the absolute values of these doses are substantially lower than the theoretical predictions.

CONCLUSIONS

For the case of low energy As and Sb implants no significant self-sputtering can be observed for doses as high as 1 10^{16} at/cm^2 whereas at high doses some profile deformation can already be observed. These results are in line with simple ion retention models in combination with sputtering. The case of B and BF_2 is more complex as more pronounced self-sputtering is observed. However literature data on the absolute magnitude of the dose loss are not consistent. The maximum retained dose and the saturation fluence both are much lower than predicted by the retention models. A possible explanation can only be found when a strong surface segregation of B is assumed which can however not be observed in the corresponding SIMS profiles. Relative differences between B and BF_2-implants on the other hand are consistent with differences in effective sputter yield per incident B or F-atom.

REFERENCES :

1. A.Agarwal, H.-J.Gossmann, D.J.Eaglesham, L.Pelaz, D.C.Jacobson, T.E.Haynes et al., Appl. Phys. Lett.71, 3141 (1997)
2. F.Schulz and K.Wittmaack, Rad. Effects 29, 31 (1976)
3. A.Agarwal, H.J. Gossmann, A.T.Fiory, V.C.Venezia and D.C.Jacobson, ECS PV 2000-9, 49 (2000)
4. M.A.Albano, V.Babaram, J.M.Poate, M.Sosnowski and D.C.Jacobson, MRS, Vol.610, B3.6.1 (2000)
5. J.Liu, X.Lu, J.Jin, Q.M.Li and W.K.Chu, Proc. 16th Int. Conf. On Applications of Accelerators in Research and Industry, eds J.L.Duggan and I.L.Morgan, AIP,-CP576, 323 (2001)
6. G.E.Coote, Nucl. Instr. Meth. B66,191 (1992)
7. H.De Witte, W.Vandervorst and R.Gijbels : Joun Appl.Phys. 89(5), 30011 (2001)
8. M.Wadsworth, D.G.Armour, R.Badheka and R.Collins, Int. J. Numerical Modelling, 3, 157 (1991)
9. J.P.Biersack and L.Haggmark, Nucl. Instr. Meth. 174, 257 (1980).
10. K.Wittmaack, Surf. Interface Anal.,24, 389 (1996)
11. W.Vandervorst, T.Janssens, R.Loo, M.Caymax, I.Peytier, R.Lindsay,J.Frühauf, A.Bergmaier and G.Dollinger, Proc. SIMS-XIII (Nara, 2001), Appl. Surf. Sci (accepted for publication)
12. K.Wittmaack Phys.Rev.B, Vol56(10),R5701-5704 (1997)

Junction and Profile Analysis using Carrier Illumination

T. Clarysse[1], W. Vandervorst[1,2], R. Lindsay[1], P. Borden[3], E. Budiarto[3], J. Madsen[3] and R. Nijmeijer[3]

[1]IMEC, Kapeldreef 75, B-3001 Leuven, Belgium
[2]Katholieke Universiteit Leuven, Electrical Engineering Department, Kardinaal Mercierlaan 92, B-3001 Leuven, Belgium
[3]Boxer Cross Inc., 125 Constitution Drive, Menlo Park, CA 94025, U.S.A.

ABSTRACT

Carrier Illumination™ (CI) is an optical technique for non-destructive in-line monitoring of post-anneal junction depth and pre-anneal PAI depth and dose with wafer mapping capabilities. This work intends to extend the use of the CI-measurements from a range-specific quantitative measurement towards a more universal quantitative analysis of junction depth, profile abruptness and implant dose. For that purpose this paper presents a systematic study of the CI response to a wide variety of post anneal implant processes, varying parameters including implant species, dose and energy, annealing condition, and surface preparation.

Samples containing B, BF_2 and As-implants with and without Ge PAI layers, with junction depths between 10-120 nm, were measured. In addition near-ideal box-like profiles (as obtained with CVD-growth) were fabricated and measured. For the abrupt CVD profiles, CI measures the junction position with sub-nm resolution independent of the CI-analysis conditions. For more graded profiles resulting from annealed implants, the correlation to the SIMS junction depth becomes a function of generation laser current (which is proportional to the applied power). As the concentration level, at which the correlation is made, can be adjusted over a concentration range of approximately $2x10^{18}$ to $2x10^{19}/cm^3$ by changing the laser current, a route towards correlating the CI measurement with profile abruptness becomes feasible.

INTRODUCTION

To yield advanced, high-performance Complementary Metal-Oxide-Semiconductor (CMOS) technologies, it is crucial to be able to characterize quickly and reliably sub-70 nm dopant structures [1]. For in-line monitoring of the involved pre- and post-anneal process steps, Carrier Illumination™ (CI) has established itself as a fast, non-contact, non-destructive tool with wafer mapping capability. For the latter the exact quantitative interpretation of the CI signal is less important as long as high repeatability and sensitivity for a particular profile can be demonstrated [2].

Previously, it has been reported [3] that for annealed structures, the CI-signal is not only dependent on the dopant junction depth but also on other parameters such as the profile steepness. Hence one can envisage that a more universal quantitative interpretation of the CI-signal expands its application towards profile analysis, eventually enabling in-line measurement where off-line tools such as spreading resistance profiling for very shallow profiles have been used [4]. In this work we therefore explore in more detail the CI signal behavior for active dopant profiles ranging from near ideal (abrupt) junctions (as can be grown by CVD) to more graded junctions as typically obtained with rapid thermal annealing (RTA) steps on low-energy implants.

BASICS OF CARRIER ILLUMINATION

In CI a pump laser (830 nm) with a 2 μm spot size is used to generate a quasi-static excess carrier profile giving rise to a depth dependent index of refraction change. The measured CI-signal is then derived from the interference of the reflected signal from a second probe laser (980 nm): the reflection originating either from the sample surface or from the region with a large change of index of refraction. The reflected probe power is given by the following theoretical formula [2]:

$$E_r^* E_r = r_s^2 E_0^2 \left\{ 1 + C(\omega, n(0)) \int_0^\infty \cos(2n(z)kz) \frac{dN(z)}{dz} dz \right\} \quad (1)$$

where E_o and E_r are respectively the incident and reflected probe field, r_s is the reflection coefficient at the surface, C is a computational coefficient, ω is the angular frequency of the probe laser, k is the field propagation constant in vacuum, and $n(z)$ and $N(z)$ are respectively the depth dependent index of refraction and excess carrier profile. The second term between the brackets represents the actual CI-signal.

For annealed structures, the shape of the induced excess carrier profile is expected to be proportional to the shape of the active dopant profile up to the depth where the excess profile becomes flat, i.e. where the dopant level reaches the induced substrate excess carrier level. Consequently, the maximum of dN/dz will in general occur at a depth where the dopant concentration level is comparable to the injection level. We will refer to the latter level as the dopant sensitivity level (DSL). The CI-signal is sensitive to the depth of this dopant sensitivity level. Only in those cases where the dopant sensitivity level corresponds with the dopant level of the underlying layer, e.g. the well region in a MOSFET structure, or for box-like profiles where the actual dopant sensitivity level is irrelevant, does one measure the "real" junction depth.

For annealed profiles that are identical, except for being shifted relative to each other with respect to the depth axis, the depth position of the maximum of dN/dz will shift accordingly. Hence, formula (1) predicts a cosine-like correlation for the CI signal versus the depth of the dopant sensitivity level. This is referred to as the CI response curve for a given dopant sensitivity (injection) level.

As the generation current (power) is increased, the injection level also increases, and consequently a change in the dopant sensitivity level is expected. For box-like structures (as can be achieved through CVD growth), this change in dopant sensitivity will result in a negligible depth change for dN/dz, due to their high profile steepness. Consequently, higher generation currents will only result in an upward (not sideward) shift of the excess profile by a given factor. Hence dN/dz and the CI signal will also be multiplied with the same factor. Therefore, the measured junction depth is independent of laser current for these box-like profiles. For more graded profiles this is, however, no longer the case as we will see in the next section.

EXPERIMENTAL DATA AND ANALYSIS

Figure 1 illustrates the experimental behavior for a series of Chemical Vapor Deposition (CVD) layers with different thicknesses grown at low temperature (680 °C) with a profile steepness of 3-4 nm/decade, i.e. close to the ideal abrupt transition case.

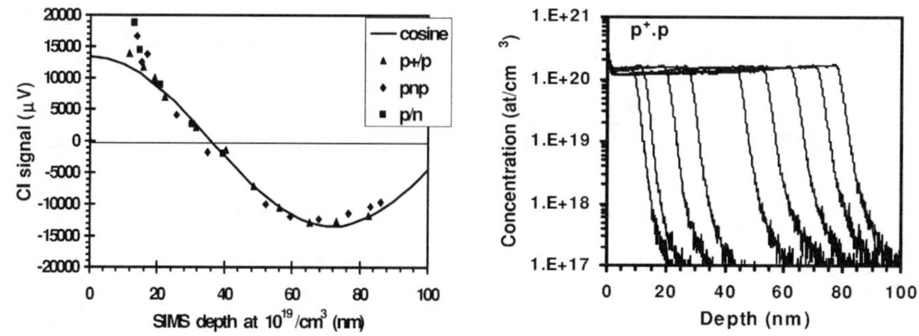

Figure 1: Left: CI-signal versus SIMS junction depth at 10^{19} /cm^3 for a series of p-type CVD layers grown on different underlying structures, Right: SIMS profiles for the p^+.p structures.

Clearly the theoretical expected cosine behavior is confirmed, except close to the surface (< 20 nm), where the signal continues to rise instead of leveling off. Preliminary simulations indicate that this deviation from theory may be due to charge crowding. As the junction approaches the surface, the depletion region occupies the entire space between the surface and the junction. A larger degree of depletion must occur within this narrower region to accommodate the same amount of charge transfer, resulting in a larger electric field. Consequently, the excess carrier profile becomes steeper, which in turn increases the CI-signal. Note that the response curve (Fig. 1) is only weakly dependent on the chosen correlation/junction level (10^{19} /cm^3) due to the high steepness of these CVD profiles. Moreover the presence of a junction (p-n transition) is irrelevant as the results for p^+-p and p^+-n structures coincide. Consequently, for junction depth monitoring applications, the data from these CVD samples can be used as a correlation table (at the selected generation current setting). Note the nearly linear response between 15 nm to 60 nm. One can determine, from Figure 1, for a given CI-signal at a fixed current level, the corresponding junction depth of a similar unknown sample with a 0.1-0.2 nm depth resolution.

For more graded junction transitions, however, the situation becomes somewhat more complex. In this work, we have measured a large database of low energy B, BF_2 and As profiles, produced by implants through 1-4 nm thick cap oxides, with or without Ge pre-amorphization implants (PAI), which were fully activated through high-temperature, rapid thermal anneals (RTA). These structures have dopant junctions (at 10^{18} /cm^3) ranging from 10 to 120 nm and profile steepness ranging from 10 to 50 nm/decade.

The underlying complexity with graded junctions is to make the correlation to a concentration level on the SIMS profile, i.e. the concentration level that needs to be plotted as "junction depth", since CI is sensitive to the depth related to the injection level, i.e. the dopant sensitivity level introduced above. The situation is illustrated in Figure 2.

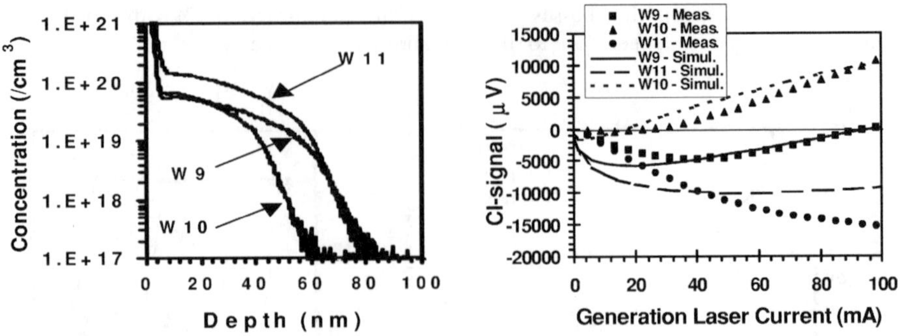

Figure 2: Left: SIMS profiles for low energy (0.5-1 keV) 2e15 B implants with 2e14 Ge PAI (30-50 keV) through 4 nm screen oxide and annealed at1050 C for 1s (wafers 10 & 11) and 1030 C for 10 s (wafer 9), Right: The corresponding experimental versus simulated CI-signal as a function of generation laser current (proportional to the applied power). Signal for simulations in a.u.

If CI is sensitive only to the 10^{19} /cm^3 level, then one would expect (based on Fig. 1) almost the same signals for wafers 9 and 11 (Fig.2/right), since the dopant profiles are less than 5 nm apart at that level (Fig. 2/left). The experimental signals are, however, different. For wafer 11 the signal is about –14000 μV (at 75 mA), in approximate agreement with Fig. 1. However, the signal of wafer 9, i.e. –2500 μV (at 75 mA), is different and would suggest erroneously that the profile depth at 10^{19}/cm^3 is less than 40 nm (based on Fig.1).

In order to better understand what is happening, let us start with a simplified interpretation of the entire current curve relative to the shape of the corresponding dopant profile. One can observe that at low current (18 mA) the signal difference between wafers 9 and 11 is quite small (less than 20 %). For higher current, the current curves move away from each other. Taking into account that a higher current (power) implies a higher excess carrier level, it is possible to correlate the low current signal with the SIMS (dopant sensitivity) level where both dopant profiles coincide, i.e. near 5×10^{18} /cm^3. The high current signal should consequently correspond to a higher SIMS (dopant sensitivity) level, where the dopant profile for wafer 9 approaches the 35 nm threshold (signal approaches zero), i.e. near 4×10^{19} /cm^3. The fact that the high current signal for wafer 9 is, however, also different from wafer 10, although they have the same depth at 4×10^{19} /cm^3, illustrates that this approach describes only a first order approximation.

A more detailed investigation involving device simulations reveals that for graded structures the derivative of the excess carrier profile is no longer a delta function (as for box-like profiles), but a broader polynomial shaped curve, with typically one positive and one negative part. The width and the relative peak heights of each part are dependent on the steepness and the peak density of the dopant profile. Consequently, the final CI-signal as given by formula (1) is also dependent on these parameters in a non-trivial way. Work addressing the additional second order effects in future modeling capabilities (to be reported elsewhere [5]) has been undertaken.

Fig. 2/right already shows a promising qualitative agreement between experimental and simulated current (power) curves. The latter were obtained starting from the corresponding SIMS dopant profiles (Fig. 2/left). These simulations in combination with the analysis of all available data indicate that for B and BF_2 the sensitivity levels for low (18 mA) and high (94 mA) generation current are respectively $5x10^{18}$ and $2x10^{19}$ /cm^3. For As, the corresponding levels have been found to be $2x10^{18}$ and $8x10^{18}$ /cm^3. Figure 3 shows the CI-signal versus depth response curves for each of these cases.

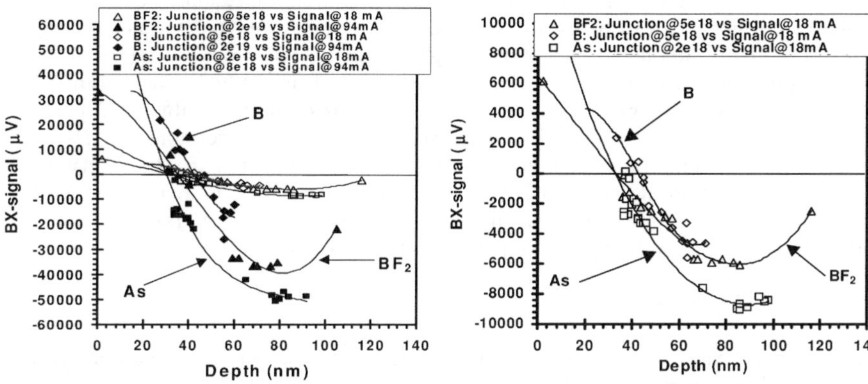

Figure 3: CI signal versus junction depth for the sensitivity levels discussed in the text. Left: Both high and low current response curves. Right: Only the low current response curves.

Several observations can be made regarding the curves shown in Fig. 3. The low-current response curves, for all species, give the best cosine-like behavior (the least scatter in absolute value), i.e. these curves are in principle best suited for the determination of the junction depth at the corresponding dopant level. They also refer to the lowest concentration level. The latter value is quite close to, for instance, the well concentrations implying that for source/drain profiles CI will be sampling the real junction depth. The B and BF_2 response curves are shifted relative to each other along the depth axis, i.e. they have a different zero-signal depth. The maximum signal and response curve steepness of the As samples is higher than for the B/BF_2 structures. Work is underway to describe the underlying physical causes of these species-related phenomena. The high current response curves have a larger scatter than the low current ones. As we have seen, the latter is directly related with the higher sensitivity of the high current signal to profile steepness (Fig. 2/right).

Given the large number of RTA annealed samples that have been investigated, the curves in Figure 3 have a universal character. Hence indications suggest that it is plausible to use the nonlinear response of the CI signal to generation laser curren to determine both the junction depth and profile steepness. In the future, as improved models and deconvolution software become available, a refinement of this procedure may be able to resolve the detailed shape of the annealed dopant profile between its low and high current sensitivity levels.

The key observations made in this study imply that sufficient care must be taken to apply the proper correlation table in junction depth monitoring applications. As long as the CI-signal-to-junction depth correlation is properly calibrated for specific implant species and dose, accurate monitoring of junction depth is ensured. Alternatively, a low laser current can be used for most implant species and conditions, in which case the CI-signal is only weakly dependent on profile steepness.

CONCLUSIONS

Carrier Illumination is a powerful technique for the non-destructive monitoring and characterization of sub-70 nm structures on whole wafers. For box-like, annealed structures, such as CVD-grown layers with a steepness of 3-4 nm/decade, it is possible to determine the junction depth with sub-nm resolution in the 15-60 nm range. For more graded structures (10-50 nm/decade), as obtained with low-energy implants and state-of-the-art high-temperature RTA, it has been shown that the CI-signal at different laser currents is sensitive to different dopant levels and that the latter correspondence depends on the species considered. Therefore, the CI-signal is sensitive to the steepness of the dopant implant, and has the potential to extract information on the shape of the dopant profile between $2x10^{18}$ and $2x10^{19}$ /cm^3. For accurate monitoring of implant and anneal module control, CI can be operated at low laser current (power). At higher laser currents, more specific calibration of the correlation table is required.

ACKNOWLEDGMENTS

The authors would like to thank Danielle Vanhaeren for the CI measurements and Luc Geenen for the SIMS measurements done at IMEC.

REFERENCES

1. International Technology Roadmap for Semiconductors 2001.
2. P. Borden, L. Bechtler, K. Lingel and R. Nijmeijer, Carrier Illumination Characterization of Ultra-Shallow Implants, in *Handbook of Silicon Semiconductor Metrology*, edited by A.C. Diebold, (Dekker Inc., New-York, 2001), 97.
3. D.C. Sing and P. Borden, *Proceedings of the Sixth International Workshop on Fabrication, Characterization and Modeling of Ultra-Shallow doping profiles in Semiconductors*, April 22-26, 2001, edited by M. Foad, (American Vacuum Society, New-York, 2001), 311
4. T. Clarysse, P. Eyben, T. Hantschel, W.Vandervorst, Material Science in Semiconductor Processing **4**, 61 (2001)
5. T. Clarysse, W. Vandervorst, R. Lindsay, P. Borden, E. Budiarto, European Materials Research Society, Spring meeting, Symposium E, June 18-21, 2002, Strasbourg, France.

Mat. Res. Soc. Symp. Proc. Vol. 717 © 2002 Materials Research Society C7.4

Calibration of Phosphorus Implantation Dose in Silicon by Radiochemical Neutron Activation Analysis

Rick L. Paul and David S. Simons
Chemical Science and Technology Laboratory
National Institute of Standards and Technology
Gaithersburg, MD 20899

ABSTRACT

A radiochemical neutron activation analysis (RNAA) procedure has been developed, critically evaluated, and shown to have the necessary sensitivity, chemical specificity, and matrix independence to certify phosphorus at ion implantation levels. ^{32}P, produced by neutron capture of ^{31}P, is chemically separated from the sample matrix, and measured using a beta proportional counter. Analysis of 12 pieces of a silicon wafer, implanted at a nominal dose of 8.5×10^{14} atoms·cm^{-2}, yielded a phosphorus concentration of $(8.30 \pm 0.16) \times 10^{14}$ atoms·cm^{-2} (expanded uncertainty), in agreement with a consensus value of $(8.23 \pm 0.33) \times 10^{14}$ atoms·cm^{-2} (1s uncertainty) determined from a round robin SIMS investigation. The relative standard deviation of < 2 % is suitable for certification of a Standard Reference Material.

INTRODUCTION

The U.S. semiconductor industry relies on secondary ion mass spectrometry (SIMS) for characterization of the depth distribution of dopants such as boron, arsenic, and phosphorus in silicon. To assist the industry in achieving high accuracy measurements, Standard Reference Materials (SRMs) of boron and arsenic implants in silicon have been developed by NIST as SIMS calibration standards.[1,2] Plans are underway to develop a phosphorus implant in silicon SRM, to be certified by radiochemical neutron activation analysis (RNAA). RNAA was chosen because other techniques lack the necessary sensitivity, chemical specificity and matrix independence to measure phosphorus at implantation levels. Preliminary measurements were carried out on six pieces of a phosphorus-implanted silicon wafer previously used in a round-robin study of SIMS measurement repeatability.[3] The measurements yielded a mean phosphorus concentration and standard deviation of $(8.35 \pm 0.20) \times 10^{14}$ atoms·cm^{-2}, in agreement with both the nominal implantation dose (8.5×10^{14} atoms·cm^{-2}) and a consensus value of $(8.23 \pm 0.33) \times 10^{14}$ atoms·cm^{-2} (1s uncertainty) determined from the SIMS investigation. In the current investigation, we set out to modify the initial procedure in order to improve precision and minimize uncertainties.

EXPERIMENTAL

Measurements were carried out on 12 pieces of the same phosphorus-implanted silicon wafer analyzed in the initial investigation. Samples were prepared for RNAA by cutting the silicon into 1 cm x 1 cm squares using a wafer saw. A micrometer was then used to measure the length and width of each piece to an accuracy of 0.0001 cm. The average surface area of the 12 samples was 0.9845 ± 0.0002 cm^2.

Blanks were prepared from two different silicon wafers which contained no implanted phosphorus. Blanks are necessary in order to correct the ^{32}P produced from neutron capture by the implanted phosphorus (near surface layer), from ^{32}P produced by neutron capture by phosphorus impurity or silicon in the bulk of the sample. Neutron capture by the silicon produces ^{32}P via the second order reaction: $\{^{30}Si\ (n, \gamma)\ ^{31}Si\ (\beta^-, 2.6\ h)\ ^{31}P\ (n, \gamma)\ ^{32}P\}$. A total of nine blanks were measured in contrast to the single blank used in the preliminary investigation. All samples and blanks were cleaned of surface contamination and sealed into clean polyethylene bags as described previously.[3]

Eight standards were prepared by depositing 50 µL aliquots of a phosphorus standard solution ((1.182 ± 0.005) mg/mL, expanded uncertainty) onto aluminum foils as described previously.[3] After drying under a lamp, each foil was sealed into a polyethylene bag. To correct the count rate of the standards for phosphorus impurity in the aluminum, a standard blank was prepared by sealing an aluminum foil into a polyethylene bag. In order to ensure that the neutron flux was uniform throughout the irradiation vessel, thin disks of iron foil, (6.35 mm diameter, mass ≈ 6.3 mg) were sealed into polyethylene bags to serve as flux monitors.

Targets were packaged into several rabbits (polyethylene irradiation vessels). Each rabbit was irradiated for 3 h in irradiation tube RT1 of the NIST 20 MW NBSR reactor at a neutron fluence rate of 1.05×10^{14} $cm^{-2} \cdot s^{-1}$. In order to ensure that the entire rabbit receive as uniform a neutron flux as possible, rabbits were flipped (rotated 180°) at the midpoint of irradiation.

Processing of samples and blanks:

The implanted silicon samples and blanks were processed using a variation of a procedure reported previously[3]:

The samples were removed from their irradiation bags, cleaned with ultrapure water and ethanol to remove surface contamination. Each sample was then transferred to a Teflon beaker and mixed with (1.979 ± 0.003 mL) of phosphorus carrier solution (3.522 ± 0.008 mg/mL P.) A mixture of 10 mL of HNO_3 and 5 mL of HF was added to the beaker, and the beaker covered with a Teflon watch glass and heated on medium heat for about 2 h in order to dissolve the silicon. The cover was then removed from the beaker and the solution evaporated to near dryness. The residue was dissolved in ~ 5 mL of $HClO_4$ and again evaporated to near dryness. This evaporation, which removes fluoride ions that could interfere with the gravimetric determination of the carrier, was not performed in the processing of the preliminary samples.

After the residue was dissolved in HNO_3, phosphorus was first precipitated as ammonium phosphomolybdate, then as magnesium ammonium phosphate, as described previously.[3] The latter precipitate was collected by suction filtration on medium porosity filter paper, washed with a 10 % solution of ammonium hydroxide and 100 % ethanol, and air dried in a clean hood for three days. The dried precipitate was weighed and transferred to a counting dish for beta counting of ^{32}P. The yield, (fraction of recovered phosphorus carrier), was determined gravimetrically, assuming a precipitate composition of $MgNH_4PO_4 \cdot 6H_2O$.

After the counting cycle was completed, the precipitate was transferred quantitatively to a 5 mL porcelain crucible (previously heated for a few hours at 700 °C, cooled in a desiccator for 1 h, then weighed on a microbalance). The crucible was heated in a muffle furnace to 650 °C for 2 h, cooled in a desiccator for 1 h. and again weighed using a microbalance. The yield was determined gravimetrically, this time assuming a precipitate composition of $Mg_2P_2O_7$. This second yield determination was omitted in the processing of the preliminary samples.

Standards processing

Aluminum foils standards were transferred to Teflon beakers, mixed with (1.979 ± 0.003 mL) of phosphorus carrier (10.08 ± 0.02) mg/mL, and each foil dissolved in a mixture of 5 mL HCl, 5 mL of HNO_3, and 1 mL of HF. Each solution was transferred quantitatively to a 50 mL polypropylene volumetric flask and diluted to 50 mL with H_2O. A (1.979 ± 0.003 mL) aliquot of each 50 mL solution was transferred to a Teflon beaker and mixed with (1.979 ± 0.003 mL) of phosphorus carrier (3.522 ± 0.008) mg/mL. (This represents an improvement over the preliminary procedure, in which a 100 µL aliquot of each 50 mL solution was used). Ten milliliters of HNO_3 and 5 mL of HF were added, along with a non-irradiated piece of non-implanted silicon (added to ensure that the chemical processing for the standard is identical to that of the samples). From here on, the sample procedure was followed.

Flux Monitors

Iron flux monitors were counted using a high purity germanium detector. The ^{59}Fe count rate was determined by integration of the 1098 keV gamma-ray peak.

Counting and Data Reduction

Counting of ^{32}P ($t_{1/2}$ = 14.28 d) was performed using a beta proportional counter. Each target was counted for 3 h. Purity of the ^{32}P from other radionuclides was ascertained by gamma-ray spectroscopy, and by monitoring beta count rates over a period of 4 half lives. Since most beta emitting nuclides also emit gamma-rays during decay (notable exceptions being ^{32}P, ^{35}S, and ^{14}C), the lack of gamma radiation in the spectrum confirms the absence of most other radionuclides. Furthermore, since no other radionuclides have a half life equal to that of ^{32}P, a plot of ln (activity) vs. time which yields a slope equal to the decay constant (ln $2/_{t1/2}$) for ^{32}P confirms the absence of other radionuclides in the target. No discernable contamination from other radionuclides was found for any of the targets analyzed.

For each target, the beta count rate at time zero (end of neutron irradiation) was determined from the beta decay curve. The count rates were then adjusted to account for differences in the amount of self absorption of beta particles in the different precipitates.[3] Corrections were made using self absorption curves (^{32}P count rate vs. milligrams of precipitate) derived from analyses of previous phosphorus standards.[4] Count rates were also adjusted for blanks.

The concentration (implantation dose per unit area) of phosphorus in each sample was calculated in atoms·cm^{-2}, using the equation:

$$[P\ (atoms \cdot cm^{-2}] = (N_{ST} \cdot DI \cdot (A_{SA}/A_{ST}) \cdot (Y_{ST}/Y_{SA}))/Ar_{SA}$$

where N_{ST} is the number of phosphorus atoms in the irradiated standard, DI is the dilution factor for the standard, A_{SA} is the ^{32}P count rate of the sample, corrected for decay time, self absorption, and blank, A_{ST} is the standard count rate, similarly corrected, Y_{ST} is the yield (fraction of recovered carrier) for the standard, Y_{SA} is the yield for the sample, and Ar_{SA} is the surface area of the sample.

RESULTS AND DISCUSSION

Results

Table 1 gives phosphorus concentrations determined for 12 samples of the implanted silicon, along with the mean value and standard deviation. The precision of the measurements, as defined by the relative standard deviation, is 1.35 %, compared with 2.35 % for the preliminary measurements. The mean value of $(8.30 \pm 0.11) \times 10^{14}$ atoms·cm^{-2} is in agreement with the consensus value of 8.23×10^{14} atoms·cm^{-2}.obtained from the round robin SIMS investigation.[3]

Evaluation of Uncertainties

In a previous publication, we performed a precursory evaluation of uncertainties for the preliminary measurements.[3] In this investigation we followed the method used in the evaluation of uncertainties for arsenic by NAA[2] to fully evaluate and compare the uncertainties for preliminary and current measurements.

Table 1. Results of analysis of phosphorus implanted silicon by RNAA

Sample	P (atoms·cm^{-2})	Sample	P (atoms·cm^{-2})
1	8.14×10^{14}	7	8.44×10^{14}
2	8.41×10^{14}	8	8.28×10^{14}
3	8.25×10^{14}	9	8.33×10^{14}
4	8.32×10^{14}	10	8.38×10^{14}
5	8.28×10^{14}	11	8.31×10^{14}
6	8.35×10^{14}	12	8.04×10^{14}
Average ± 1s	**$(8.30 \pm 0.11) \times 10^{14}$**		
Relative standard deviatio (precision)	**1.35 %**		

Type A uncertainties evaluated included those arising from counting uncertainties and measurement replication, measurement of surface area, and blank correction. A significant improvement over the preliminary measurements was noted in the uncertainty due to measurement of surface area. The current samples were cut to precisely rectangular geometry and the dimensions measured using a micrometer. By contrast, the preliminary silicon samples were cleaved to approximately rectangular geometry, and the surface area calculated from the mass, thickness, and the tabulated density for silicon.[3]

Type B uncertainties evaluated included those arising from neutron self shielding corrections, irradiation geometry, determination of carrier yields, beta decay curve fitting, beta self absorption corrections, and measurement of standards.. Significant procedural improvement was made in the evaluation of uncertainties due to carrier yield determination. For the current investigation, uncertainties in the determination of the carrier yield were evaluated by comparing yields obtained from gravimetric determination of phosphorus as $MgNH_4PO_4{\cdot}6H_2O$ with determination as $Mg_2P_2O_7$.[4] For the preliminary investigations, this second yield determination was not performed, and uncertainties were estimated from $MgNH_4PO_4{\cdot}6H_2O$ and $Mg_2P_2O_7$ data obtained in previous investigations Improvement was also noted in the standards uncertainty. Precision was improved by processing 2 mL (1.979 ± 0.003 mL, calibrated precisely) aliquots of each 50 mL diluted standard, rather than the 100 µL aliquots used in the preliminary investigation. The use of the smaller aliquots seemed to introduce additional uncertainty, as the precision obtained for measurement of six samples (relative standard deviation) was no better than 1.5 %, while the precision for eight standards measured in this investigation was 1.2 %.

Expanded uncertainties were evaluated using guidelines set by the International Organization for Standardization (ISO).[5] All type A and B uncertainties were added in quadrature to obtain a combined uncertainty of 0.95 % for the measurements (Table 2). Using a coverage factor of 2, the relative expanded uncertainty was calculated as 1.89 %, a significant improvement over the 3.18 % uncertainty calculated for the preliminary measurements. The final value for the phosphorus concentration in the silicon was determined to be (8.30 ± 0.16) atoms cm^{-2}.

CONCLUSIONS

Improvements in the procedure for determination of phosphorus in silicon have resulted in significantly better precision for measurement of both samples and standards. The most significant improvements in the procedure were the use of a wafer saw to cut the silicon pieces to precise dimensions, the addition of a perchloric acid evaporation after dissolution to destroy fluorides which could interfere with gravimetric yield determination, the introduction of a second gravimetric determination as $Mg_2P_2O_7$, and a smaller dilution factor for the standards. Additionally, since the activity of the standard foils was significantly higher than that of the silicon samples, samples and standards were processed in separate rooms in order to minimize the possibility of cross contamination. Using the improved procedure it will be possible to certify phosphorus in a silicon SRM with an expanded uncertainty of better than 2 %, which is acceptable for the SRM.

Table 2. Uncertainty evaluation for determination of phosphorus by RNAA. Preliminary analyses are compared with the current investigation. Relative uncertainties are expressed as a percentage of the mean value.

	Preliminary values	This investigation
Number of replications (n)	6	12
Mean ± 1 standard deviation (s)	**(8.35 ± 0.20) x 10^{14} atoms• cm^{-2}**	**(8.30 ± 0.11) x 10^{14} atoms• cm^{-2}**
Precision (rel. std. dev.)	**2.35 %**	**1.35 %**
Type A Uncertainty		
measurement replication ($s/\sqrt{n}$)	0.98 %	0.37 %
measurement of area	0.80 %	0.02 %
blank correction	n. d.	0.06 %
Combined Type A	1.27 %	0.38 %
Type B uncertainty		
beta decay curve fitting	0.3 %	0.3 %
beta self absorption / attenuation	0.05 %	0.05 %
yield determination	0.34 %	0.28 %
self-shielding correction	0.024 %	0.024 %
irradiation geometry	0.09 %	0.09 %
standard sensitivity	0.84 %	0.72 %
Combined Type B	0.96 %	0.87 %
Combined Uncertainty	1.59 %	0.95 %
Coverage Factor	2	2
Expanded Uncertainty	3.18 %	1.89 %
Final Value	**(8.35 ± 0.27) x 10^{14} atoms• cm^{-2}**	**(8.30 ± 0.16) x 10^{14} atoms• cm^{-2}**

References

1. NIST SRM 2137, Boron Implant in Silicon Standard for Calibration of Concentration in a Depth Profile.
2. R. R. Greenberg, R. M. Lindstrom, and D. S. Simons, *J. Radioanal. Nucl. Chem*, **245,** 57 - 63 (2000).
3. R. L. Paul and D. S. Simons, in *Characterization and Metrology for ULSI Technology*, edited by D. G. Seiler, A. C. Diebold, T. J. Shaffner, R. McDonald, W. M. Bullis, P. J. Smith, and E. M. Secula, (American Institute of Physics, AIP Conference Proceedings **550**, Melville, NY, 2001) pp. 677 – 681.
4. R. L. Paul, *J. Radioanal. Nucl. Chem.*, **245,** 11 – 15 (2000).
5. *Guide to the Expression of Uncertainty in Measurement,* ISBN 92-67-10188-9, 1st edn. ISO, Switzerland, (1993).

Mat. Res. Soc. Symp. Proc. Vol. 717 © 2002 Materials Research Society C7.7

Pulsed Force-Scanning Spreading Resistance Microscopy (PF-SSRM) for high spatial resolution 2D-dopant profiling.

P. Eyben, M. Fouchier, P. Albart, J. Charon-Verstappen and W. Vandervorst[a]
IMEC vzw, Kapeldreef 75, 3001 Leuven, Belgium
[a]also : KULeuven, INSYS, Belgium

ABSTRACT

Scanning Spreading Resistance Microscopy (SSRM) is now widely used for two-dimensional doping profiling with high spatial resolution. The need for a high force between the tip and the sample in order to obtain a good electrical contact, leads to a fast degradation of the tip (and the sample) while scanning. Tip damage is mainly due to the shear force occurring while scanning in contact mode at high forces leading to breakage (cleavage) of sharp tips or a rapid increase of tip radius (wear). The latter adversely affects the accuracy of the electrical measurements, as the contact radius is a determining parameter for quantification. The strong abrasive force also necessitates the use of tips composed of very hard material such as doped diamond, which has however a limited resistivity, and so far prevented the use of metallic probes. In addition the high force also prevents the simultaneous acquisition of high quality topography data. The solution to these problems is obtained by implementing the Modulated Force Principle (MFP). The latter consists of applying a variable (for instance pulsed) force while scanning, reducing the force during the lateral movement of the tip and synchronizing the electrical measurements with the high force periods. The latter results in lower lateral forces and introduces a quasi *multi point contact* mode. MFP also allows to obtain a better topography image by synchronizing the topography measurement with the low force part of the force cycle. The MFP leads to a drastic reduction of the surface and probe damage while maintaining high quality electrical data. The implementation of multiplexed detectors within the force cycle further enables the simultaneous acquisition of spreading resistance and topography during one scan, and/or the combination with multiple linear current detectors, capacitance sensors or tunneling current measurements.

INTRODUCTION

Scanning Spreading Resistance Microscopy (SSRM) is a powerful electrical characterization technique, based on Atomic Force Microscopy (AFM), which has been developed in recent years to probe the two-dimensional resistivity (and carrier) distribution in semiconductor devices [1][2]. In SSRM, a very small conductive tip is used to measure the local spreading resistance, which is intimately linked (directly proportional) to the local resistivity [3]. Scanning a cross section of the sample provides a two-dimensional map of the local spreading resistance with a spatial resolution set by the tip radius (typically 5-15nm). The main advantages of SSRM ly in its robustness, as it is far less sensitive to surface preparation than for instance Scanning Capacitance Microscopy (SCM) [4], its excellent dynamic range covering the entire dopant range of interest (10^{14}-10^{20} at/cm^3) with constant sensitivity and its high spatial resolution (set by the tip radius only) combined with very accurate junction delineation capabilities. Indeed different from SCM, no bias induced junction displacements occur and a correct measurement of the junction depth can be obtained [5]. A crucial advantage is also its ease of quantification as a straight conversion of spreading resistance (SR) to local resistivity (ρ) using the ideal formula $R=\rho/4a$ (where "a" is the tip radius) or a calibration curve already provides high quantification accuracy [3]. A typical example of a CMOS transistor characterized with SSRM is shown in fig.1 highlighting the SSRM-capability to characterize the source/drain regions as well as the (small) V_t adjust and halo-implants. Obviously SSRM is one of the few techniques able to probe these subtle dopant profile variations in detail. Crucial information such as the effective channel length of a MOS transistor or the internal emitter size of a bipolar transistor can also easily be extracted.

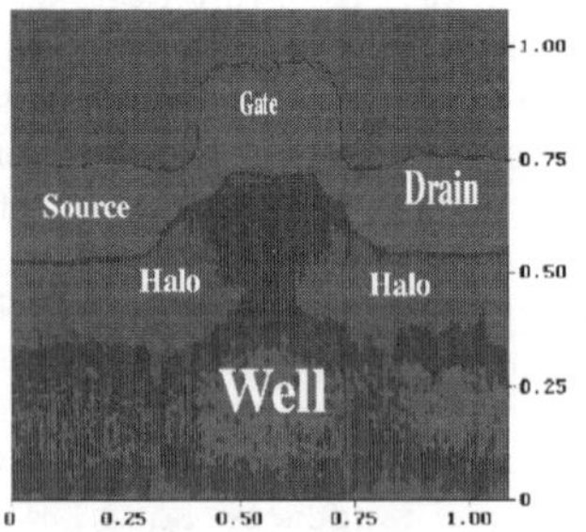

Figure 1 : SSRM image of 0.25 μm CMOS-transistor. The colour coding is linked to the different

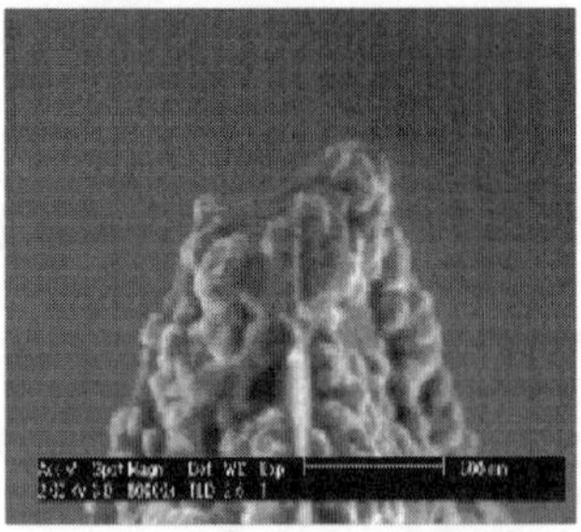

Figure 2. SEM picture of a diamond coated Si probe after use for a few scans at typical SSRM forces (~GPa). Probe has been broken due to the huge shear force during the SSRM scan.

Obtaining a reproducible electrical contact between the tip and the sample, requires the creation of the β-tin transformation in the probe-Si interface region [6]. As such the applied force needs to be high enough to induce in the contact region a pressure as high as 5-10 GPa. The latter is true for probing concepts such as SSRM and Scanning Spreading Resistance Spectroscopy (SSRS) or Nanopotentiometry [7].

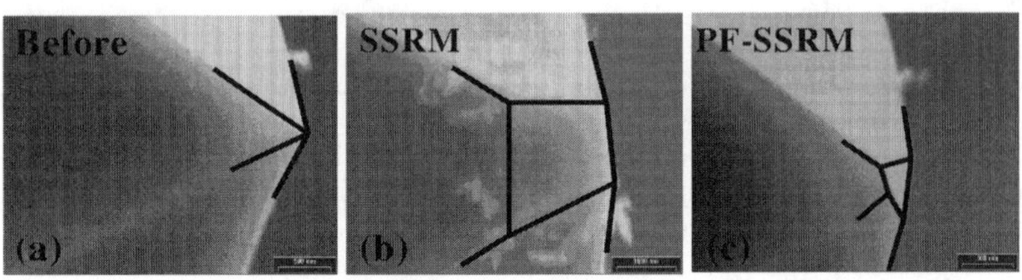

Figure 3. SEM picture of full Ni-P probes (a) before use (b) after use for a few scans with SSRM (c) after use for a few scans with PF-SSRM. The wearing of the probe is drastically reduced with PF-SSRM.

The use of this high force unfortunately leads to a degradation of the tip due to the high lateral force exerted while scanning in contact mode. Tip damage manifests itself as breakage or cleavage of high aspect ratio Si-tips (even when coated with diamond, figure 2) or as a rapid increase of tip radius for full diamond and metallic tips (see figure 3). Needless to say that the latter affects the accuracy of the electrical measurements, as the contact radius is a determining parameter in the quantification, as well as the spatial resolution. The high force also excludes the simultaneous acquisition of topography data as no sensitivity is left and furthermore leads to rapid degradation of the initial topography (see figure 4).

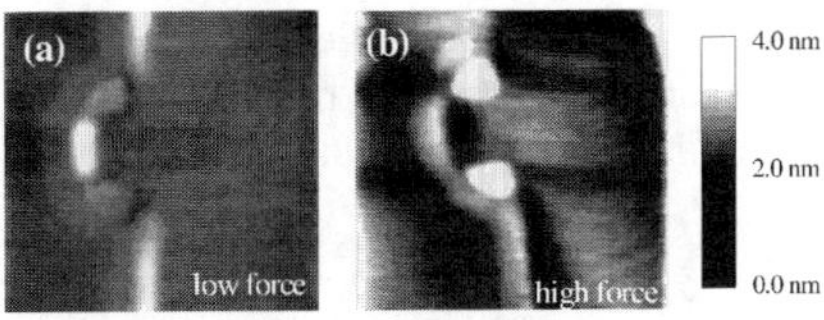

Figure 4. Comparison between topography measured at low force (left) and at high force (right).

EXPERIMENTAL

<u>Principle and realization of the Pulsed Force SSRM</u>

The Modulated Force Principle and multiplexing

In order to overcome these shortcomings, the Modulated Force Principle (MFP) is introduced which consists of applying a variable force (for example pulsed or sinusoidal force) whereby during a low force period the actual movement of the tip occurs and electrical measurements are done during a (short) high force period. In this respect the scratching of the contact mode is replaced by a quasi *multi point contact* mode (figure 5). It is important to note that during the displacement the tip remains in contact with the sample thereby maintaining the feedback control on the applied force.

The implementation of the MFP requires a synchronization of the electrical measurements (SSRM, SSRS or Nanopot) with the force peaks and of the topography measurement with the low force part of the force cycle. Once such a synchronization and timing is realized it is clear that a more general multiplexing of detection sensors becomes available as well. Hence one might consider the combined use of SSRM (high force) with SCM (low force), SSRM with TUNA (tunneling current detection at low force), etc...Obviously the entire system

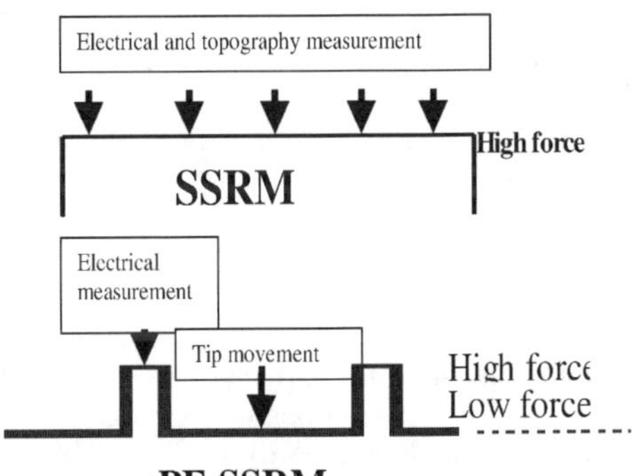

Figure 5. Comparison between SSRM (top) and Pulsed Force SSRM (bottom). The constant high force is replaced by a low force interrupted by short high force peaks. In PF-SSRM the electrical measurements in PF-SSRM is confined to the high force period, the topography measurement to the low force period. In SSRM both are performed simultaneously.

Realization of the Pulsed Force SSRM

In the atomic force microscope used (Digital Instruments 3000 with SSRM-module), the force between the tip and the sample is determined by a piezoelectric system. By using the SAM (Signal Access Module) of the microscope, a PC and an I/O card, we are able to control this force and to synchronize the electrical and topography measurements with its modulation. The overall equipment is controlled with a Labview software combined with a Visual C++ program to visualize the measurements.

RESULTS

Limitations of the technique

We have realized tests to check the pulse shape conformity and positioning control, the the piezoelectric time response and the amplifier integration time. Excellent registration of the tip movement is maintained and no loss in spatial resolution could be observed. However since the time available for the electrical measurements is now drastically reduced, speed of the electrical sensors is becoming the limiting factor. In particular in lowly areas (low currents) the integration time needed by the logarithmic amplifier imposes an upper limit to the scan speed. For 256 points/line and a pulse ratio of 10% (what means that 90% of the displacement happens at low force), the scan speed is at present limited to 5 seconds per scan line. Obviously the development of faster electronics can resolve this limitation.

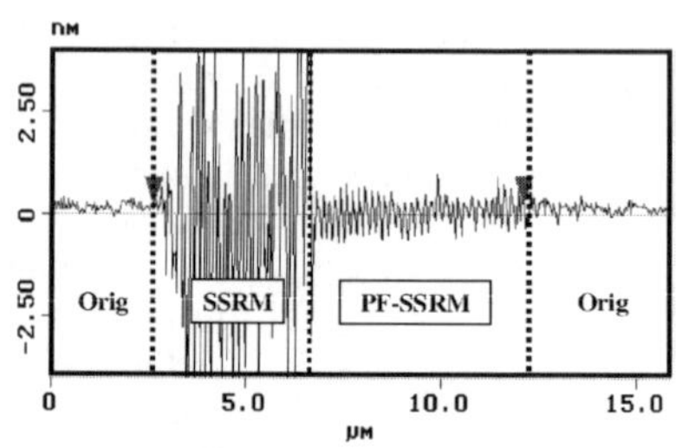

Figure 7. Topography section. The latter shows that the sample damaging is reduced when using PF-SSRM

Probe and sample damage

One of the most important advantages pursued in this concept is the reduction in sample/tip damage. The success of that can be judged in fig.7 by comparing the surface damage induced by classical SSRM and PF-SSRM. In figure 7, a line scan through a topography image is shown where the sample was partly scanned in classical SSRM mode and partly in PF-SSRM mode. The reduction of the sample damage is clearly visible.

With respect to probe damage a similar reduction is observed and cleavage of diamond coated probes occurs far less frequently. Moreover metallic probes wear much less rapidly (see figure 3c). It however remains impossible to realize PF-SSRM measurements with classical metallic probes (Ni). This is most likely due to the fact that those probes are softer than Si and thus are unable to punch through the native oxide. We are now developing new metallic probes using harder metals as IrO2 or TiN which are harder than Si and SiO2, that may allow electrical measurements with metallic probes in the near future.

Electrical performance of PF-SSRM measurements

The staircase test-structures [8] have been used to test the actual electrical performance of PF-SSRM. As shown in fig.8 the PF-SSRM results are completely similar to classical SSRM results and no loss in dynamic range or sensitivity is observed.

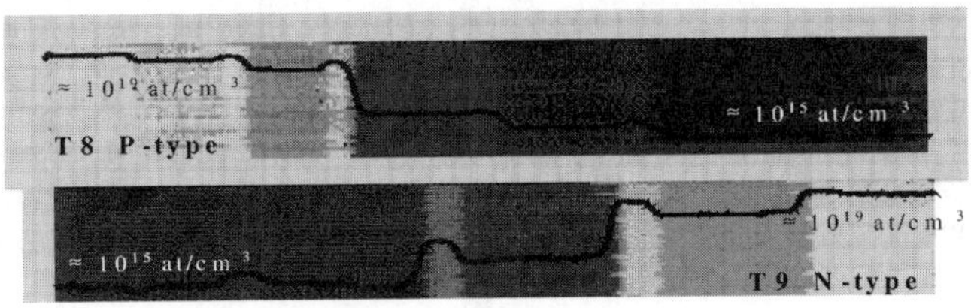

Figure 8. PF-SSRM pictures of the classical T8 (top) and T9 (bottom) staircase test-structures. The average section has been superimposed and clearly shows all the different layers (7 for T8 and 5 for T9).

As a final test measurements on a 0.5 μm PMOS (figure 9a) and 0.18 μm NMOS (figure 9b) transistor have been realized. The measurement on the PMOS-transistor were done with a commercial diamond coated Si-probe (Nanosensors) and the one on NMOS-transistor with our own full diamond probe [9]. Both experiments demonstrate that PF-SSRM is completely functional and provides the same sensitivity and spatial resolution as standard SSRM. Again it is very easy to observe the LDD-structure in the NMOS-case. However the accompanying topography image in PF-SSRM provides extra information on the structural details of the devices and aids in determining the dopant registration (size of the gates, presence of nitride, silicided region, …) .

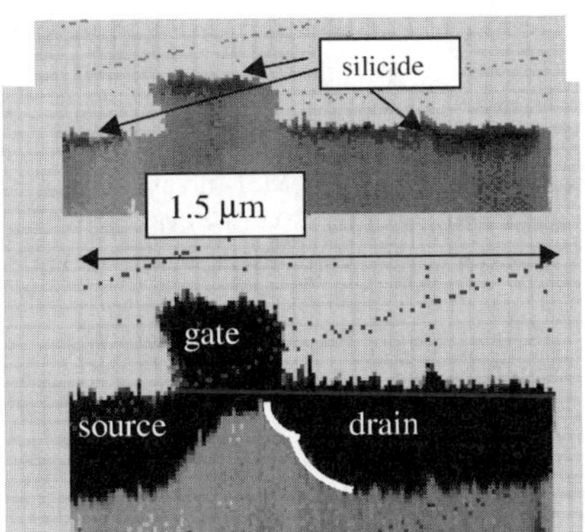

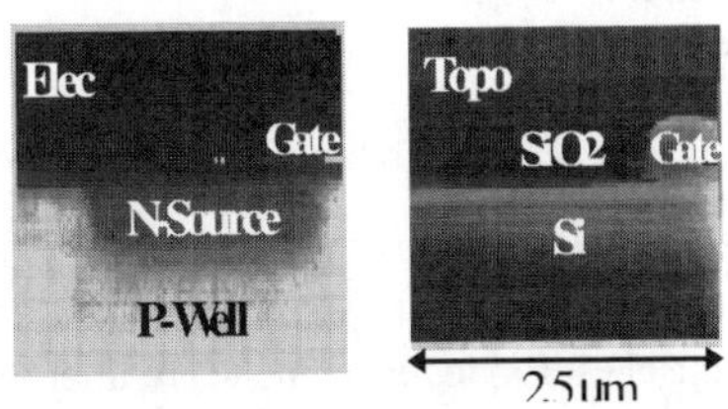

Figure 9b (top). PF-SSRM picture (left) electrical, (right) topography of a 0.5 PMOS transistor.

Figure 9a (left). PF_SSRM picture (left) electrical, (right) topography of a 0.18 NMOS transistor.

CONCLUSIONS

A modulation of the force during scanning and during measurements has been implemented in a commercial AFM-instrument equipped with a SSRM-module. The strongly reduced force during tip movement greatly reduces the sample and tip damage without any loss in spatial resolution. This implementation of the pulsed force SSRM concept preserves all the advantages of the SSRM technique in terms of resolution robustness and ease of quantification and offers a lot of complementary advantages as simultaneously a high quality topography image can be obtained. The reduced tip degradation also has positive impact on the quantification of the SSRM-results are the electrical radius of the probe remains longer constant.

The synchronization of the electrical measurements to a specific time interval has opened the way towards multiplexing of several techniques. Hence the combination in one scan of high and low force electrical AFM-based techniques as SSRM and TUNA will also be possible.

An additional advantage of the reduced tip damage is the potential use of metallic tips which have a smaller resistance (<100ohms) and thus enable to measure extremely highly doped areas. Alternatively one might consider the use of higher aspect ratio tips witch typically have a smaller electrical radius, and thus provide increased lateral resolution for the different techniques.

ACKNOWLEDGEMENTS

P. Eyben is indebted to the Flemish Institute for Support of Scientific Research (IWT) for his Ph. D fellowship.

REFERENCES

1. P. De Wolf, T. Clarysse, W. Vandervorst, J. Snauwaert, L. Hellemans, J. Vac. Sci. tech B14(1996)380-385
2. P. De Wolf, "Two-dimensional carrier profiling of semiconductor structures with nanometer resolution", Ph.D., KUL, May 1998.
3. T. Clarysse, P. De Wolf, H. Bender, and W. Vandervorst, J. Vac. Sci. Technol. B 14(1), 1996
4. V.V.Zavyalov, J.S.mcMurray and C.C.Williams, Proc. 1998 Int.Conf. on Characterization and Metrology for ULSI Technology (Gaithersburg).Eds. D.G.Seiler, A.C.Diebold, W.M.Bullis, T.J.Schaffner, R.McDonald and E.J.Walters,, AIP Conf Proc-449, 753(1998)
5. P.Eyben, N.Duhayon, T.Clarysse and W.Vandervorst, J.Vac.Sci.Technol. B (accepted for publication, 2002)
6. P.De Wolf, J.Snauwaert, T.Clarysse, W.Vandervorst and L.Hellemans, Appl. Phys. Lett. 66(12), 1530 (1995)
7. T.Trenkler, R.Stephenson, P.Jansen and W.Vandervorst, J.Vac.Sci.Technol. B18(1), 586 (2000)
8. T.Clarysse, M.Caymax, P.De Wolf, T.Trenkler, W.Vandervorst, J.McMurray, J.Kim, C.C.Williams, J.Clark and G.Neubauer, J.Vac.Sci.Technol. B16, 394 (1998)
9. T.Hantschel, P.Niedermann, T.Trenkler and W.Vandervorst, Appl. Phys. Lett. 76, 1603 (2000)

Mat. Res. Soc. Symp. Proc. Vol. 717 © 2002 Materials Research Society C7.8

Damage and dopant profiles produced by ultra-shallow boron and arsenic ion implants into silicon at different temperatures characterised by medium energy ion scattering.

J. A. van den Berg*, D. G. Armour*, S. Zhang, S. Whelan*, M. Werner*, E. H. J. Collart**, R. D. Goldberg**, P. Bailey*** and T. C. Q. Noakes***.
*Joule Physics Laboratory, School of Sciences, University of Salford, Salford, M5 4WT, UK.
**Applied Materials UK Ltd, Foundry Lane, Horsham, W. Sussex, RH13 5PX.
***CLRC Daresbury Laboratory, Daresbury, Cheshire, WA4 4AD, UK.

ABSTRACT

Medium energy ion scattering (MEIS), operated at sub-nm depth resolution in the double alignment configuration, has been used to examine implant and damage depth profiles formed in Si(100) substrates irradiated with 2.5 keV As^+ and 1 keV B^+ ions. Samples were implanted at temperatures varying between −150 °C, and 300 °C to doses ranging from 3×10^{14} to 2×10^{16} cm^{-2}. For the As implants the MEIS studies demonstrate the occurrence of effects such as a dopant accommodation linked to the growth in depth of the damage layer, dopant clustering, as well as damage and dopant movement upon annealing. Following epitaxial regrowth at 600°C, approximately half of the As was observed to be in substitutional sites, consistent with the reported formation of As_nV complexes ($n \leq 4$), while the remainder became segregated and became trapped within a narrow, 1.1 nm wide layer at the Si/oxide interface.

MEIS measurements of the B implants indicate the formation of two distinct damage regions each with a different dependence on implant dose, the importance of dynamic annealing for implants at room temperature and above, and a competing point defect trapping effect at the Si/oxide interface. B^+ implantation at low temperature resulted in the formation of an amorphous layer due to the drastic reduction of dynamic annealing processes.

Notably different dopant distributions were measured by SIMS in the samples implanted with As at different temperatures following rapid thermal annealing (RTA) up to 1100 °C in an oxidising environment. Implant temperature dependent interactions between defects and dopants are reflected in the transient enhanced diffusion (TED) behaviour of As.

INTRODUCTION

The industrial scale production of very shallow, highly doped layers of silicon by ion implantation of species such as B and As requires the use of high current, low energy ion implanters in conjunction with rapid thermal processors. The effects of Si interstitial damage formation, clustering and associated transient enhanced diffusion (TED) during annealing has been investigated for well over a decade [1,2]. The way in which the implanted atoms are accommodated during the implant stage and the nature and distribution of the damage existing at the end of this stage are known to determine the extent to which dopant redistribution and activation occur during post-implant annealing. For the high doses of interest for the production of source and drain extensions, both the implant species and damage distributions are strong functions of dose due to complex processes such as dynamic annealing and collision induced redistribution that occur during implantation. At very low ion energies considered, the situation is further complicated by the proximity of the surface which exacerbates the contributions of range shortening and sputtering to the retained dose saturation and acts as either a sink or source for defects and significantly influences dopant redistribution and segregation effects.

The complexity of these processes is reflected in the behaviour of As and B implanted into Si at energies of 2.5 keV and 1 keV, respectively. High depth resolution, medium energy ion scattering (MEIS) studies of the build up and redistribution of radiation damage and, for the case of As, the implanted atoms, have demonstrated the critical dependence of the annealing behaviour on the condition of the Si matrix at the end of the implant stage [3, 4]. A further insight into the mechanisms responsible for the redistribution has been obtained by varying the implant temperature over a range that covers the temperatures at which simple defects such as interstitials, vacancies and di-vacancies become mobile.

In the present work, the implant temperature was varied from –150 to 300 °C and the As and B doses covered the range from 3×10^{14} to 2×10^{16} cm^{-2}. The double alignment MEIS measurements used scattering geometries for the As and B implants that provided sub-nanometer depth resolutions. The As studies showed that the initial distribution of the implanted atoms was strongly dependent on dose and that significant clustering and movement of As to substitutional sites as well as segregation to the Si / oxide interface occurred during annealing. For B, the implant temperature had a more dramatic effect in that at the lowest temperature, a continuous amorphous layer was formed for doses above 1×10^{15} cm^{-2}. At the higher temperatures, the damage build up as a function of dose was complex, initially comprising a near-surface disordered region with the development of a second, deeper damaged region at doses exceeding 1×10^{15} cm^{-2}.

The MEIS study of the annealing behaviour of the samples implanted at different temperatures was complemented by secondary ion mass spectrometry (SIMS) measurements of the redistributed dopant profiles. The significant influence of the implant temperature clearly demonstrated the importance of the as-implanted condition of the silicon.

EXPERIMENTAL

The ion implantations were carried out using the ultra-low energy implanter at Salford [5]. This machine covers the energy range from 10 eV to 30 keV and allows the pressure in the sample environment to be maintained at below about 1×10^{-8} mbar during implantation. Dose rates up to 1×10^{14} ions sec^{-1} cm^{-2} were used. For the B^+ and As^+ implants of present interest, the temperature of the Si(100) target was varied between -150 °C and 300 °C using liquid nitrogen cooling and radiative heating, respectively. All implants were performed with the native oxide present. MEIS analysis was carried out at the Daresbury facility using 100 and 200 keV He^+ ions in combination with the $[\bar{1}\bar{1}1]$ channelling direction for both species and the [111] blocking direction for the B implants and the [332] direction for the As implants. These conditions result in depth resolutions of 0.7 and 0.9 nm respectively. The slightly reduced depth resolution for As implants was chosen in order to ensure an improved mass resolution that results in a good separation between the As peaks and the Si edge. Using the mean energy approximation, MEIS spectra may straightforwardly be converted to depth profiles [6]. The SIMS analysis was carried out in a Cameca 4f using a 2 keV 0_2^+ primary beam and oxygen flooding.

RESULTS AND DISCUSSION

The implantation of 2.5 keV As^+ over a wide dose range demonstrates the occurrence of a number of complex accommodation, damage and redistribution phenomena. The ability of MEIS to measure the depth distributions quantitatively with high depth resolution enables the effects of dose on the as-implanted distributions to be investigated. Figure 1 shows the experimental As profiles for doses of 1×10^{14}, 1.5×10^{15} and 6×10^{15} cm^{-2} as well as profile simulations using TRIM [7] and the IMPETUS codes [8]. It shows that the As peak position shifts to greater depth and that the profile spreads out towards the surface with increasing dose. This dependence on dose cannot be explained on the basis of range shortening and sputtering alone. The shape of the high dose profiles can be explained by the IMPETUS model in which it is assumed that in addition to the above effects, the rate at which As atoms are redistributed by collisional mixing is dependent on the local As concentration [9]. Implicit in this assumption is the fact that as the implanted atom concentration builds up, significant changes in the way in which the atoms are accommodated in what is amorphous or very highly damaged silicon occur. MEIS observations of the Si damage peak (not show) demonstrate that for As doses from 1×10^{14} cm^{-2} upwards the damage layer grows from the surface to greater depth, apparently trapping As at, on average, smaller depths that would be expected on the basis of the range distribution in virgin Si, as shown by the TRIM results. This has implications not only for relocation during implantation but also for subsequent annealing.

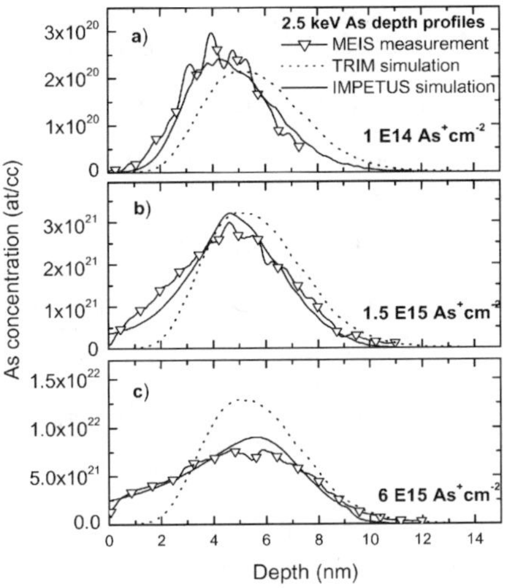

Figure 1. Depth profiles of 2.5 keV As^+ implanted into Si as measured by MEIS and simulated by TRIM [7] and IMPETUS [8].

As indicated, a measure of the extent of the damage produced by such room temperature implants can be obtained from MEIS data. The energy spectra shown in Figure 2 refer to the implantation of 2.5 keV As^+ to a dose of 1.5×10^{15} cm^{-2}. For reference the spectra for virgin and amorphous Si samples are also included. Depth scales associated with the As implant and Si damage peaks are indicated in the spectra. In the as-implanted sample, there is an amorphous layer in which all the retained As is accommodated. Rapid thermal annealing at 500 ^oC for 20 s does not affect the As profile but reduces both the height and width of the silicon damage peak. This implies that annealing, even at this comparatively low temperature, produces some crystalline material in the initially amorphous layer and removes some end of range damage. However, an increase in the annealing temperature to 600 ^oC has a very significant effect on both the damage and As profiles. Not only has a major part of the damage annealed out via solid phase epitaxial regrowth, but at the same time the shape and area of the As peak have changed dramatically. Random Rutherford backscattering and SIMS measurements on this sample have confirmed that there has been virtually no loss of As so the change in area is a measure of the amount of As that has taken up substitutional positions in the regrown silicon and are thus not visible to the He^+ ions. The As that is still visible in MEIS spectra, approximately 30 % of the original dose, has segregated in a 1.1 nm thick layer at the oxide interface, apparently trapped under the oxide layer which itself is about 1.7 nm thick, as can be determined from the O peak at 153 keV. Thus it appears that the part of the As that can not be accommodated in the regrown Si, is 'snow-ploughed' in front of the moving amorphous / crystalline interface. Annealing at 900 ^oC results in a slight reduction in the amount of damage and further segregation of As near to the surface. Due to a slight growth in thickness of the oxide layer it appears that, following the higher temperature anneal, the As in this region has moved slightly deeper into the bulk. After this anneal in excess of 50 % of the implanted As has become invisible to the ion beam. This observation is consistent with the formation of an As-vacancy

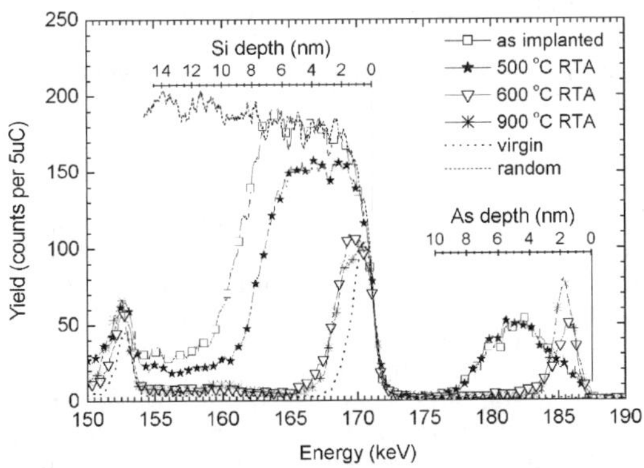

Figure 2. MEIS spectra of 2.5 keV As ions implanted into Si to 1.5×10^{15} cm^{-2} and following RTA for 10 s.

cluster, As_nV ($n \le 4$) proposed on the basis of ab-initio calculations and EXAFS and standing wave experiments [10,11]. For n = 4 this structure yields a tetragonal configuration in which the As atoms lie along two mutually perpendicular <110> rows adjacent to the vacancy which thus accommodates the larger size of the As atoms. Due to its structure this substitutional As is electrically inactive. The ejection of interstitials in the formation of As_nV clusters is also in agreement with the observation of a 3 nm thick surface damage peak that is substantially deeper than the 1.6 nm surface peak measured for the virgin sample. It is ascribed to the accumulation of interstitials at the oxide interface [4].

The effects discussed above demonstrate that complex defect-defect and defect-dopant interactions occur during both implantation and annealing. The sensitivity of these processes to the actual nature of the damage can be seen in Figure 3 where the effect of implant temperature on the As redistribution during annealing is shown. These SIMS spectra were taken after 10 s RTA treatments (in a N_2 / 10 % O_2 ambient) of the samples implanted at the three different temperatures. As the anneal temperature is increased, not only does the amount of enhanced diffusion increase but the dependence on the implant temperature also becomes more marked. At 875 °C, the small amount of movement is independent of implant temperature (only 300 °C case shown), while for the higher anneal temperatures, the implant temperature is a significant factor. For both the higher temperature anneals, there is more diffusion in the sample implanted at the lowest temperature than for the higher temperatures. It is important to note that the above effects are only observed if annealing is carried out in an oxidising environment to ensure retention of the high concentrations of As required for the diffusion enhancement to occur. However, as evidenced by the sample annealed in a N_2 + 0.3% O_2 environment, the actual oxygen content is also important due to interstitial injection effects during oxidation.

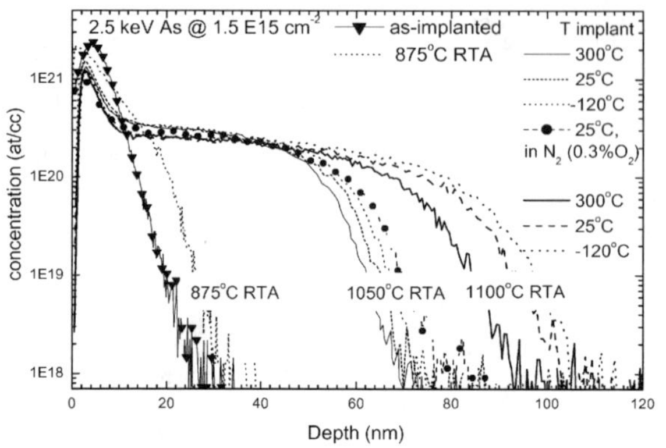

Figure 3. SIMS profiles of 2.5 keV As implanted to a dose of 1.5×10^{15} cm^{-2} at temperatures of –120, 25 and 300 °C, followed by RTA for 10 s at 875, 1050 and 1100 °C.

The effects of the implantation of B into Si are fundamentally different to those associated with As due to the significantly different energy deposition density. In contrast to As bombardment that produced amorphous material at doses above 1.5×10^{15} cm^{-2} for all the implant temperatures studied, B implantation only produced a well-defined amorphous layer at −150 °C. One consequence of this inability to amorphise is the development of a sequence of complex defect structures during dose build up [3]. This is shown in the series of MEIS spectra shown in Figure 4 which refers to the bombardment with 1 keV B^+ ions at 25 °C. Over the dose range from 3×10^{14} to 2×10^{16} cm^{-2} the damage changes from a very shallow near-surface region containing a low concentration of displaced silicon atoms to a considerably deeper amorphous layer. The figure shows that for doses below about 1×10^{15} cm^{-2} the damage is characterised by a simple broadening of the surface peak of the virgin sample. All this damage is located within the first 3 – 4 nm under the surface. This is less than the mean projected range of 5.3 nm for 1 keV B^+ and implies that there is a preferential accumulation of displaced atoms (interstitials) in the near-surface region [4]. At higher doses, a second, deeper, damage region is formed with a peak in the

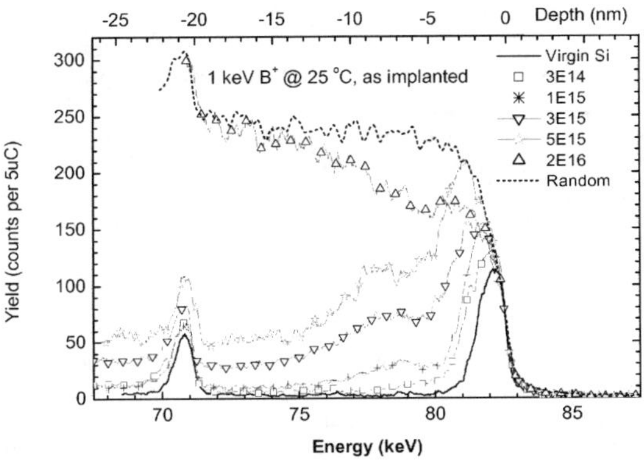

Figure 4. MEIS spectra showing the accumulation of damage in Si as a function of dose for 1 keV B^+ ion implantation.

displaced atom density at a depth of approximately 7 nm. The scattering peak corresponding to this damage moves to slightly greater depths and the surface peak height increases as the dose is increased to 5×10^{15} cm^{-2}. At this dose the surface peak actually reaches the random level suggesting that the anomalous near-surface damage region has amorphised.

The double peak damage distributions can be explained in terms of the dynamic annealing of point defects whereby a fraction of the vacancies and interstitials produced along the projectile ion track recombine, interstitials migrate to the oxide interface and become trapped and defect-implanted atom clusters grow by acquisition of migrating point defects [2, 3, 4,].

The complex nature of the damage build-up at room temperature and the critical role of mobile defects suggests that the behaviour of B implants will be strongly influenced by the bombardment temperature. That this is indeed the case is shown in Figure 5 where MEIS spectra of 1 keV B^+ implanted to a dose of 3×10^{15} cm^{-2} at target temperatures of –150, 25 and 300 °C are compared. The spectrum for the highest temperature implant, in common with that at room temperature, exhibits a two peak structure while the damage produced at the lowest temperature is characterised by a single damage peak that envelopes the region containing both peaks in the higher temperature samples. By considering the TRIM predicted B range distribution (B) and the peaks of the distributions of vacancies and interstitials (V and I) as shown in Figure 5, the detailed features of the MEIS spectra and the differences between the depths of the damage peaks can be explained. In the region where the V and I distributions have their maxima (3 - 3.5 nm) the highest dynamic annealing rates would be expected to occur and this reduces the accumulation of damage at room temperature and above. The higher dynamic annealing at 300 °C implant temperature is the result of the increased point defect mobility compared to the 25 °C implant and it is likely that for this implant, vacancies have largely been annihilated in this region. The increased near-surface damage peak for the 25 °C implant is the result of the higher fraction of non-annihilated interstitials that again are trapped in this region. The occurrence of the single damage peak for the -150 °C implant

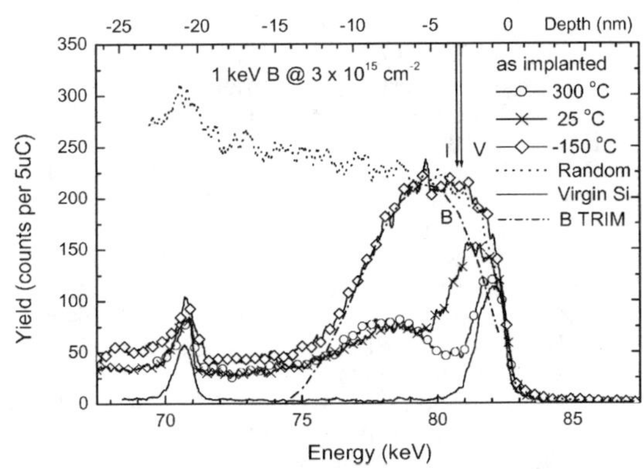

Figure 5. MEIS spectra of 1 keV B^+ showing the dependence of the damage profile on the implant temperature.

reaching random height is believed to be due the observation that at this temperature point defects have become immobile [12] largely eliminating dynamic annealing processes during implantation. Damage accumulates over the whole of the range of the B implant to depth of ~15 nm and an amorphous layer to a depth of ~8 nm is formed as confirmed by XTEM [4]. The coincidence of the damage profile with the B distribution indicates a close connection between the two and suggests that displaced atoms are associated with the presence of B containing clusters. It should finally be mentioned that that solid phase epitaxial regrowth of this layer at 600 °C is found arrested at the mean projected range (R_p) by clusters containing B atoms [4].

Conclusions

High depth resolution MEIS studies of the damage distributions produced by 2.5 keV As^+ and 1 keV B^+ ion implantation into Si, to doses between 3×10^{14} and 3×10^{15} cm^{-2}, and of the dopant distributions for the case of As, have given an insight in the effects of dopant accommodation and clustering in addition to the movement of the damage and dopant following RTA. The As peak position and shape was found to be substantially dependent on the implant dose. Following RTA at 600 °C, which leads to solid phase epitaxial regrowth of the Si matrix amorphised by the As implantation, ~50% of the dopant is observed to be in substitutional lattice sites, in agreement with the reported formation of As_nV clusters ($n\leq4$), while the remainder segregated within a narrow, 1.1 nm wide layer trapped under the Si/oxide interface. SIMS studies of As implanted at different temperatures (–120 °C, 25 °C and 300 °C) show an increased amount of TED for the low temperature implants following RTA up to 1100°C in an oxidizing ambient. MEIS studies of B implanted samples demonstrate that the occurrence of dynamic annealing for implants at room temperature and above results in the formation of two distinct damage regions, each with a different dependence on the implant dose. The near-surface damage region grows due to the migration and trapping of point defects at the Si/oxide interface that escape dynamic annealing. B implantation at –150 °C produces a single amorphous region.

References

1. P. Stolk, H. J. Gossman, D. J. Eaglesham, D. C. Jacobson, C. S. Rafferty, G. H. Gilmer, M. Jareiz, J. M. Poate, H. S. Luftman and T. E. Hayes, J. Appl. Phys **81,** 6031 (1997) and refs.
2. N. E. B. Cowern and C. S. Rafferty, MRS Bulletin **25**, 39 (2000).
3. J. A. van den Berg, S. Zhang, S. Whelan, D. G. Armour, R. D. Goldberg, E. J. H. Collart, P. Bailey and T. C. Q. Noakes, Nucl. Instrum. and Meths **B 183**, 154 (2001).
4. J. A. van den Berg, D.G. Armour, S. Zhang, S. Whelan, D. G. Armour, T.-S. Wang, A. G. Cullis, R. D. Goldberg, E. J. H. Collart, P. Bailey and T. C. Q. Noakes, J.Vac. Sci. Technol. B (to be published).
5. R. D. Goldberg, D. G. Armour, J. A. van den Berg, N. Knorr, H. Ohno, S. Whelan, S. Zhang, C. E. A. Cook and M. A. Foad, Rev. Sci. Instr. **71** (2), p. 1032, (2000).
6. R.M. Tromp, in Practical Surface Analysis, Vol. 2: Ion and Neutral Spectroscopy, D. Briggs and M.P. Seah (eds.), 1992, Wiley, p.577; W. K. Chu, J W Mayer, M Nicolet, "Backscattering Spectrometry", (Acad. Press, New York, 1978).
7. J. P. Biersack, and L. G. Haggmark, Nucl. Instr. and Meth. **174,** 257 (1980).
8. R. Badheka, M. Wadsworth, D. G. Armour, J. A. van den Berg, and J. B. Clegg. Surface. Interf. Anal. **15**, 550 (1990).
9. R. Badheka and D. G. Armour, S. Whelan and J. A. van den Berg, (unpublished results).
10. M. A. Berding and A. Sher, Phys. Rev. **B 58,** 3853 (1998).
11. P. M. Rousseau, P. B. Griffin, W. T. Fang and J. D. Plummer, J. Appl. Phys. **84,** 3593 (1998).
12. K. Nordlund, P. Partyka, Y. Zhong, I. K. Robinson, R. S. Averback and P. Ehrhart, Nucl Instr. and Meth. **B 147**, 399 (1999).

AUTHOR INDEX

SUBJECT INDEX